Introduction to Modeling Convection in Planets and Stars

Princeton Series in Astrophysics

Edited by David N. Spergel

Introduction to Modeling Convection in Planets and Stars

Magnetic Field, Density Stratification, Rotation

Gary A. Glatzmaier

PRINCETON UNIVERSITY PRESS

PRINCETON AND OXFORD

Published by Princeton University Press, 41 William Street,
Princeton, New Jersey 08540
In the United Kingdom: Princeton University Press, 6 Oxford Street,
Woodstock, Oxfordshire OX20 1TW

press.princeton.edu

Cover ilustration © Gary A. Glatzmaier. A simulated magnetic field maintained by 3D rotating convection.

Library of Congress Cataloging-in-Publication Data

Glatzmaier, Gary A., 1949–
Introduction to modeling convection in planets and stars : magnetic field, density stratification, rotation / Gary A. Glatzmaier.
    pages cm. — (Princeton series in astrophysics)

    Summary: "This book provides readers with the skills they need to write computer codes that simulate convection, internal gravity waves, and magnetic field generation in the interiors and atmospheres of rotating planets and stars. Using a teaching method perfected in the classroom, Gary Glatzmaier begins by offering a step-by-step guide on how to design codes for simulating nonlinear time-dependent thermal convection in a two-dimensional box using Fourier expansions in the horizontal direction and finite differences in the vertical direction. He then describes how to implement more efficient and accurate numerical methods and more realistic geometries in two and three dimensions. In the third part of the book, Glatzmaier demonstrates how to incorporate more sophisticated physics, including the effects of magnetic field, density stratification, and rotation. Featuring numerous exercises throughout, this is an ideal textbook for students and an essential resource for researchers. Describes how to create codes that simulate the internal dynamics of planets and stars. Builds on basic concepts and simple methods. Shows how to improve the efficiency and accuracy of the numerical methods. Describes more relevant geometries and boundary conditions. Demonstrates how to incorporate more sophisticated physics." — Provided by publisher.

Includes index.
ISBN 978-0-691-14172-5 (hardback)—ISBN 978-0-691-14173-2 (paperback)
1. Convection (Astrophysics)—Computer simulation. 2. Convection (Astrophysics)—Mathematical models. 3. Planets—Atmospheres. 4. Stars—Atmospheres. I. Title.
QB462.3.G53 2014
    523.4—dc23
                          2013010051
British Library Cataloging-in-Publication Data is available

This book has been composed in Times

Printed on acid-free paper. ∞

Typeset by S R Nova Pvt Ltd, Bangalore, India

Printed in the United States of America

10 9 8 7 6 5 4 3 2 1

To Tracy, my best friend,
and to all of our wonderful pets, past and present

Contents

Preface

As stated in the title, this is a book for people interested in modeling convection in planets and stars. It begins with the basics of computer modeling and assumes the reader has no previous computer modeling experience but does have at least a basic understanding of classical physics, vector calculus, partial differential equations, and simple computer programming. The book is a compilation of my lecture notes for teaching students at the University of California Santa Cruz how to write their own computer programs to simulate time-dependent thermal convection, internal gravity waves, and magnetoconvection. I have taught Part 1 of this book as a side project in my graduate and undergraduate courses on fluid dynamics and have included Chapter 11 in my courses on magnetohydrodynamics (MHD). In this way students gain experience in and appreciation for the art of computer modeling, while gaining a much better understanding of the fluid dynamics. In addition, I have taught Parts 2 and 3 to all the graduate students I have supervised at UCSC. Being able to write and debug their own convection programs has become a "rite of passage" for my graduate students to work toward a PhD. The focus of Part 1 and most of Parts 2 and 3 is two-dimensional (2D) models because most numerical methods can be implemented in either 2D or 3D, because 2D models are simpler to write, and because 2D models require far fewer computational resources. With this preparation, some of my students have gone on to write their own, more sophisticated, computer programs to produce original research for their PhD theses; others have chosen to study and modify existing computer programs for their thesis research. By teaching this material over the years I have learned the many subtle issues students typically need to have carefully explained, issues too detailed to be mentioned in research papers. I have made an effort to include these explanations throughout the book.

Part 1, *The Fundamentals*, reviews the concepts and equations of thermal convection and then describes, step by step, how to design a computer program that employs basic numerical methods for solving these equations to simulate convection in a 2D cartesian box of fluid heated from below. Internal gravity waves can be simulated by simply reversing the thermal boundary conditions. By prescribing a stable thermal stratification in part of the fluid domain and an unstable stratification in another part one can simulate a combination of gravity waves and convection. Double-diffusive convection is also a combination of gravity waves and convection; it occurs when buoyancy is due to perturbations in both temperature and composition with the secondary constituent of the fluid being much less diffusive than temperature. The numerical method presented in this part to simulate these types of dynamics is spectral in the horizontal direction and finite difference in the vertical direction, which introduces the reader to these two very different methods. The linear stability problem is first addressed, which provides readers a way to check their time-dependent linear program. Then the nonlinear terms are added to produce numerical simulations. I have chosen the Galerkin method to calculate nonlinear terms so readers gain a better understanding of how energy cascades between spatial scales; a more efficient (spectral-transform) method is described in Part 2. Graphical analyses of the simulated data, including making movies, is also discussed in Part 1.

Part 2, *Additional Numerical Methods*, describes alternative numerical methods that improve accuracy and efficiency and provide more realistic geometry. For example, semi-implicit, instead of explicit, time integration schemes are presented; fully finite-difference and fully spectral methods are discussed; the spectral-transform method for calculating nonlinear terms, instead of the Galerkin method, is described; and a local cartesian geometry is converted into global 2D, 2.5D, and 3D spherical-shell geometries. For efficiency, I include magnetic, density stratification, and rotational terms and equations in the discussion of the numerical methods for 2.5D and 3D spherical-shell convection in Section 10.6 before formally introducing these physical effects in Chapters 11–13. The reader could simply choose to ignore these extra terms until they are needed in Part 3 of this book or could read the introductions to Chapters 11–13 before proceeding through Section 10.6.

Part 3, *Additional Physics*, reviews the effects of magnetic fields, density stratification, and rotation. Chapter 11 (Magnetic Field) does not require any of the material in Part 2; therefore readers who are interested in adding magnetic field to the model described in Part 1 can go directly to Chapter 11. The linear analyses and nonlinear simulations with these additional physical effects are described in detail for 2D simulations; they are also described for 2.5D and 3D in Chapter 13, including two standard benchmarks for global 3D convective dynamos. In the final section (13.7) I list several more sophisticated computer modeling features of planetary and stellar convection that are beyond the scope of this book.

This book could be used by anyone with a basic background in physics, mathematics, and computer programming as a self-study guide to learn how to develop computer programs that simulate convection or other fluid dynamics. Parts of this book could also be taught as supplemental material in courses on classical fluid dynamics, magnetohydrodynamics, stellar structure and dynamics, planetary science, geodynamics, physical oceanography, and atmospheric science. Alternatively, the book could be used for a dedicated, one- or two-semester course on computer modeling of convection in any subset of these fields. Part 1 is presented in a very fundamental way with many details carefully explained to help readers who have had no previous experience in computer modeling. Parts 2 and 3 cover more advanced material with the assumption that the reader has mastered the material in Part 1. *Exercises* at the end of each chapter ask the reader, for example, to derive various mathematical results presented in the text. *Computational projects* are also listed that require computer programming to produce, for example, simulations modified relative to those presented in the text.

I have written the computer programs and have run the simulations for all the examples described and displayed in this book. Online copies of computer graphical movies of some of these are available via the book's Web page at *http://press.princeton.edu/titles/10158.html*. However, I have decided not to include copies of most of the computer programs because the point of the book is to inspire and encourage readers to learn how to design and write programs themselves. I feel too many scientists today rely on computer programs written by a small subset of their scientific community, and run these programs as "black boxes" with little understanding of the approximations that have been made to the equations or the details and limitations of the numerical methods employed to solve the equations. I have, however, included via the book's Web page copies of the basic programs described in Part 1 for those who may need help debugging their programs. Copies of the various subroutines printed in the appendixes can also be downloaded via the Web page.

Although I do cite several papers and books that describe computer modeling studies of convection in planets and stars, the list is far from complete. This book is not meant to be a review of such computer modeling studies; many excellent review papers and books have been written on those topics. I have not described parameterized convection such as that employed in Mixing Length theory (within the astrophysics and planetary communities), in hydrostatic circulation models (within the atmospheric and ocean communities), or in Mean Field models (within the geodynamo and solar dynamo communities). This book is also not a description of the latest and most sophisticated numerical or programming methods for modeling convection in planets and stars; more advanced reviews and books very adequately describe such methods. Instead, this book is meant to be a tutorial for those wishing to learn the basics of writing and using computer models of convection. Hopefully, all readers will gain an appreciation and excitement for computer modeling and some will go on to improve existing computer programs or write completely new ones.

I wish to thank the three reviewers for their helpful suggestions. I also want to thank my former and current students who have given me very useful feedback; I originally developed these 2D models as teaching tools for them. It is my hope that several of my students will find this book useful for teaching their students someday.

Gary A. Glatzmaier
Santa Cruz, California
2012

PART 1

The Fundamentals

Chapter One

A Model of Rayleigh-Bénard Convection

There are two basic types of fluid flows within planets and stars that are driven by thermally produced buoyancy forces: thermal convection and internal gravity waves. The type depends on the thermal stratification within the fluid region. The Earth's atmosphere and ocean, for example, are in most places *convectively stable*, which means that they support internal gravity waves, not (usually) convection (but see Chapter 7). On warm afternoons, however, the sun can heat the ground surface, which changes the vertical temperature gradient in the troposphere and makes the atmosphere *convectively unstable*; the appearance of cumulus clouds is an indication of the resulting convective heat (and moisture) flux. Thermal convection likely also occurs in the Earth's liquid outer core, which generates the geomagnetic field, and, on a much longer time scale, in the Earth's mantle, which drives plate tectonics and, on a much shorter time scale, initiates earthquakes and volcanic eruptions. Thermal convection is seen on the surface of the sun and likely occurs in the outer 30% of the solar radius, where solar magnetic field is generated. Below this depth buoyancy likely drives internal gravity waves. Rotation strongly influences the style of the convection and waves in all of these examples except the mantle, which is dominated by viscous forces.

Computer simulation studies, over the past few decades, have significantly improved our understanding of these phenomena. Some studies, like those for the atmospheres of the Earth and sun, have provided physical explanations and predictions of the observations. Others, like those for the deep interiors of the Earth and sun, have provided detailed theories and predictions of the dynamics that cannot be directly observed. As computers continue to improve in speed and memory, computer programs are able to run at greater spatial and temporal resolutions, which improves the quality of and confidence in the simulations. Numerical and programming methods have also improved and need to continue to improve to take full advantage of the improvements in computer hardware.

1.1 BASIC THEORY

We begin with a simple description of the fundamental dynamics expected in a fluid that is convectively stable and in one that is convectively unstable. Then we review the equations that govern fluid dynamics based on conservation of mass, momentum, and energy.

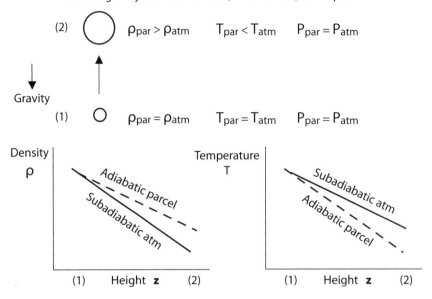

Figure 1.1 A schematic of a test parcel that is raised adiabatically from position (1) to position (2) in an atmosphere that has *subadiabatic* temperature and density stratifications.

1.1.1 Thermal Convection and Internal Gravity Waves

The thermal stability of a fluid within a gravitational field is determined by its horizontal-mean (i.e., ambient) vertical temperature gradient. The classic way of describing this is to consider a fluid in *hydrostatic equilibrium*, i.e., the weight of the fluid above a given height (per cross-sectional area) is supported by the pressure at that height. Therefore, the vertical pressure gradient is negative. (As usual, "vertical" here and throughout this book refers to the direction of increasing height or radius, opposite to that of the gravitational acceleration.) In the interiors of planets and stars the horizontal-mean density and temperature also decrease with height. The question is how does the vertical temperature gradient of this fluid (atmosphere) compare with what an adiabatic temperature gradient would be.

Consider a small (test) parcel of fluid (Fig. 1.1) that, at its initial position (1), has the same pressure, density, and temperature as the surrounding atmosphere at that position. Imagine raising the parcel to a new height (2), fast enough so there is no heat transfer between it and the surrounding atmosphere but slowly enough that it remains in pressure equilibrium with its surroundings; that is, its upward velocity is much less than the local sound speed. Assuming this process is reversible and also adiabatic since there is no heat transfer, the parcel's entropy remains constant while rising; that is, this is an isentropic process. However, since it remains in pressure equilibrium with the surroundings, its density and temperature both decrease as it

Thermal convection in an unstable (superadiabatic) atmosphere

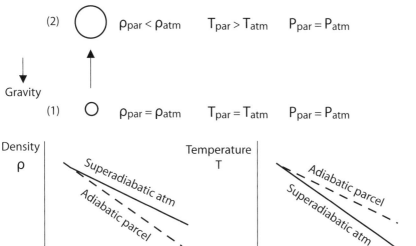

Figure 1.2 A schematic of a test parcel that is raised adiabatically from position (1) to position (2) in an atmosphere that has *superadiabatic* temperature and density stratifications.

rises because the decrease in pressure causes it to expand. If, when reaching its new higher position (2), its temperature has decreased more than the temperature of the surrounding atmosphere has decreased over that change in height, its density there will be greater than the density of the surrounding atmosphere there (assuming a typical coefficient of thermal expansion). Therefore, the parcel will be antibuoyant and, when no longer externally supported, will fall. As the parcel falls its temperature increases faster than the surrounding temperature and when it passes the initial position its temperature exceeds the temperature of surrounding atmosphere, causing the now buoyant parcel to eventually stop falling and then to start rising. This process of accelerating downward when it is above the initial position and accelerating upward when below the initial position is called an *internal gravity wave* and the surrounding atmosphere is said to be *convectively stable*. Recall that this occurs when the surrounding temperature decreases less rapidly with height than an adiabatic temperature profile would, since the test parcel moves adiabatically. That is, the surrounding temperature gradient is *subadiabatic*. The temperature stratification would be extremely stable if the surrounding temperature increased with height. In reality, thermal and viscous diffusion cause internal gravity waves to decay with time unless they are continually being excited.

Now consider the case for which the changes in the parcel's temperature as it moves up and down are less than that of the surrounding atmosphere (Fig. 1.2). That is, consider a surrounding atmosphere with a *superadiabatic* temperature gradient.

In this case, when reaching its new higher position (2), the parcel's temperature will be higher than the temperature of the surroundings there and so its density will be less than that of the surroundings. This makes the parcel buoyant and so, when no longer externally supported, it continues to rise. Likewise, if the parcel were initially lowered, it would continue to sink. Typically, a rising parcel will eventually encounter a cold impermeable top boundary where it gives up heat by conduction, contracts, and becomes antibuoyant. This causes it to fall until it encounters a hot impermeable bottom boundary where it absorbs heat by conduction, expands, becomes buoyant, and rises. This process is called *thermal convection* and in this case the surrounding atmosphere is said to be *convectively unstable*. In reality, thermal and viscous diffusion would cause this process to decay with time unless the excess temperature drop maintained across the region (compared to the adiabatic temperature drop) is larger than a critical value (Section 3.3). That is, if thermal diffusion (i.e., conduction) is not efficient enough at transferring heat upward through the atmosphere, thermal convection will occur, which will transfer heat as *convective heat flux*.

Thermal convection exists in planets and stars and takes on a variety of forms. Thermal convection likely occurs throughout the mantles of terrestrial planets, like the Earth, and also throughout most if not all of their liquid cores. However, because of their very different viscosities, the time scale for mantle convection is typically a hundred million times longer than that for core convection and, whereas the style of mantle convection is unaffected by planetary rotation (Coriolis forces), core convection is dominated by rotation. The interactions between the Earth's mantle and its fluid core are discussed in Buffett (2007). Where, within the interior of stars and giant planets, thermal convection occurs depends on the interior structure. Most one-dimensional (1D) evolutionary models of giant gas planets predict convection throughout their liquid/gas interiors. Thermal conduction in stars is by radiative transfer, which is less efficient at heat transfer in regions where atomic excitation and ionization occur, i.e., where the opacity is large and the adiabatic gradient is less steep, respectively. Convection typically occurs in these regions because the temperature gradient would need to be steeper than the adiabatic gradient to conduct all of the heat upward (i.e., outward). For example, the sun has a convection zone in roughly the outer 30% of its radius; radiative transfer is sufficient to carry the heat flux within the inner 70% of its radius, where the gas is fully ionized. Lower mass stars, which are cooler than the sun and therefore have a larger fraction of un-ionized gas, have much deeper convection zones; stars with less than 30% of a solar mass are fully convective. Stars more massive (and so hotter) than the sun have very shallow outer convection zones, if any, and much larger stars have convection within their central cores where the strong temperature dependence of the nuclear energy generation rate maintains a sufficiently steep temperature gradient.

Computer modeling has made and will continue to make significant contributions to our understanding of the interior dynamics of planets and stars. Our study of computer modeling begins with simple models in this Part 1. In Chapters 1–5 we focus on modeling thermal convection. In Chapter 6 we discuss the relatively simple changes to the model needed to simulate internal gravity waves. A combination of convection and gravity waves occurs in double-diffusive convection, which we describe in Chapter 7.

1.1.2 Equations of Motion

The fluid dynamics and thermal dynamics of these processes are governed by the classical conservation laws for mass, momentum, and energy. However, since we are considering a continuous fluid, these laws are written in terms of the densities of mass, momentum, energy, and force.

The mass conservation equation,

$$\frac{\partial \rho}{\partial t} = -\nabla \cdot \rho \mathbf{v}, \tag{1.1a}$$

says that the local (Eulerian) time rate of change of mass density (ρ) is determined by the convergence (i.e., negative divergence) of mass flux ($\rho \mathbf{v}$) at that location and time (t); $\mathbf{v}$ is the fluid velocity. Note that the Lagrangian time derivative of density ($d\rho/dt$), which is the rate at which the density of a fluid parcel changes as it moves with the flow, is the sum of the Eulerian time derivative ($\partial \rho/\partial t$) and the advection of density ($\mathbf{v} \cdot \nabla)\rho$. (The Lagrangian derivative is also called the material derivative.) Therefore, Eq. 1.1a can also be written as

$$\frac{d\rho}{dt} = -\rho \nabla \cdot \mathbf{v}. \tag{1.1b}$$

Newton's Second Law of motion applied to a fluid describes momentum conservation: mass density times acceleration equals the net force density on a fluid parcel as it moves. This equation,

$$\rho \frac{d\mathbf{v}}{dt} = -\nabla p + \nabla \cdot \boldsymbol{\sigma} + \mathbf{g}\rho, \tag{1.2}$$

is called the *Navier-Stokes equation* after Claude-Louis Navier and George Gabriel Stokes.

The first two terms on the right side are the macroscopic representation of the effects due to molecules (or atoms). The first is the negative pressure gradient, a force density from high to low pressure, p, which is due to static normal stress. The second is the divergence of the viscous stress tensor, $\boldsymbol{\sigma}$ or "σ_{ij}", to indicate that it is a tensor. Unless noted, we assume a *Newtonian fluid*; that is, viscous stress is proportional to the rate of strain of the fluid:

$$\sigma_{ij} = 2\rho v \left(e_{ij} - 1/3\, e_{kk}\, \delta_{i,j} \right) \tag{1.3}$$

where

$$e_{ij} \equiv 1/2 \left(\partial v_i/\partial x_j + \partial v_j/\partial x_i \right) \tag{1.4}$$

is the rate of strain tensor (for $i = 1, 2, 3$ or x, y, z in cartesian coordinates), v is the viscous diffusivity (also called the kinematic shear viscosity), and the kronecker delta function $\delta_{i,j}$ is one for $i = j$ and zero if not. Note, $e_{kk} = \nabla \cdot \mathbf{v}$. As is usually done for subsonic convection problems, we have neglected the small contribution to the viscous stress due to bulk viscosity: $\lambda e_{kk}\delta_{ij}$, where λ is called the of kinematic bulk viscosity. If the dynamic shear viscosity ρv were constant in space, the divergence of the viscous stress tensor (when neglecting the bulk viscosity) would reduce to

$$\nabla \cdot \boldsymbol{\sigma} = \rho \nu \left(\nabla^2 \mathbf{v} + 1/3 \nabla (\nabla \cdot \mathbf{v}) \right) . \qquad (1.5)$$

The last term on the right of Eq. 1.2 is the gravitational force density, $\mathbf{g}$ being the gravitational acceleration.

By doing a Taylor expansion about both position and time, the left side of Eq. 1.2 can be written in the Eulerian form as $\rho(\partial \mathbf{v}/\partial t + (\mathbf{v} \cdot \nabla)\mathbf{v})$, which by using Eq. 1.1 also equals $\partial \rho \mathbf{v}/\partial t + \nabla \cdot (\rho \mathbf{v} \mathbf{v})$. Above we called $\rho \mathbf{v}$ mass flux; here we call it momentum density. The Reynolds stress tensor, $\rho \mathbf{v} \mathbf{v}$, is the momentum flux due to the flow. That is, it states how each of the three components of momentum is being transported in each of the three directions. For example, $\rho v_x v_z$ is the rate that the x-component of momentum is being transported in the z-direction, which is also the rate that the z-component of momentum is being transported in the x-direction. The divergence of this tensor is a vector equal to the net rate that each of the three components of momentum is diverging at the given position and time.

A few more words may be appropriate about the Eulerian and Lagrangian time derivatives. In an Eulerian representation we ask how the properties of the fluid are changing in time on, for example, a set of grid points in space, without keeping track of where the current fluid parcels at these locations originated. In a Lagrangian representation, on the other hand, there are no set grid points in space. Instead, we ask how the properties *and* the coordinate locations of a given set of fluid parcels change with time. The Eulerian approach is preferred for a continuous fluid that fills a defined volume. The Lagrangian approach is preferred for a discontinuous set of particles interacting within an otherwise empty volume of space. We are adopting the Eulerian approach.

The first law of thermodynamics describes internal energy conservation: the rate of change of the internal energy of a fluid parcel plus the rate the fluid parcel does work equals the rate it absorbs heat. Note that "work" and "heat transfer" are *process* functions, not properties of the fluid. However, internal energy conservation can also be described in terms of state functions. The rate the fluid does work per mass is pressure times the rate of change of the volume per mass (i.e., specific volume, which is $1/\rho$); therefore, using Eq. 1.1b, the rate fluid does work per volume is $p \nabla \cdot \mathbf{v}$. The rate the fluid absorbs heat per volume can also be written in terms of state variables as $\rho T \, dS/dt$, where S is specific entropy (i.e., entropy per mass). Therefore, conservation of internal energy density is

$$\rho \frac{de}{dt} + p \nabla \cdot \mathbf{v} = \rho T \frac{dS}{dt} = \nabla \cdot (k \nabla T) + Q , \qquad (1.6)$$

where e is internal energy per mass (i.e., specific internal energy). The first heating term on the far right side of Eq. 1.6 is the convergence of diffusive heat flux, $-k \nabla T$, where T is temperature, $k = c_p \rho \kappa$ is thermal conductivity, c_p is specific heat capacity at constant pressure, and κ is thermal diffusivity. The remaining term, Q, represents viscous and ohmic heating and any other heating or cooling, e.g., nuclear.

This relationship between process and state functions is valid within the very good approximation of local thermodynamic equilibrium (LTE). That is, since "fluid dynamics" is a macroscopic description of the state and evolution of a fluid averaged over length and time scales large compared to the molecular structure and processes, state variables like temperature, pressure, density, specific internal energy, and specific entropy are defined as continuous functions of space and time that usually vary slowly enough on macroscopic length and time scales that thermodynamic equilibrium can be assumed in small neighborhoods around every location and time within the domain of study. For example, although temperature can vary in space (and therefore drive a diffusive heat flux), around any point within the fluid there is a small neighborhood in which the velocities of the particles have a well-defined Maxwellian distribution defined by the local temperature.

In addition to the internal energy density equation (Eq. 1.6), a useful equation is one that describes the rate of change of kinetic and gravitational potential energy densities. This is obtained by taking the dot product of fluid velocity and the momentum equation (Eq. 1.2), which gives

$$\rho \left(\frac{\partial}{\partial t} \left(\tfrac{1}{2} v^2 \right) + (\mathbf{v} \cdot \nabla) \left(\tfrac{1}{2} v^2 \right) \right) = -\mathbf{v} \cdot \nabla p + \mathbf{v} \cdot (\nabla \cdot \boldsymbol{\sigma}) - \rho \mathbf{v} \cdot \nabla \Phi, \tag{1.7}$$

where we have written the gravitational acceleration as $\mathbf{g} = -\nabla \Phi$, Φ being the gravitational potential energy per mass. Assuming Φ is time independent and using Eq. 1.1, the gravitational work term in Eq. 1.7 can be written as

$$-\rho \mathbf{v} \cdot \nabla \Phi = -\frac{\partial}{\partial t} (\rho \Phi) - \nabla \cdot (\rho \Phi \mathbf{v}).$$

Equation 1.1 can also be used to write the Lagrangian time derivative on the left side of Eq. 1.7 as

$$\rho \left(\frac{\partial}{\partial t} \left(\tfrac{1}{2} v^2 \right) + (\mathbf{v} \cdot \nabla) \left(\tfrac{1}{2} v^2 \right) \right) = \frac{\partial}{\partial t} \left(\tfrac{1}{2} \rho v^2 \right) + \nabla \cdot (\tfrac{1}{2} \rho v^2 \mathbf{v}).$$

It is also convenient to write the pressure work term in Eq. 1.7 as

$$-\mathbf{v} \cdot \nabla p = -\nabla \cdot (p \mathbf{v}) + p \nabla \cdot \mathbf{v}.$$

Also it can be shown (e.g., Batchelor, 1967) that part of the work done by viscous forces goes into viscous heating and the remaining part into the convergence of viscous energy flux. That is,

$$\mathbf{v} \cdot (\nabla \cdot \boldsymbol{\sigma}) = -2\rho v \left(e_{ij} e_{ij} - 1/3 (\nabla \cdot \mathbf{v})^2 \right) + \nabla \cdot (\mathbf{v} \cdot \boldsymbol{\sigma}). \tag{1.8}$$

Therefore, substituting these expressions into Eq. 1.7 gives the mechanical energy density equation, i.e., the rate of change of the sum of the kinetic and gravitational potential energy densities:

$$\frac{\partial}{\partial t} \left(\tfrac{1}{2} \rho v^2 + \rho \Phi \right) = -\nabla \cdot \left[(\tfrac{1}{2} \rho v^2 + \rho \Phi + p) \mathbf{v} - \mathbf{v} \cdot \boldsymbol{\sigma} \right]$$
$$+ p \nabla \cdot \mathbf{v} - 2\rho v \left(e_{ij} e_{ij} - 1/3 (\nabla \cdot \mathbf{v})^2 \right). \tag{1.9}$$

Combining this equation with the internal energy density equation (Eq. 1.6) and using Eq. 1.1 to write

$$\rho \frac{de}{dt} = \frac{\partial \rho e}{\partial t} + \nabla \cdot (\rho e \mathbf{v}),$$

and setting Q to just the viscous heating rate (i.e., the negative of the first term on the right of Eq. 1.8) gives the rate of change of the internal, kinetic, and potential energy densities:

$$\frac{\partial}{\partial t}\left(\rho e + \tfrac{1}{2}\rho v^2 + \rho\Phi\right) = -\nabla\cdot\left[(\rho e + \tfrac{1}{2}\rho v^2 + \rho\Phi + p)\mathbf{v} - k\nabla T - \mathbf{v}\cdot\boldsymbol{\sigma}\right].$$
(1.10)

Note that the "PdV" work and the viscous heating, which occur in Eq. 1.6, cancel those in Eq. 1.9. Equation 1.10 says that the local rate of change of total energy density equals the convergence of the fluxes of enthalpy density ($\rho h = \rho e + p$), kinetic energy density ($\rho v^2/2$), and gravitational potential energy density ($\rho\Phi$) plus the thermal diffusive heat flux ($-k\nabla T$) and the viscous energy flux ($-\mathbf{v}\cdot\boldsymbol{\sigma}$). For stress-free impermeable boundaries that do not move or deform, which we will usually assume, the volume integral of Eq. 1.10 over the domain of the fluid (i.e., the rate of change of the total energy in the fluid) reduces to minus the surface integral over the boundary of the fluid of the thermal heat flux out of the domain. Note, additional energy and energy flux terms would be included if magnetic fields exist within the fluid (Eq. 11.16); however, if the fluid were rotating, Coriolis forces (measured within a rotating frame of reference) would do no work because they are always perpendicular to the local flow.

In this Part 1 we make a very common and traditional approximation to these conservation equations, the *Boussinesq approximation*, which simplifies these equations to a form very similar to that of an incompressible fluid. In Part 3 we discuss a more realistic approximation for planets and certainly stars, the *anelastic approximation*, which accounts for the effects of density stratification. For a review of the fully compressible equations, briefly described here, and of their Boussinesq and anelastic approximations see, for example, Braginsky & Roberts (2007).

1.2 BOUSSINESQ EQUATIONS

Since density, temperature, and pressure increase with depth in the atmospheres and interiors of planets and stars, understanding the conditions under which the Boussinesq equations represent a valid approximation for these compressible fluids is critical. The Boussinesq approximation to the fluid flow equations (Boussinesq, 1903; Spiegel & Veronis, 1960) assumes that the vertical extent of the modeled domain is small relative to the hydrostatic scale heights of density, temperature, and pressure. A *density scale height*, for example, at a given location within the domain is $-(d\ln\rho/dr)^{-1}$ and is roughly the distance over which density decreases by a factor of e. This assumption is fairly valid for convection in an ocean, mantle, or liquid core of a terrestrial planet. It may also be valid near the center of a giant planet or star, where the local density scale height is much greater than the radius of the body. However, it is not valid in the outer part of the interior or in the atmosphere of a giant planet or star where the density scale height is much smaller.

The Boussinesq and anelastic (Chapter 12) approximations filter out sound waves, which typically represent even smaller perturbations in density and pressure. Sound waves usually travel much faster than the fluid flows in planetary and stellar interiors and would therefore require much smaller computational time steps to resolve. This huge reduction in the number of computational time steps needed to simulate convection (or gravity waves) compared to what would be required for a compressible model is the main reason for employing the Boussinesq (or anelastic) approximation in numerical models. Effectively, the speed of sound is assumed to be infinite. That is, a computational time step in a Boussinesq (or anelastic) simulation is assumed to be long compared to the time it would take for changes in pressure to be communicated throughout the modeled domain. Therefore, another condition for the Boussinesq (and anelastic) approximation to be valid is that the fluid velocity be small relative to the local sound speed (Eq. 12.8); that is, the Mach number (the ratio of the fluid velocity to the local sound speed) needs to be less than, say, 0.1. If, on the other hand, the fluid velocity for a particular problem were comparable to the local sound speed, a fully compressible model would be needed.

The objective here is to describe how to develop a model and the corresponding code that can be run on a computer to simulate thermal convection in a nearly incompressible liquid within a uniform gravitational field, heated on the bottom boundary and cooled on the top boundary, i.e., *Rayleigh-Bénard convection*. Laboratory experiments of Rayleigh-Bénard convection were first done by Henri Bénard in 1900 and later the linear stability analysis (Section 3.4) was described by Rayleigh (1916).

In this book, "model" refers to the equations, numerical methods, and the assumptions and approximations upon which these are based; "code" refers to the computer program that translates the model into computer language; and "simulation" refers to the numerical results obtained when the code is run on a computer.

As usual, the independent variables, for this Eulerian representation, are the time, t, and the cartesian spatial coordinates, x, y, and z. The gravitational acceleration, $\mathbf{g} = -g_o\hat{z}$, is directed downward (i.e., g_o is positive and $\hat{z}$ is the unit vector in the positive vertical direction).

For simplicity, we make the Boussinesq approximation; that is, when the change in hydrostatic density across the domain is small relative to the volume-averaged density, the background density (ρ_o) is taken to be constant in space and time. Therefore, to first order, the mass conservation equation, 1.1b, is simply

$$\nabla \cdot \mathbf{v} = 0. \qquad (1.11)$$

This does not imply that density is exactly constant in space and time, but only that the amplitude of its rate of change is small, i.e., of the order of the relative change in hydrostatic density across the domain, compared to the amplitudes of the three individual contributions to the divergence of velocity.

The local effects of pressure on density are also assumed to be small. That is, density perturbations (ρ) are assumed to be produced only by temperature perturbations (T) according to the equation of state:

$$\rho = -\rho_o \alpha T, \qquad (1.12)$$

where α is the constant coefficient of thermal expansion.

These perturbations produce buoyancy forces that drive the convection according to the momentum equation 1.2, which to first order within the Boussinesq approximation reduces to

$$\frac{\partial \mathbf{v}}{\partial t} = -(\mathbf{v}\cdot\nabla)\mathbf{v} - \rho_o^{-1}\nabla p + \alpha g_o T\hat{z} + \nu\nabla^2\mathbf{v}. \tag{1.13}$$

This equation was obtained from Eq. 1.2 by first subtracting from it the momentum equation of the hydrostatic background state:

$$0 = -\frac{dp_o}{dz} - g_o\rho_o.$$

Density, in the remaining terms, is, to first order, the constant background density ρ_o.

The viscous force in Eq. 1.13 simplifies to a constant viscous diffusivity, ν, times the Laplacian of velocity. This can be thought of as the convergence of a molecular momentum flux that is modeled as being proportional to the negative gradient of fluid velocity. That is, viscous forces tend to smooth out gradients in velocity by transporting momentum from regions of high momentum to those of low momentum.

Updating density perturbations via the equation of state 1.12 instead of the mass conservation equation 1.1 is why sound waves are filtered out in a Boussinesq model. That is, the divergence of Eq. 1.13 removes the time derivative because of Eq. 1.11; therefore the resulting pressure Poisson equation determines the pressure everywhere within the domain at each time step given the updated velocity and temperature perturbations.

The rate of change of internal energy density, $\rho de/dt$, for a perfect gas (Section 12.1.1) is $\rho c_v dT/dt$, where c_v is specific heat capacity at constant volume. The internal energy density equation 1.6 within the Boussinesq approximation reduces to

$$\frac{\partial T}{\partial t} = -(\mathbf{v}\cdot\nabla)T + \kappa\nabla^2 T \tag{1.14}$$

when only retaining the heating due to the convergence of diffusive heat flux. (Viscous heating is usually neglected when using the Boussinesq approximation because it is argued that it should be small (Spiegel & Veronis, 1960); mantle convection may be an exception (Jarvis & McKenzie, 1980).) Note that the work term, $p\nabla\cdot\mathbf{v}$, in Eq. 1.6 has not been neglected here because $\nabla\cdot\mathbf{v}$ is the same order as the other terms in this equation. Spiegel & Veronis (1960) show that (for a perfect gas) the two terms on the left of Eq. 1.6 add, to first order, to simply $\rho c_p dT/dt$ (i.e., c_v has been changed to c_p) when T is taken to be the temperature perturbation (here including the depth-dependent horizontal mean) relative to an adiabatic temperature profile (Section 1.1.1). The adiabatic temperature gradient (Eq. 12.10) and the parameters c_v, c_p, and κ are all assumed to be constants. Equation 1.14 also works for a perfectly incompressible liquid, in which case $c_p = c_v$ and there is no "PdV" work since $\nabla\cdot\mathbf{v}$ vanishes.

A simple way to summarize the Boussinesq approximation is that it ignores variations in density except in the buoyancy force and in the equation of state.

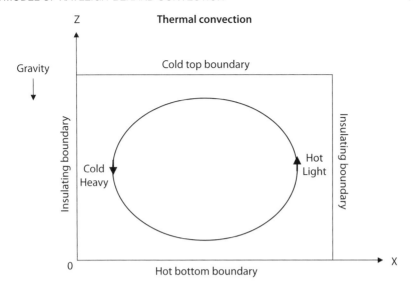

Figure 1.3 A schematic of a simple one-cell thermal convection pattern in the fluid box
         domain of unstably stratified fluid.

1.3 MODEL DESCRIPTION

Here in Part 1 we define the fluid domain as a rectangular region bounded by walls
that are impermeable and stress-free (Fig. 1.3). One could think of this rectangular
box as a small region within a global planet or star. Modifications made to the
geometry and physics of this problem later in Parts 2 and 3 build upon this first
scenario, improving the physical realism of the simulations.

The bottom and top boundaries of our fluid box are maintained at constant
temperatures by external heaters and coolers; the bottom boundary temperature
is higher than the top boundary temperature by an amount ΔT. Heat diffuses in
through the bottom boundary and out through the top boundary. The side boundaries are set to be thermally insulating; that is, there is no heat flow through them.

To keep the problem more manageable, we allow fluid flow and gradients only
in two directions, the horizontal (x-direction) and the vertical (z-direction). Therefore, $\partial/\partial y = 0$ and the y-component of the fluid velocity, v_y, vanishes. In this two-
dimensional (2D) problem the linear terms have eight spatial derivatives on the fluid
velocity and pressure, four in z and four in x. Therefore, eight boundary conditions
are formally required to maintain a unique solution of the velocity and pressure.
We apply all eight on the velocity, none on the pressure perturbation, because the
amplitude and gradient of the pressure perturbation are both expected to vary with
time and location on the impermeable boundaries.

Let's consider the velocity boundary conditions more carefully. Impermeable
means fluid cannot pass through the boundaries. Therefore, the vertical component
of the flow, v_z, has to vanish on the top ($z = D$) and bottom ($z = 0$) boundaries.

Likewise, the horizontal component of the flow, v_x, vanishes on the side boundaries ($x = 0$ and L). Stress-free means the rates of tangential strain vanish at the boundaries. That is, the fluid slips without resistance along the boundaries instead of the more physical condition that requires the fluid at the boundary not to move relative to the boundary. Our stress-free condition allows fluid at the boundary to flow parallel to the boundary but requires the gradient, normal to the boundary, of this parallel flow to vanish at the boundary. Therefore, at the top and bottom boundaries, $\partial v_x/\partial z$ vanishes and at the side boundaries $\partial v_z/\partial x$ vanishes.

The heat equation has four spatial derivatives on the temperature perturbation, two in z and two in x, and therefore four boundary conditions are required. The isothermal top and bottom boundaries are forced by setting the temperature perturbation to be zero at $z = D$ and at a prescribed value, ΔT, at the bottom boundary, $z = 0$. No heat flows through the insulating side boundaries. There is no advective heat flux through them since they are impermeable. The diffusive (conductive) heat flux through a side boundary is proportional to the horizontal gradient of the temperature there, which is set to zero at the side boundaries to prevent any diffusive heat flux through them. That is, the other two boundary conditions are that $\partial T/\partial x$ vanishes at $x = 0$ and $x = L$.

At this point readers need to decide if they wish to use dimensional or nondimensional variables. Nondimensional variables, which are dimensional variables scaled by values representative of the chosen problem, have traditionally been used so the equations and solutions can easily be characterized and readily be applied to many different physical problems. On the other hand, variables represented in centimeters, grams, seconds (CGS units) or meters, kilograms, seconds (MKS units) are easily monitored without first having to divide each variable by its chosen scaling parameter. In addition, using dimensional variables facilitates testing and debugging. For example, if a term in an equation is supposed to contain the radius of a planet squared and one mistakenly had the radius cubed, it would be relatively easy to spot if the radius were written in centimeters instead of scaled to something near unity. Also, the characteristic nondimensional numbers that would appear in the nondimensional equations can easily be constructed and presented along with the dimensional values of the simulation. What often happens, however, when people present just their nondimensional results for a study of the internal dynamics of a particular planet or star is that the degree of agreement (or disagreement) with observations is not obvious to the readers (or audience).

In this Part 1 of the book we use nondimensional variables and equations (for the most part) because the problems described are relatively simple and not meant to be realistic simulations of any particular type of planet or star. Later in the book we use dimensional variables and equations, for writing codes and presenting results, when simulating a particular problem with more realistic geometry and physics.

As a side note, it is interesting that researchers in some communities, like those in the Earth sciences, have traditionally used MKS units; whereas those in other communities, like astronomy and astrophysics (which deal with much greater masses and lengths), have traditionally used CGS units (which are smaller mass and length units). Planetary science seems to be a mixed bag. Those studying planetary atmospheres and surfaces tend to use MKS units, probably because they started in the Earth sciences; whereas many studying the dynamics of the deep interiors of giant gas planets tend to use CGS units because they also study stellar interiors.

In this chapter we choose a traditional set of scales for the problem. Length, time, and temperature are scaled by the depth of the box (D), the thermal diffusion time (D^2/κ), and the temperature drop across the depth (ΔT), respectively. For convenience, we choose $\rho_o \kappa^2/D^2$ as the pressure scale. Then Eq. 1.11 is multiplied by D^2/κ, Eq. 1.13 by D^3/κ^2 and Eq. 1.14 by $D^2/\kappa \Delta T$. This results in the following nondimensional versions of these equations:

$$\nabla \cdot \mathbf{v} = 0 \,, \tag{1.15}$$

$$\frac{\partial \mathbf{v}}{\partial t} = -(\mathbf{v} \cdot \nabla)\mathbf{v} - \nabla p + \mathrm{RaPr}\, T \hat{z} + \mathrm{Pr}\, \nabla^2 \mathbf{v} \,, \tag{1.16}$$

$$\frac{\partial T}{\partial t} = -(\mathbf{v} \cdot \nabla)T + \nabla^2 T \,. \tag{1.17}$$

All variables are now nondimensional and $0 \leq z \leq 1$ and $0 \leq x \leq a$, where $a = L/D$, the aspect ratio of the box.

This scaling results in two nondimensional numbers, which characterize the type of flow based on prescribed fluid properties and boundary conditions:

$$\mathrm{Ra} \equiv \frac{g_o \alpha \Delta T D^3}{\nu \kappa} \,, \tag{1.18}$$

$$\mathrm{Pr} \equiv \frac{\nu}{\kappa} \,. \tag{1.19}$$

The *Rayleigh number*, Ra, is a measure of the convective driving; the terms in the numerator promote convection, whereas those in the denominator inhibit convection. The *Prandtl number*, Pr, is the ratio of viscous to thermal diffusion; a small Pr usually means flow structures are smaller scale than thermal structures and vice versa for large Pr.

Note that an alternative choice would have been to scale the time by the viscous diffusion time, D^2/ν, which would have resulted in a slightly different arrangement of nondimensional numbers in the equations. However, neither of these scalings are particularly appropriate for simulations that are strongly driven by buoyancy because the resulting convective velocities are typically much greater than the thermal and viscous diffusion velocities, κ/D and ν/D, respectively.

SUPPLEMENTAL READING

Batchelor (1967)

EXERCISES

1. *Viscous force density*
   Derive Eq. 1.5 starting from Eqs. 1.3 and 1.4 assuming a constant dynamic viscosity.

2. *Nondimensional equations*

   Demonstrate how scaling length by the depth of the box (D), time by the thermal diffusion time (D^2/κ), temperature by the temperature drop across the depth (ΔT), and pressure by $\rho_o \kappa^2/D^2$ transforms Eqs. 1.11, 1.13, and 1.14 into Eqs. 1.15, 1.16, and 1.17, respectively.

3. *An alternative set of nondimensional equations*

   Find another set of nondimensional equations by again scaling length by the depth of the box (D) and temperature by the temperature drop across the depth (ΔT) but time by the *viscous* diffusion time (D^2/ν) and pressure by $\rho_o \nu^2/D^2$.

4. *The Boussinesq equation of state for a perfect gas*

   Explain how the Boussinesq approximation to the equation of state, 1.12, can be appropriate for a perfect gas, assuming the domain spans much less than a density scale height. See Spiegel & Veronis (1960).

5. *The Boussinesq internal energy equation for a perfect gas*

   Explain how the Boussinesq approximation to the internal energy equation, 1.14, can be appropriate for a perfect gas, assuming the domain spans much less than a density scale height. See Spiegel & Veronis (1960).

Chapter Two

Numerical Method

Now we describe a numerical method for solving these equations on a computer. The vorticity-streamfunction formulation is introduced as a means of conserving mass. This formulation was used for this problem by Nigel Weiss and his collaborators (e.g., Moore et al., 1973; Weiss, 1981a,b). To introduce the reader to two very different spatial discretizations, the vertical derivatives are approximated with a local (finite-difference) method and the horizontal derivatives with a global (spectral) method. The nonlinear terms are computed in spectral space; a more efficient spectral-transform method is introduced in Chapter 10. The time integration is an explicit Adams-Bashforth scheme; an improved semi-implicit scheme is described in Chapter 8.

2.1 VORTICITY-STREAMFUNCTION FORMULATION

There are several ways to solve this system of equations. Typically the solution is evolved in time via computational time steps (i.e., a long series of snapshots), each step requiring an update of all the variables in space. For some problems, like the one described here, it is convenient to first update for the vorticity, $\omega \equiv \nabla \times \mathbf{v}$, and then solve for the fluid velocity, $\mathbf{v}$, each time step. This is the approach presented here.

Recall that for this 2D problem $v_y = 0$ and $\partial/\partial y = 0$. Therefore,

$$\boldsymbol{\omega} = \left(\frac{\partial v_x}{\partial z} - \frac{\partial v_z}{\partial x} \right) \hat{\mathbf{y}} \equiv \omega \hat{\mathbf{y}}.$$

An equation for this vorticity is obtained by taking the curl of the equation for momentum conservation (1.16):

$$\nabla \times \frac{\partial \mathbf{v}}{\partial t} = -\nabla \times ((\mathbf{v} \cdot \nabla)\mathbf{v}) - \nabla \times \nabla p + \mathrm{RaPr} \nabla \times T \hat{\mathbf{z}} + \mathrm{Pr} \nabla \times \nabla^2 \mathbf{v}. \qquad (2.1)$$

One might wonder at this point how this could actually lead to a more convenient method of solution. Since the curl commutes with the Eulerian time derivative, the term on the left is $\partial \omega / \partial t \, \hat{\mathbf{y}}$. Also, since the curl commutes with the Laplacian operator, the viscous term on the far right is $\mathrm{Pr} \nabla^2 \omega \, \hat{\mathbf{y}}$. The buoyancy term is simply $-\mathrm{RaPr} \partial T / \partial x \, \hat{\mathbf{y}}$ and, since the curl of a gradient always vanishes, the pressure term drops out. This leaves the curl of the nonlinear advection term. Using the standard vector identity for $\nabla (\mathbf{A} \cdot \mathbf{B})$ one can see that

$$(\mathbf{v} \cdot \nabla)\mathbf{v} = \tfrac{1}{2} \nabla v^2 - \mathbf{v} \times (\nabla \times \mathbf{v}); \qquad (2.2)$$

then taking the curl of these terms it can be shown that this reduces to simply

$$\nabla \times ((\mathbf{v} \cdot \nabla)\mathbf{v}) = (\mathbf{v} \cdot \nabla)\omega \hat{\mathbf{y}}. \tag{2.3}$$

Therefore, since all the terms in the vorticity equation are in the y-direction, the scalar equation for the y-component of vorticity is

$$\frac{\partial \omega}{\partial t} = -(\mathbf{v} \cdot \nabla)\omega - \mathrm{RaPr}\frac{\partial T}{\partial x} + \mathrm{Pr}\nabla^2 \omega. \tag{2.4}$$

This equation states that the Eulerian time derivative of vorticity (on the left side) is, at a given point in time and space, determined by the sum of the three terms on the right side: advection of vorticity by the flow, vorticity generation by buoyancy, and vorticity diffusion by viscosity, respectively. The advection term simply means that when fluid is flowing toward a given location from a place where its vorticity is greater than it is at the given location, for example, the vorticity at the given location will increase with time. The viscous diffusion term always tries to smooth the vorticity by reducing its extreme values. That is, when and where the second derivative in space of vorticity is positive (e.g., a minimum value of the function) the viscous term is positive, which tries to make vorticity increase with time. Likewise, when and where the second derivative is negative (e.g., a maximum value) this term tries to decrease vorticity. The buoyancy "torque" term says that when the temperature perturbation increases in the x-direction, as in Fig. 1.3, fluid tends to sink on the left and rise on the right. This counterclockwise circulation represents a component of vorticity directed in the negative y-direction, which is why, with the minus sign, this term would generate negative vorticity. Likewise, a negative $\partial T/\partial x$ would drive a clockwise circulation, i.e., generate vorticity in the positive y-direction.

Of course now, having updated the vorticity, one needs to update the fluid velocity. This is accomplished by defining a streamfunction, ψ, such that

$$\mathbf{v} \equiv \nabla \times (\psi \hat{\mathbf{y}}) = -\frac{\partial \psi}{\partial z}\hat{\mathbf{x}} + \frac{\partial \psi}{\partial x}\hat{\mathbf{z}} \tag{2.5}$$

and recognizing that

$$v_x = -\frac{\partial \psi}{\partial z} \quad \text{and} \quad v_z = \frac{\partial \psi}{\partial x}. \tag{2.6a,b}$$

Notice that this automatically satisfies the mass conservation equation (1.11), $\nabla \cdot \mathbf{v} = 0$, and, when substituted into the definition of vorticity ($\boldsymbol{\omega} = \nabla \times \mathbf{v}$), provides the equation needed to solve for the streamfunction,

$$\omega = -\nabla^2 \psi, \tag{2.7}$$

using the updated vorticity. Then the two components of velocity can be calculated via Eqs. 2.6.

Notice also that $\nabla \psi \cdot \mathbf{v}$ vanishes with the help of Eqs. 2.6; therefore, contours of constant ψ are tangent to the local $\mathbf{v}$. In addition, $\nabla \psi$ has the same amplitude as the velocity; so the local density of a set of contours of ψ (that differ by a constant increment) is proportional to the amplitude of the local fluid velocity. Therefore, plots showing contours of ψ are instantaneous "streamlines" of the flow.

2.2 HORIZONTAL SPECTRAL DECOMPOSITION

The problem is now described by Eqs. 2.4, 2.7, and 1.17. There are several ways to solve this system. We choose a spectral method in the horizontal direction and a finite-difference method in the vertical direction. The spectral method is based on Fourier expansions in x. That is, instead of solving for the time-dependent values of a function on a finite set of grid points in x, we will approximate the function as a finite series of sines or cosines in x and solve for the time-dependent and z-dependent coefficients of these sines or cosines.

Usually one would begin by formally expanding all three dependent variables, T, ω, and ψ, in both sines and cosines. However, the chosen set of side boundary conditions allows us to expand the temperature in cosines only,

$$T(x,z,t) = \sum_{n=0}^{N_n} T_n(z,t)\, \cos(n\pi x/a), \tag{2.8a}$$

and the vorticity and streamfunction in sines only,

$$\omega(x,z,t) = \sum_{n=1}^{N_n} \omega_n(z,t)\, \sin(n\pi x/a), \tag{2.8b}$$

$$\psi(x,z,t) = \sum_{n=1}^{N_n} \psi_n(z,t)\, \sin(n\pi x/a). \tag{2.8c}$$

Here n is the mode number and N_n is the chosen truncation level in the expansions; the higher the value of N_n the better the spatial resolution in the horizontal direction. Note that the temperature expansion begins with $n = 0$ because this first term does not vanish for cosines. It represents the x-averaged value of T at the given z and t. The vorticity and streamfunction, on the other hand, have no $n = 0$ contributions because the four boundaries are impermeable and fixed in space and the fluid is incompressible, $\nabla \cdot \mathbf{v} = 0$.

Note also that the $n = 1$ modes represent a half wavelength spanning the length of the box. One could double the length of the box (i.e., double the value of the aspect ratio a) and change π to 2π in the arguments of the sines and cosines to effectively add on the mirror image of the original solution. However, if this were desired, it could be accomplished much more efficiently via graphics instead of doing the additional amount of computational work to solve for the mirror image of the solution.

One could also define a wavenumber $k \equiv n\pi/a$, which would be the number of radians per unit length spanned by the Fourier mode with wavenumber k. This would simplify the arguments of the sines and cosines to be just kx; but it is more convenient for the subscripts of their coefficients to be integers n instead of real numbers k.

Now, with these spectral expansions, consider the boundary conditions at $x = 0$ and a. The insulating condition, $\partial T/\partial x = 0$, is automatically satisfied, for each mode n because T is expanded in only cosines (Eq. 2.8a). Likewise, each mode satisfies the impermeable condition, $v_x = -\partial\psi/\partial z = 0$, and the stress-free condition, $\partial v_z/\partial x = \partial^2\psi/\partial x^2 = 0$. Note that since v_x vanishes for all z on the side boundaries, so does $\partial v_x/\partial z$; therefore both ω and ψ need to vanish on the side boundaries, as they do with the chosen sine series (2.8b,c).

One of the reasons for expanding in Fourier functions is that they are orthogonal. That is, for n_1 and n_2 not both zero,

$$\int_0^a \sin(n_1\pi x/a)\,\sin(n_2\pi x/a)\,dx = \begin{cases} \pm a/2 & \text{if } n_1 = \pm n_2 \\ 0 & \text{if } |n_1| \neq |n_2| \end{cases}, \qquad (2.9a)$$

$$\int_0^a \cos(n_1\pi x/a)\,\cos(n_2\pi x/a)\,dx = \begin{cases} a/2 & \text{if } n_1 = \pm n_2 \\ 0 & \text{if } |n_1| \neq |n_2| \end{cases}. \qquad (2.9b)$$

Substituting the spectral expansions for T, ω, and ψ into Eqs. 1.17, 2.4, and 2.7 and employing the orthogonality of Fourier functions allows one to write the following set of equations for each individual mode n:

$$\frac{\partial T_n}{\partial t} = -[(\mathbf{v}\cdot\nabla)T]_n + \left(\frac{\partial^2 T_n}{\partial z^2} - \left(\frac{n\pi}{a}\right)^2 T_n\right), \qquad (2.10)$$

$$\frac{\partial\omega_n}{\partial t} = -[(\mathbf{v}\cdot\nabla)\omega]_n + \text{RaPr}\left(\frac{n\pi}{a}\right)T_n$$
$$+ \text{Pr}\left(\frac{\partial^2\omega_n}{\partial z^2} - \left(\frac{n\pi}{a}\right)^2\omega_n\right), \qquad (2.11)$$

$$\omega_n = -\left(\frac{\partial^2\psi_n}{\partial z^2} - \left(\frac{n\pi}{a}\right)^2\psi_n\right). \qquad (2.12)$$

Equation 2.10 is for $n = 0 \rightarrow N_n$ but Eqs. 2.11 and 2.12 are for $n = 1 \rightarrow N_n$ because $\sin(0) = 0$.

Recall that the solutions to these equations are the coefficients in the Fourier expansions of the temperature, vorticity, and streamfunction and that these coefficients are functions of only z and t. The "$[\ ]_n$" terms are nonlinear; that is, they involve the product of two unknown quantities and, as will be shown, they depend on contributions from many other modes. The nonlinear terms are more challenging to compute; but they make the solutions more interesting and realistic. Before describing how they can be computed each computational time step, we consider in Chapter 3 the other terms in these equations, which are linear.

However, the same boundary conditions are imposed for both the linear and nonlinear problems. Having ensured the boundary conditions on the side boundaries by our particular choice of sine and cosine expansions, we need to enforce the boundary conditions on the top ($z = 1$) and bottom ($z = 0$) boundaries. Isothermal boundary conditions are imposed on the mean (x-independent) temperature perturbation, T_0 (i.e., for $n = 0$, not to be confused with the constant background temperature); so $T_0 = 0$ at $z = 1$ and $T_0 = 1$ at $z = 0$. For $n > 0$, $T_n = 0$ at both the top and bottom boundaries. The impermeable boundary condition means that v_z vanishes on the top and bottom boundaries; therefore, by Eqs. 2.6, $\psi_n = 0$ at $z = 0$ and 1. We also want the boundaries to be stress-free, i.e., $\partial v_x / \partial z = 0$. Again according to Eqs. 2.6, this means that $\partial^2 \psi / \partial z^2 = 0$ on both the top and bottom boundaries; and, by Eq. 2.12 and the above impermeable condition, this also means that $\omega_n = 0$ on these boundaries. Therefore, all three Fourier coefficients, T_n, ω_n, and ψ_n, for $n > 0$ vanish on the top and bottom boundaries. Note that the full variables, $\omega(x, z, t)$ and $\psi(x, z, t)$, vanish on all four boundaries.

2.3 VERTICAL FINITE-DIFFERENCE METHOD

There are several ways to solve this set of partial differential equations. Here we choose a simple method that we can continue to employ for the nonlinear problem: a second-order accurate finite-difference method to represent derivatives with respect to z and t. The basic idea is to use the values of a variable at neighboring locations to approximate the derivatives of that variable at a given location. The more "neighbor" values used the higher the order of accuracy. A second-order accurate method means that the local error is proportional to the cube of the grid spacing, Δz^3, for spatial derivatives; the global error is proportional to Δz^2.

Finite-difference methods are obtained by expanding the variable in Taylor series based on powers of the grid spacing. Consider a function, $f(z)$, on a discrete set of grid points, $z_k = (k - 1)\Delta z$ for $k = 1 \rightarrow N_z$, where N_z is the number of vertical grid points with uniform grid spacing $\Delta z \equiv 1/(N_z - 1)$. Let f_k represent the value of f at z_k. To obtain an approximation to the first derivative of f with respect to z at z_k, represent f_{k+1} as a Taylor expansion of f about z_k. That is,

$$f_{k+1} = f_k + \left(\frac{\partial f}{\partial z} \right)_k \Delta z + \frac{1}{2} \left(\frac{\partial^2 f}{\partial z^2} \right)_k \Delta z^2 + O(\Delta z^3). \tag{2.13}$$

Likewise, represent f_{k-1} as a Taylor expansion about z_k:

$$f_{k-1} = f_k - \left(\frac{\partial f}{\partial z} \right)_k \Delta z + \frac{1}{2} \left(\frac{\partial^2 f}{\partial z^2} \right)_k \Delta z^2 + O(\Delta z^3). \tag{2.14}$$

Subtracting Eq. 2.14 from 2.13 and dropping terms of order Δz^3 and smaller gives the centered finite-difference approximation to the first derivative,

$$\left(\frac{\partial f}{\partial z} \right)_k = \frac{f_{k+1} - f_{k-1}}{2\Delta z}. \tag{2.15}$$

Likewise, adding Eqs. 2.13 and 2.14 and again dropping terms of order Δz^3 and smaller gives the finite-difference approximation to the second derivative,

$$\left(\frac{\partial^2 f}{\partial z^2} \right)_k = \frac{(f_{k+1} - 2f_k + f_{k-1})}{(\Delta z)^2}. \tag{2.16}$$

A simple way to understand Eq. 2.15 is to think of the first derivative of f at a given grid point, z_k, as being simply the linear slope between the values of the function on either side of the grid point. The second derivative is then represented as the finite difference of the first derivatives at $z_k \pm \Delta z/2$:

$$
\begin{aligned}
\left(\frac{\partial^2 f}{\partial z^2} \right)_k &= \left[\left(\frac{\partial f}{\partial z} \right)_{k+1/2} - \left(\frac{\partial f}{\partial z} \right)_{k-1/2} \right] \Big/ \Delta z \\
&= \left[\frac{(f_{k+1} - f_k)}{\Delta z} - \frac{(f_k - f_{k-1})}{\Delta z} \right] \Big/ \Delta z \\
&= \frac{(f_{k+1} - 2f_k + f_{k-1})}{(\Delta z)^2}.
\end{aligned}
$$

Therefore, for this central finite-difference method, both the first and second derivatives require information from only the two nearest neighbors. This would require a modification at the top and bottom boundaries if Eqs. 2.10–2.12 needed to be solved on these boundaries. However, for our problem the conditions $T_n = \omega_n = \psi_n = 0$ are forced to be satisfied on these boundaries instead of the fluid equations.

A finite-difference method is typically much simpler to program than a spectral method. However, a spectral method, for a problem with relatively simple boundaries and geometry, usually converges much more rapidly to the actual solution as the number of modes increases compared to the rate a finite-difference method converges as the number of grid levels increases. A spectral method is said to have "spectral accuracy," i.e., if the actual solution is infinitely differentiable a spectral method with N modes is effectively Nth-order accurate and, for a given number of modes, the error depends only on the smoothness of the actual solution. An approximate "rule" for many situations is that one needs at least about $2N$ grid levels per dimension with a finite-difference method to obtain the accuracy of a spectral method with N modes per dimension (e.g., Orszag, 1971b). This is a "rough" estimate because it depends on how one defines accuracy, e.g., based on spatial and temporal averages or on the local amplitude and phase of the numerical solution. It also depends on the smoothness in space and time of the actual solution. Tests I have performed on simulations of fairly laminar convection suggest that indeed a factor of at least 2 is needed for a fourth-order finite-difference method and a factor of roughly 4 for a second-order finite-difference method.

2.4 TIME INTEGRATION SCHEME

Next consider the Eulerian time derivatives in Eqs. 2.10 and 2.11. There are many different time integration schemes, which fall into two major categories: explicit and implicit. An explicit scheme uses information at the current and previous time steps to update a variable to the new time step. This is usually simpler than implicit schemes but has more severe constraints on how large the computational time step can be. That is, although an explicit time step requires fewer computational resources than an implicit step, an explicit scheme typically requires many more time steps to evolve the solution over the same amount of simulated time. In Chapter 8 we describe a semi-implicit scheme, which uses information from both the current and the new time steps to update a variable to the new time step.

In this Part 1, however, we choose a second-order accurate explicit scheme, the Adams-Bashforth time integration scheme. The idea is to advance a variable from the current time step to the new step using an approximation to the time derivative midway between these two steps, as in Eq. 2.15. Note that although the local error (i.e., for one time step) is formally proportional to Δt^3, the global error (i.e., accumulated over the span of many time steps) is proportional to Δt^2. The time derivative is a function of the variable itself and likely other variables in the problem, all of which are calculated only at the discrete time steps. This scheme uses the variables at the current time step and the previous one to extrapolate the time derivative to the midway point between the current and new steps, assuming the time derivative is changing approximately linearly over these three (small) time steps. A simple analysis shows that the time derivative at the midway point is estimated to be the time derivative at the current step weighted by 3/2 minus the time derivative at the previous step weighted by 1/2. For example, let G represent the left side of Eq. 2.10 for a given Fourier mode number n at time $t + \Delta t/2$, where Δt is the length of the computational time step. Then Eq. 2.10, dropping the subscript n for the moment, is

$$\left(\frac{\partial T}{\partial t}\right)_{t+\Delta t/2} = G_{t+\Delta t/2},$$

which is approximately

$$\frac{T_{t+\Delta t} - T_t}{\Delta t} = 3/2\, G_t - 1/2\, G_{t-\Delta t}. \tag{2.17}$$

Therefore, the Adams-Bashforth scheme is

$$T_{t+\Delta t} = T_t + \Delta t/2\, (3G_t - G_{t-\Delta t}). \tag{2.18}$$

Since this time integration scheme is explicit and Eqs. 2.10 and 2.11 have thermal and viscous diffusion terms, respectively, Δt needs to be less than the time needed for a thermal perturbation or shear flow of the size of a grid cell to diffuse between two adjacent grid points, i.e., Δz. The thermal diffusion velocity in the z-direction for a grid-cell size perturbation is $\kappa/\Delta z$; therefore, the time needed for a perturbation to diffuse over a distance Δz is $(\Delta z)^2/\kappa$. The actual constraint is somewhat more severe (see, for example, Ferziger & Perić, 1997). In our nondimensional variables, it is

$$\Delta t < (\Delta z)^2/4 . \tag{2.19}$$

If a Δt larger than this limit is chosen, a numerical instability quickly develops, which can easily be detected because the amplitudes of the variables grow out of control.

If Pr is greater than one, i.e., viscous diffusivity greater than thermal diffusivity, then it too appears in the denominator of the term on the right in expression 2.19 because the limit is based on the greatest diffusion velocity. The diffusive time constraint in the x-direction can be approximated by calling $\Delta x = a/N_n$. However, typically this is chosen to be larger than Δz because the spectral method is more accurate than the finite-difference method.

2.5 POISSON SOLVER

Now that the solution method for the prognostic equations, 2.10 and 2.11, has been outlined to update $T_n(z, t)$ and $\omega_n(z, t)$, respectively, we need to describe how the Poisson equation, 2.12, can be solved at each time step for the updated streamfunction, $\psi_n(z, t)$, given the updated vorticity, $\omega_n(z, t)$. An updated $\psi_n(z, t)$ is needed to compute the updated velocities via Eqs. 2.6, which are required at the next time step for the advection terms in Eqs. 2.10 and 2.11.

Equation 2.12 written for mode n at the kth z-level, using Eq. 2.16 to approximate the second derivative, is

$$\omega_k = -\left(\frac{(\psi_{k+1} - 2\psi_k + \psi_{k-1})}{(\Delta z)^2} - \left(\frac{n\pi}{a} \right)^2 \psi_k \right), \tag{2.20}$$

where here the subscript n has been dropped. It should be obvious now that the way to solve this for ψ_k is to use a tridiagonal solver. For example, a simple Fortran subroutine that solves a tridiagonal matrix equation of rank N_z is listed in Appendix A. In this subroutine, the inputs are the matrix operator (i.e., only its nonzero elements, which are stored in arrays sub, dia, and sup) and the right-hand side of the equation, rhs. There are also two work arrays, wk1 and wk2, that the subroutine uses. The output solution vector is sol. All of these are one-dimensional arrays with N_z elements.

For example, if $N_z = 5$, the matrix equation would be

$$
\begin{bmatrix}
\text{dia}(1) & \text{sup}(1) & & & \\
\text{sub}(2) & \text{dia}(2) & \text{sup}(2) & & \\
& \text{sub}(3) & \text{dia}(3) & \text{sup}(3) & \\
& & \text{sub}(4) & \text{dia}(4) & \text{sup}(4) \\
& & & \text{sub}(5) & \text{dia}(5)
\end{bmatrix}
\begin{bmatrix}
\text{sol}(1) \\
\text{sol}(2) \\
\text{sol}(3) \\
\text{sol}(4) \\
\text{sol}(5)
\end{bmatrix}
=
\begin{bmatrix}
\text{rhs}(1) \\
\text{rhs}(2) \\
\text{rhs}(3) \\
\text{rhs}(4) \\
\text{rhs}(5)
\end{bmatrix}. \tag{2.21}
$$

The indices, $k = 1$ to N_z, correspond to the z-levels (vertical grid points), where the bottom boundary is at $k = 1$ and the top boundary is at $k = N_z$. Each internal row (i.e., for $k = 2$ to $(N_z - 1)$) corresponds to a different k in Eq. 2.20. That is,

$$\text{sub}(k) = -\frac{1}{(\Delta z)^2},$$

$$\text{dia}(k) = \left(\frac{n\pi}{a} \right)^2 + \frac{2}{(\Delta z)^2},$$

$$\text{sup}(k) = -\frac{1}{(\Delta z)^2},$$

$$\text{rhs}(k) = \text{omg}(k, n),$$

and on output $sol(k)$ will be $psi(k, n)$. Rows $k = 1$ and N_z correspond to the bottom and top boundary conditions on ψ, respectively; that is, $\psi_1 = \psi_{N_z} = 0$. Therefore,

$$dia(1) = 1, \qquad sup(1) = 0, \qquad rhs(1) = 0, \qquad (2.22a,b,c)$$

$$sub(N_z) = 0, \qquad dia(N_z) = 1, \qquad rhs(N_z) = 0. \qquad (2.22d,e,f)$$

Note, since we have chosen impermeable boundaries, which force ψ to vanish on the boundaries at all times, we could reduce the order of the matrix equation to $(N_z - 2)$ by eliminating the bottom and top rows. However, since N_z is usually large compared to 2, the savings would be minimal. Also, if we later wish to implement a condition that allows ψ to be time-dependent on the boundary, instead of constant, the entire N_z rows would be needed in the matrix.

After writing a general tridiagonal solver one should first test it by choosing N_z to be something like 100 and prescribing artificial arrays for sub, dia, sup, and sol using random numbers. Make the absolute values of the elements in the dia array greater than those in the sub and sup arrays; that is, the diagonal needs to be dominant (as they are for our problem) for a tridiagonal solver to be numerically stable. Then multiply (via an inner product) the matrix (composed of sub, dia, and sup) and the solution vector (sol) to get the right-hand side vector (rhs). Then, with the chosen matrix operator and rhs, check that the tridiagonal solver produces a sol that agrees with the originally chosen solution. The subroutine for this tridiagonal solver, which we use in this chapter to solve the Poisson equation for the streamfunction, is also employed in later chapters for improving the numerical method and modifying the problem.

Also, note that the first loop over z-levels in Appendix A, which computes the work arrays wk1 and wk2, does not involve rhs. Therefore, wk1 and wk2 could be computed and saved before the time integration begins instead of being computed every time step. Then wk1 and wk2 would be input arrays and sub, dia, and sup would not be needed for this routine. Only the second set of loops over z-levels in Appendix A, which does involve rhs, would be needed in this subroutine. Doing this would save computational time but would require additional memory to save the wk1 and wk2 arrays, which would then be two-dimensional arrays, i.e., over both z-levels k and mode numbers n.

SUPPLEMENTAL READING

Boyd (2001)
Canuto et al. (1988)
Ferziger & Perić (1997)
Peyret (2002)

EXERCISES

1. *Curl of advection*
   Show in detail how Eqs. 2.2 and 2.3 are obtained.

2. *No x-averaged vorticity or streamfunction*

Show why the vorticity and streamfunction have no $n = 0$ contributions when the four boundaries are impermeable and fixed in space and the fluid is incompressible.

3. *Adams-Bashforth time integration scheme*

Show why the time derivative midway between the current time step and the new time step can be approximated as the time derivative at the current step weighted by 3/2 minus the time derivative at the previous step weighted by 1/2.

Chapter Three

Linear Stability Analysis

In this chapter we describe a linear stability analysis (i.e., solving for the critical Rayleigh number and mode) so readers can check their linear codes against the analytic solution. Dropping the nonlinear terms in Eqs. 2.10 and 2.11 not only simplifies the problem but also redefines the problem. For this linear analysis, each Fourier mode n can be considered a separate and independent problem since the linear terms in Eqs. 2.10–2.12 involve only a single value of n. The question being asked now is under what conditions, i.e., what values of Ra, Pr, and a, will the amplitude of the linear solution grow with time for a given mode n. In other words, this is a *linear stability* problem.

3.1 LINEAR EQUATIONS

By examining the linear versions of Eqs. 2.10–2.12 one sees that although the temperature perturbation, T, is needed in the vorticity equation (2.11), the temperature equation (2.10), which determines the evolution of T, is independent of both the vorticity and the streamfunction. It is only a diffusion equation, which would simply cause any initial temperature perturbation to decay away. This, in turn, would cause the vorticity and streamfunction to also decay away, which would not be very exciting. Therefore, for this linear problem, we consider the x-independent part of the temperature (i.e., the temperature averaged in the x-direction) to be a prescribed background (i.e., reference) state that depends on z but not on time, t. This couples the three equations and allows a stability analysis (e.g., Chandrasekhar, 1961). The idea is that we are interested in how an initial temperature and flow pattern would grow or decay, with the prescribed background temperature profile and the chosen values of the nondimensional parameters, before becoming so large that the nonlinear terms would become important.

Therefore, we first define a simple time-independent and x-independent *conductive* reference state temperature, $T_0(z)$. The subscript 0 means this is the $n = 0$ coefficient in the cosine expansion of the temperature. Recall that ω_0 and ψ_0 are zero everywhere at all times; therefore, since this T_0 is independent of x, the linear parts of Eqs. 2.11 and 2.12 are satisfied for $n = 0$. All terms in the linear part of Eq. 2.10 for $n = 0$ also vanish since T_0 is also prescribed to be time-independent. Therefore, the thermal diffusion term in that equation is simply

$$\frac{\partial^2 T_0}{\partial z^2} = 0.$$

The solution to this equation that satisfies the temperature boundary conditions at the top and bottom boundaries is simply

$$T_0(z) = 1 - z. \tag{3.1}$$

Consequently,

$$\frac{\partial T_0}{\partial z} = -1.$$

This x and t independent background temperature gradient is assumed, for this linear problem, to be large relative to the $n > 0$ perturbations in the temperature gradient. Therefore, when solving for the $n > 0$ linear temperature perturbations, the nonlinear (advection) term in Eq. 2.10 is approximated as a linear term by using only the $n = 0$ temperature gradient part in that term, which, as seen above, is just -1. This reduces the advection term in Eq. 2.10 to simply the cosine coefficient of v_z. That is, for this linear problem, temperature at a given location tends to increase with time when hot fluid flows up from below (because of Eq. 3.1) and decreases with time when cold fluid flows down from above. The upflows and downflows (v_z) are obtained using Eqs. 2.6 for a given mode n:

$$(v_z)_n = \left(\frac{\partial \psi}{\partial x}\right)_n = \left(\frac{n\pi}{a}\right)\psi_n, \tag{3.2}$$

which is the coefficient of the $\cos(n\pi x/a)$ term in the spectral expansion of the temperature equation.

With this linear approximation to the temperature advection term Eq. 2.10 is now coupled to the linear versions of Eqs. 2.11 and 2.12. Therefore, for a given mode $n > 0$, the linear equations describing the z- and t-dependent Fourier coefficients of temperature, vorticity, and streamfunction are

$$\frac{\partial T_n}{\partial t} = \left(\frac{n\pi}{a}\right)\psi_n + \left(\frac{\partial^2 T_n}{\partial z^2} - \left(\frac{n\pi}{a}\right)^2 T_n\right), \tag{3.3}$$

$$\frac{\partial \omega_n}{\partial t} = \text{RaPr}\left(\frac{n\pi}{a}\right)T_n + \text{Pr}\left(\frac{\partial^2 \omega_n}{\partial z^2} - \left(\frac{n\pi}{a}\right)^2 \omega_n\right), \tag{3.4}$$

$$\omega_n = -\left(\frac{\partial^2 \psi_n}{\partial z^2} - \left(\frac{n\pi}{a}\right)^2 \psi_n\right). \tag{3.5}$$

The boundary conditions, as described above, are

$$T_n = \omega_n = \psi_n = 0 \text{ for } z = 0 \text{ and } 1, \quad n > 0. \tag{3.6}$$

The next thing needed for this linear stability problem is to set the initial conditions. Typically, the initial fluid velocity is set to zero, i.e., for each mode n and for all z

$$\omega_n = \psi_n = 0 \text{ at } t = 0.$$

A simple initial condition for the temperature perturbation that satisfies the boundary condition (Eq. 3.6) is

$$T_n = \sin(\pi z) \text{ at } t = 0.$$

Recall that this is for $n > 0$ since the $n = 0$ mode has already been prescribed (Eq. 3.1). Note that since this is a linear problem the amplitude of the initial temperature perturbation is arbitrary as far as the resulting time dependence is concerned.

3.2 LINEAR CODE

When the tridiagonal matrix solver for the system 2.21 is working the main code can be designed and written. (Note the words "code" and "program" are used interchangeably.) Instead of providing a sample computer code that would solve this problem, we only provide guidance; the programming language and the particular design of the code are up to the reader/modeler.

The parameters representing the number of z-levels, N_z, and the number of Fourier modes, N_n, need to be chosen and set. One-dimensional real arrays for z, sub, dia, sup, wk1, and wk2 need to be dimensioned as $(1:N_z)$. In addition, we need two-dimensional real arrays $(1:N_z,0:N_n)$ for psi, omg, and tem. Note there is no dimension for time since we plan to overwrite the values of these variables every time step instead of keeping a history of them, which would obviously require a large amount of storage. We also need three-dimensional real arrays $(1:N_z,0:N_n,1:2)$ for domgdt and dtemdt. These arrays store the time derivatives of the vorticity and temperature, respectively, with the third dimension indicating that the value is for the previous time step (1) or for the current step (2). Recall that these two time levels are needed for the Adams-Bashforth time integration scheme, Eq. 2.18.

Typically a code like this begins by specifying or asking the user for the values of the defining parameters, Ra and Pr, the aspect ratio, a, the size of the time step, dt, and the number of time steps to be run, nsteps. Also, combinations of constants that appear in the equations and are used each time step should be calculated once and stored. For example, pi $=$ 4.*atan(1.), which is an easy way of setting the value of π to machine roundoff. Also define a constant c=pi/a and the uniform spacing between z-levels, dz $=$ 1./(N_z-1), and a constant oodz2=1/dz^2. Set the value of a computational time step, dt, according to the constraint 2.19 or this divided by Pr if Pr > 1; to be safe, set dt to something like 90% of this limit. In addition, since the sub and sup parts of the tridiagonal matrix never change, they can be set once at the beginning of the code. Note sup(N_z) and sub(1) are never used.

The initial conditions for omg and psi, for all z and n, are usually set to zero, i.e., no initial velocity. A simple initial condition for tem, which satisfies its boundary conditions, is tem(k, n)=sin(pi*z(k)), where the mode number $n = 1$ to N_n and the z-level z(k) $= (k - 1)$*dz with $k = 1$ to N_z. The values of dtemdt$(k, n, 1)$ and domgdt$(k, n, 1)$ will be needed for the first time step. The easiest way to deal with this is to just set these to zero since the initial conditions are arbitrary anyway.

The rest of the code is a loop over the time steps. The first part of each step involves computing dtemdt$(k, n, 2)$ and domgdt$(k, n, 2)$ in a nested loop over all internal z-levels ($k = 2$ to $N_z - 1$) and all nonzero mode numbers ($n = 1$ to N_n). These are updated according to Eqs. 3.3 and 3.4 using the finite-difference approximation, Eq. 2.16, for the second derivatives in z.

Recall that each n is an independent stability problem since we have dropped the nonlinear terms. Therefore, instead of doing all N_n problems simultaneously, we could write the code to do one at a time. However, since we will need to compute all modes simultaneously later when we do include the nonlinear terms, we recommend also calculating all modes simultaneously for this linear problem. Typically though N_n is chosen to be much smaller for the linear problem, something like twice the aspect ratio.

Now, update tem and omg to the new time step with another nested loop over k and n using their respective time derivatives at the current and previous time steps via Eq. 2.18. After updating these, update the streamfunction, psi, with a loop over n that first constructs the array dia, which depends on n, and then calls the tridiagonal matrix solver subroutine with the updated omg being the rhs on input and psi being the sol on output.

To prepare for the next time step, make another nested loop over k and n to copy the current $(k, n, 2)$ values of the time derivatives into the previous location $(k, n, 1)$ of these arrays. This can be avoided by using an even number of time steps (nstep) and keeping track of which value in the third dimension represents the current and previous steps.

Finally, some output is needed to let the user know how the amplitudes of tem, omg, and psi are changing in time. An easy way to do this is to print out (to the screen or to a file) the values of these three variables for all n but only for one z-level, at, for example, the mid-depth or one-third depth.

3.3 CRITICAL RAYLEIGH NUMBER

The plan now, for this linear stability problem, is to find the value of the Rayleigh number, Ra, for a given aspect ratio, a, and Prandtl number, Pr, that gives a solution that neither grows nor decays with time, i.e., the conditions for "marginal stability." This is called the critical Rayleigh number, $Ra_{crit}(n)$, for mode n. That is, if one sets $Ra > Ra_{crit}(n)$ the result will be a "supercritical" solution with amplitudes $T_n(z, t)$, $\omega_n(z, t)$, and $\psi_n(z, t)$ that increase exponentially with time. Likewise, setting $Ra < Ra_{crit}(n)$ will produce an exponentially decaying "subcritical" solution. Different modes, n, have different $Ra_{crit}(n)$. The mode number with the smallest $Ra_{crit}(n)$ is the critical mode number, n_{crit}, and $Ra_{crit}(n_{crit})$ is the overall critical Rayleigh number for the chosen Pr and a. Actually, n_{crit} and Ra_{crit} are independent of Pr for this problem.

The idea is to choose a Pr and an a and then try many values of Ra in an attempt to bracket the $Ra_{crit}(n)$. The closer one gets to the critical Rayleigh number the more time steps will be needed to decide if the solution is increasing or decreasing. In practice, one estimates the critical Rayleigh number by setting it to the value midway between an Ra that produces a solution that just barely increases with time and one that gives a solution that just barely decreases with time.

Note that in either case the absolute value of tem will initially decrease and that of omg and psi will increase as thermal energy converts to kinetic energy; nstep therefore needs to be large enough (typically several thousand) for the solution, i.e., tem, omg, and psi, to evolve beyond this initial transient and all be either increasing or decreasing with time at the same (relative) exponential rate. To monitor the rate one could print the amplitudes of tem, omg, and psi at, for example, one-third depth for each mode n. A more convenient set of numbers to print would be

$$\ln(|\text{tem}(N_z/3, n)_{current}|) - \ln(|\text{tem}(N_z/3, n)_{previous}|)$$

and likewise for omg and psi. Here, "previous" means the value saved from the previous time this was checked, which the user needs to specify; for example, this might be checked once every 500 computational time steps. (Note, one could just print the ratios of the previously checked to current values; however, often truncation errors prevent accurate estimates of these ratios when the function values become very (exponentially) small or large.) These three natural logarithm values should converge to the same time-independent value, which if positive would indicate an exponentially growing solution and if negative, an exponentially decaying solution. The closer the Ra is to the critical value the more time steps will need to be run to determine if the solution is increasing or decreasing. For example, one may need to run 100,000 numerical time steps or more with $\Delta t = 10^{-5}$.

In addition, since all modes n are computed simultaneously, it is possible to choose a value for Ra that is supercritical for one or more modes but subcritical for the others. That is, the amplitudes of some of the modes may be increasing while those for the other modes are decreasing. As mentioned, the first mode to increase with time as Ra is increased is the critical mode number for the chosen aspect ratio.

Another important point is that for many cases, depending on the spatial resolution and the chosen Ra, Pr, and a, single precision arithmetic, i.e., using 4-byte real numbers, may not work for this problem. It is highly recommended that all computations suggested in this book be performed in double precision, i.e., with 8-byte real numbers. Integers, on the other hand, usually need only 4 bytes. Also, if the ratios mentioned above do not converge to the same constant value for a given Ra and a (and if the code is correctly written), the problem could be that more resolution in the z-direction is needed, i.e., N_z would need to be increased, although usually $N_z = 100$ should be adequate.

3.4 ANALYTIC SOLUTIONS

Now we show that the linear stability problem can be solved much more efficiently using an analytic (Rayleigh 1916) approach instead of employing the numerical method described in Section 3.3. Of course, if the linear problem were our only objective, we would not have bothered to develop the linear code that simulates a solution's time dependence. Our intent here though is to check the accuracy of the linear code by comparing the critical Rayleigh numbers and mode numbers it predicts with the analytic results before adding the nonlinear terms to the code.

To find an expression for the value of $\text{Ra}_{crit}(n)$ for a given aspect ratio and mode number set the time derivatives in Eqs. 3.3 and 3.4 to zero since we wish to find the Ra for which the n-mode solution is neither increasing nor decreasing with time.

Note, we are assuming that, for our defined problem, convection at Ra = $\text{Ra}_{crit}(n)$ occurs as a steady-state solution, i.e., not with an oscillatory time dependence as can occur when there is double-diffusion (Section 7.3) or rotation (Section 12.4.1). In addition, instead of using a finite-difference approximation for the second-order derivatives in z, Fourier expand the temperature, vorticity, and streamfunction coefficients in z. This is now a fully spectral method. Since all three of these vanish at both boundaries, only the sine functions survive; so we have

$$T_n(z, t) = \sum_{m=1}^{N_m} T_{nm}(t) \, \sin(m\pi z), \qquad (3.7a)$$

$$\omega_n(z, t) = \sum_{m=1}^{N_m} \omega_{nm}(t) \, \sin(m\pi z), \qquad (3.7b)$$

$$\psi_n(z, t) = \sum_{m=1}^{N_m} \psi_{nm}(t) \, \sin(m\pi z), \qquad (3.7c)$$

where these coefficients depend on the x-mode number n, the z-mode number m, and time.

Next substitute these expressions into Eqs. 3.3–3.5 with the time derivatives set to zero and the z-derivatives now done analytically. The orthogonality property of sine functions allows one to examine each nm-mode separately. That is, there is a set of three algebraic equations and three unknowns (T_{nm}, ω_{nm}, ψ_{nm}) for each set of mode numbers n and m. This set can easily be solved, resulting in the following expression for the critical Rayleigh number:

$$\text{Ra}_{crit}(n, m) = \left(\frac{\pi}{a}\right)^4 \frac{\left(n^2 + (am)^2\right)^3}{n^2}. \qquad (3.8)$$

It should be noted that this convenient analytic solution exists because of the simple geometry and convenient boundary conditions we have chosen. In general an eigenvalue problem needs to be solved numerically to find Ra_{crit} as a function of mode numbers (see, for example, Glatzmaier & Gilman, 1981a).

For our problem, however, Eq. 3.8 is quite useful. For example, a square box ($a = 1$) with a single convection cell ($n = m = 1$) has $\text{Ra}_{crit} = 2^3\pi^4 = 779.27$. Note, for a given aspect ratio a and x-mode number n, the z-mode number m with the smallest Ra_{crit} is $m = 1$. That is, the most unstable convective pattern is one for which the vertical trajectories extend continuously from the bottom to the top boundary, which maximizes the upward convective heat flow between these boundaries. Therefore, Eq. 3.8 for $m = 1$ can be used to check if a linear numerical code is working correctly. Notice that for this linear stability problem the critical Rayleigh number does not depend on the Prandtl number Pr; however, the style of the nonlinear supercritical solutions (Chapter 4) does depend on the Pr.

The Rayleigh number is a measure of how strong convective driving is relative to viscous and thermal diffusion, which hinder the convection. It is related to the ratio of the convective heat flux to the conductive (i.e., diffusive) heat flux before the onset of convection. The conductive heat flux is the rate thermal energy is transported, per unit time per cross-sectional area, due to thermal diffusion, i.e., molecular motions (or unresolved subgrid-scale turbulence) transporting energy from hot to cold regions. In the vertical direction, it is computed as $-k\,\partial T/\partial z$, where the constant of proportionality, k, is the thermal conductivity, equal to $c_p\rho_o\kappa$; c_p is the specific heat capacity at constant pressure. The negative sign simply means that heat diffuses "down gradient", i.e., from hot to cold. An estimate of the amplitude of this flux is therefore $c_p\rho_o\kappa\,\Delta T/D$. The convective heat flux is the amount of thermal energy transported upward, per unit time per cross-sectional area, by convection, i.e., by hot fluid rising and cold fluid sinking. To estimate this term we need an estimate of what a typical convective velocity will be. One way to do this is to assume the buoyancy and viscous forces will be comparable. That is, $\rho g_o = \nu\rho_o\partial^2 v_z/\partial z^2$. Then, using $\rho_o\alpha\,\Delta T$ to estimate the amplitude of the density perturbation, ρ, and using u/D^2 to estimate the amplitude of $\partial^2 v_z/\partial z^2$, the estimate of the convective velocity is $u \approx \alpha\Delta T g_o D^2/\nu$. Using this to estimate a typical vertical velocity, v_z, and using ΔT to estimate a typical temperature perturbation, T, the convective heat flux, $c_p\rho_o v_z T$, would be estimated to be $c_p\rho_o\alpha\,\Delta T^2 g_o D^2/\nu$. This, together with the above estimate for the conductive flux, makes the ratio of these two estimated heat fluxes equal to what is defined in Eq. 1.18 as the Rayleigh number, Ra. However, this description is more relevant for the nonlinear problem for which amplitudes are meaningful. Also, it applies to the bulk of the fluid where there is a significant convective heat flux.

To find the critical mode number, n_{crit}, for a given aspect ratio, a, set the derivative with respect to n of Eq. 3.8 to zero, which gives, for $m = 1$,

$$n_{crit} = \text{nearest integer to } \frac{a}{\sqrt{2}}. \tag{3.9}$$

The most unstable horizontal mode number is approximately equal to the aspect ratio because the preferred number of convection cells in the box depends on the shape of the box. Circulation cells that extend from the bottom to the top of the box ($m = 1$) are typically more unstable, i.e., grow faster, because buoyancy forces act over more of the fluid trajectory than they would if there were multiple cells in the z-direction, which would require additional horizontal flow driven by pressure gradients instead of buoyancy forces. Also, in general, the cells that have a more circular shape are the easiest to excite because the flow has less curvature and so less viscous resistance to buoyancy. In addition, a circular flow minimizes diffusion of temperature perturbations, which are needed to provide buoyancy forces. On the other hand, in order to limit the number of upwellings and downwellings, the preferred aspect ratio of a single cell tends to be slightly greater than unity. Therefore, for cells that stretch from the bottom to the top, the most unstable mode number, n, is expected to be slightly less than the aspect ratio of the box, a, as stated in Eq. 3.9. For example, if the length of the box were five times its height ($a = 5$), one would expect slightly fewer than five convection cells in the x-direction ($n_{crit} = 4$). For more discussion and details on this linear stability problem see, for example, Chandrasekhar (1961).

Note that an aspect ratio of $a = \sqrt{2}$ makes n_{crit} exactly 1 (Eq. 3.9) and gives the smallest critical Rayleigh number:

$$\mathrm{Ra}_{crit} = \frac{27\pi^4}{4} = 657.5. \tag{3.10}$$

In summary, to find the overall Ra_{crit} for a given aspect ratio, first use Eq. 3.9 to find n_{crit} and then use this value of n in Eq. 3.8 (with $m = 1$). These analytic predictions can then be compared with estimates of the critical Rayleigh numbers obtained using the linear code. Assuming there are no bugs in the code, better agreement will be obtained by using better resolution in the z-direction, i.e., a higher value of N_z, which will reduce artificial (numerical) diffusion.

SUPPLEMENTAL READING

Chandrasekhar (1961)

EXERCISES

1. *Critical Rayleigh number*
   Derive the analytic expression for the critical Rayleigh number, Eq. 3.8, by substituting Eqs. 3.7a–c into Eqs. 3.3–3.5 and setting all time derivatives to zero.
2. *Critical Rayleigh numbers vs. aspect ratio and horizontal mode number*
   Using Eq. 3.8 with $m = 1$, plot Ra_{crit} as a function of the aspect ratio, $0.1 < a < 10$, for horizontal mode numbers $n = 1, 3$, and 10; and plot Ra_{crit} as a function of the horizontal mode number (an integer), $1 < n < 10$, for aspect ratios $a = 0.1$, $\sqrt{2}$, and 10.

COMPUTATIONAL PROJECTS

1. *Critical Rayleigh numbers via a linear convection code*
   Compare the critical Rayleigh numbers for $\mathrm{Pr} = 0.3$, $a = 5$ and $n = 1, 3$, and 10 obtained using a linear convection code to those obtained using Eq. 3.8 with $m = 1$. Demonstrate, using the linear convection code, that the critical Rayleigh numbers are independent of the value of Pr.
2. *Overall critical Rayleigh number via a linear convection code*
   Compare the critical Rayleigh number, Ra_{crit}, at the critical mode number, n_{crit}, for $\mathrm{Pr} = 0.3$ and $a = 5$ obtained using a linear convection code to that obtained using Eqs. 3.9 and 3.8 with $m = 1$.

Chapter Four

Nonlinear Finite-Amplitude Dynamics

Now the nonlinear terms are added to produce finite-amplitude simulations. Here we choose to calculate the nonlinear terms using a Galerkin method in spectral space; this is replaced with the more efficient spectral-transform method in Chapter 10. However, a Galerkin method provides a clear understanding for how nonlinear terms disperse energy among the available modes and what is meant by spectral aliasing. For a review of spectral methods see, for example, Canuto et al. (1988), Boyd (2001), and Peyret (2002).

4.1 MODIFICATIONS TO THE LINEAR MODEL

The linear solution approximates only the initial growth of supercritical convection that begins with small temperature perturbations. Very quickly the amplitudes become large enough that the nonlinear terms become significant and stop the exponential growth. This is what is called a *finite amplitude* simulation since the amplitudes of the variables are no longer arbitrary. For example, doubling all amplitudes of a solution would no longer satisfy the set of equations because the amplitudes of the linear terms would increase by 2, whereas those of the nonlinear terms would increase by 4. For values of Ra below a certain limit the nonlinear solution evolves to a steady state, i.e., the time derivatives go to zero and all variables become constant in time. A more realistic and interesting situation occurs with larger values of Ra, which produce time-dependent solutions.

The set of coupled nonlinear equations 2.10–2.12 require expressions for the Fourier coefficients of the nonlinear terms, which describe the advection of temperature and vorticity. The x-independent ($n = 0$) part of the temperature perturbation is now also part of the solution, instead of being prescribed as was done for the linear problem. That is, the z-dependent horizontal mean of the temperature now can evolve away from its initial condition (except for its constant boundary values). Recall that at the bottom and top boundaries T_n, ω_n, and ψ_n are zero for $n > 0$. Our isothermal boundary conditions require the temperature for $n = 0$, T_0, to remain one at the bottom boundary and zero at the top boundary.

Since the full advection of temperature is now part of the temperature equation, including the advection of $T_0(z, t)$, we no longer need the linearized version of this term that we added for the linear stability problem. An alternative approach would be to retain this term and consider $T_0(z, t)$ to be the horizontal mean of the temperature perturbation relative to the time-independent conductive profile, $(1 - z)$, in which case this $T_0(z, t)$ would also vanish at the two boundaries.

However, since there is no advantage in doing this, we continue to define $T_0(z, t)$ to be the horizontal mean of the temperature perturbation relative to a constant (in time and space) background temperature. Therefore, the term $(n\pi/a)\psi_n$, which appears in the linear equation 3.3, needs to be removed in the numerical code that is now being modified for the nonlinear problem.

4.2 A GALERKIN METHOD

Now we describe how to add the nonlinear terms to the code. Note that the nonlinear terms involve products of two variables, each of which are expanded in either sines or cosines in x. However, what we need are expressions that involve just sines for Eqs. 2.11 and 2.12 and just cosines for Eq. 2.10. The following set of trigonometric identities provides exactly what is needed:

$$\sin(a)\sin(b) = (\cos(a - b) - \cos(a + b))/2, \tag{4.1a}$$

$$\cos(a)\cos(b) = (\cos(a - b) + \cos(a + b))/2, \tag{4.1b}$$

$$\sin(a)\cos(b) = (\sin(a + b) + \sin(a - b))/2. \tag{4.1c}$$

Consider the term representing the advection of vorticity:

$$-(\mathbf{v} \cdot \nabla)\omega = -\left[v_x \frac{\partial \omega}{\partial x} + v_z \frac{\partial \omega}{\partial z} \right]$$

$$= -\left[-\frac{\partial \psi}{\partial z} \frac{\partial \omega}{\partial x} + \frac{\partial \psi}{\partial x} \frac{\partial \omega}{\partial z} \right].$$

Using the expressions 2.6 for v_x and v_z in terms of derivatives of ψ and writing ω and ψ in terms of their Fourier expansions, Eqs. 2.8b,c, makes the advection of vorticity a double summation over two mode numbers, which we will call n' for the Fourier expansion of ω and n'' for the Fourier expansion of ψ. Each product of a sine and cosine in this double summation is then written as a sum of two sines using Eq. 4.1c. The result is

$$-(\mathbf{v} \cdot \nabla)\omega = -\sum_{n'=1}^{N_n} \sum_{n''=1}^{N_n}$$

$$\times \left[-\left(\frac{n'\pi}{2a} \right) \frac{\partial \psi_{n''}}{\partial z} \omega_{n'} \left(\sin((n'' + n')\pi x/a) + \sin((n'' - n')\pi x/a) \right) \right.$$

$$\left. + \left(\frac{n''\pi}{2a} \right) \psi_{n''} \frac{\partial \omega_{n'}}{\partial z} \left(\sin((n'' + n')\pi x/a) - \sin((n'' - n')\pi x/a) \right) \right], \tag{4.2}$$

which nicely gives us only sines like the other terms in the vorticity equation. However, what we need are the Fourier coefficients in a single summation over sines for this nonlinear term, i.e.,

$$-(\mathbf{v} \cdot \nabla)\omega = -\sum_{n=1}^{N_n} [(\mathbf{v} \cdot \nabla)\omega]_n \sin(n\pi x/a). \tag{4.3}$$

By comparing Eqs. 4.2 and 4.3 one can see that

$$
-[(\mathbf{v} \cdot \nabla)\omega]_n = -\frac{\pi}{2a} \sum_{n'=1}^{N_n} \sum_{n''=1}^{N_n} \left[\left(-n' \frac{\partial \psi_{n''}}{\partial z} \omega_{n'} + n'' \psi_{n''} \frac{\partial \omega_{n'}}{\partial z} \right) \delta_{n''+n',n} \right.
$$
$$
\left. - \left(n' \frac{\partial \psi_{n''}}{\partial z} \omega_{n'} + n'' \psi_{n''} \frac{\partial \omega_{n'}}{\partial z} \right) (\delta_{n''-n',n} - \delta_{n'-n'',n}) \right], \qquad (4.4)
$$

where the kronecker delta function $\delta_{i,j}$ is one for $i = j$ and zero if not.

This strategy of computing the contribution to the nonlinear terms for each mode due to the binary interactions of many other modes is called a *Galerkin* method. That is, the spectral coefficients for the nonlinear terms are computed in spectral (n) space without ever going to physical (x) space. (Note the method here is always in physical space in terms of the z coordinate.) Chapter 10 describes a more efficient method for calculating the nonlinear terms called a spectral-transform method, for which variables are transformed to physical space to compute the nonlinear terms each time step. However, working with the Galerkin method should provide an understanding of how the interaction of two mode numbers n' and n'' via advection produces contributions to mode numbers $(n' + n'')$ and $\pm(n' - n'')$.

Now consider the other nonlinear term, the advection of temperature:

$$
-(\mathbf{v} \cdot \nabla)T = -\left[v_x \frac{\partial T}{\partial x} + v_z \frac{\partial T}{\partial z} \right]
$$
$$
= -\sum_{n=0}^{N_n} [(\mathbf{v} \cdot \nabla)T]_n \cos(n\pi x/a). \qquad (4.5)
$$

The Fourier coefficients for this term are evaluated in a manner similar to those for the advection of vorticity. One difference is that now we hope to end up with a series of cosines (Eq. 4.5), like the other terms in the temperature equation. We will again call the ψ mode numbers n''; those for the temperature will be n'. Recall that the cosine expansion of temperature includes $n' = 0$. We treat this contribution separately so the rest of the n' contributions can be treated in a similar way and at the same time as those for the vorticity advection. The result is

$$
-[(\mathbf{v} \cdot \nabla)T]_n = -\frac{n\pi}{a} \psi_n \frac{\partial T_0}{\partial z} - \frac{\pi}{2a} \sum_{n'=1}^{N_n} \sum_{n''=1}^{N_n} \left[\left(-n' \frac{\partial \psi_{n''}}{\partial z} T_{n'} + n'' \psi_{n''} \frac{\partial T_{n'}}{\partial z} \right) \delta_{n''+n',n} \right.
$$
$$
\left. + \left(n' \frac{\partial \psi_{n''}}{\partial z} T_{n'} + n'' \psi_{n''} \frac{\partial T_{n'}}{\partial z} \right) (\delta_{n''-n',n} + \delta_{n'-n'',n}) \right], \qquad (4.6)
$$

where the $(\delta_{n''-n',n} + \delta_{n'-n'',n})$ should be replaced by just $\delta_{n''-n',0}$ for $n = 0$; otherwise each of the $n' = n''$ cases would be counted twice. Notice that the first term on the right of Eq. 4.6 is the $n' = 0$ contribution from the double summation, for which the only nonzero contribution occurs when $n'' = n$. This term is the same as the linearized advection term used in Section 4.2, except that now $\partial T_0/\partial z$ is allowed to evolve.

This Galerkin method works nicely here because of the simple geometry and convenient boundary conditions we have chosen. A Galerkin method could also be used to calculate the nonlinear terms in, for example, a 3D spherical convection problem for which the horizontal structures are expanded in spherical harmonics (Section 10.6); but the process would be even more difficult and less efficient.

4.3 NONLINEAR CODE

Constructing a nonlinear code by adding the nonlinear terms to a linear code requires planning and care. There is no one way to do this; the goal should be to produce a clearly organized code that runs efficiently.

4.3.1 Initial Conditions

At the beginning of the nonlinear code the input parameters could be provided in, for example, a *namelist* or *module* format if using Fortran. Besides those already mentioned for the linear case, a logical parameter should be input that tells the code if this run should start from random initial conditions or by opening an input data file and reading in the data from a previous run. If the former is the case, again it would be good to set the initial values for $omg(k, n)$ and $psi(k, n)$ to zero so the total integrated momentum is zero and will remain zero since the boundaries are stress-free. (Note, the momentum would need to be integrated over the convecting box and its mirror image by doubling the value of a and replacing $\pi nx/a$ with $2\pi nx/a$ for the arguments for the sines and cosines.) The initial conditions for $tem(k, n)$ are no longer needed for every mode number n because the nonlinear terms will quickly distribute energy among the other modes. However, if only modes 0 and, say, 10 were initialized (all others having zero amplitude), then only mode numbers that are multiples of 10 would be amplified. Therefore, it is important to initialize at least $n = 1$ or, alternatively, two mode numbers that differ by one. The initial profile for $tem(k, 0)$ could be the same as what was prescribed for it in the linear case, $(1 - z)$, which satisfies the boundary conditions; it could also be $(1 - z)$ to some power. Likewise, the nonzero mode numbers that are initialized as small perturbations could have a $\sin(\pi z)$ vertical profile, multiplied by a random number between -1 and 1 and then by a constant much less than unity since now the amplitudes are important. That is, the initial nonzero values represent small temperature perturbations relative to the order-one value of the x-averaged initial temperature profile.

4.3.2 Nonlinear Terms

There are several ways to gather the contributions to $-[(\mathbf{v} \cdot \nabla)\omega]_n$ and $-[(\mathbf{v} \cdot \nabla)T]_n$ for each mode n at each z-level and time step. However, one does not want to actually do the double summations over both n' and n'' for each n because this would involve N_n^3 "if" statements for every z-level and time step, especially since most of the kronecker delta functions would be zero. In fact, for a given n and, say, n' there will be no more than three values of n'' that make a nonzero contribution.

So, here is one way to add these nonlinear terms, starting with a copy of the linear code. The loop over the nstep time steps begins, as before, by computing the linear parts of dtemdt$(k, n, 2)$ and domgdt$(k, n, 2)$ in a nested loop over all internal z-levels ($k = 2$ to $N_z - 1$) and all mode numbers ($n = 0$ to N_n); the loop over n should be the inner loop. Note that now $n = 0$ is included; so one might change all loops that started with $n = 1$ in the linear code to now start with $n = 0$ for the nonlinear code. However, only dtemdt gets a nonzero contribution for the $n = 0$ mode. Also, as mentioned above, the linearized advection term in Eq. 3.3 is dropped here.

After this loop over n, but still inside the loop over k, we want to again loop over n to add in the nonlinear terms. First we take care of $n = 0$, which applies only to dtemdt and only for $n'' = n'$. So in the code add

$$-\frac{\pi}{2a} \sum_{n'=1}^{N_n} n' \left(\frac{\partial \psi_{n'}}{\partial z} T_{n'} + \psi_{n'} \frac{\partial T_{n'}}{\partial z} \right)$$

to the current dtemdt$(k, 0, 2)$ again using Eq. 2.15 to approximate the z-derivatives.

Next we take care of the other values of n, from 1 to N_n, while still inside the loop over k. Inside this loop over n we again need to loop over n'; but we treat $n' = 0$ separately since it applies only to dtemdt and only for $n'' = n$. That is, in the code add

$$-\frac{n\pi}{a} \psi_n \frac{\partial T_0}{\partial z}$$

to dtemdt$(k, n, 2)$. Then we loop over the others, $n' = 1$ to N_n. Within this loop we have just three values of n'' (for a given n and n') that contribute to both dtemdt and domgdt according to the expressions inside the double summations of Eqs. 4.4 and 4.6 because of the kronecker delta functions. These are $n'' = n - n'$, $n'' = n + n'$, and $n'' = n' - n$. The appropriate parts of these expressions need to be added to dtemdt$(k, n, 2)$ and domgdt$(k, n, 2)$. Note for each of these three cases the value of n'' needs to be checked to make sure that it is greater than or equal to 1 and less than or equal to N_n; otherwise there is no contribution.

With the nonlinear terms now computed, the loop over z-levels (k) is complete, as are the time derivatives of ω and T for the current time step. Next, as in the linear code, ω_n and T_n are updated for all n at all z-levels using the time derivatives at the current and previous time steps. Then, since Eq. 2.12 is linear, ψ_n is updated the same way it is in the linear code using the tridiagonal solver. Also, as in the linear code, set dtemdt$(k, n, 1)$=dtemdt$(k, n, 2)$ and similarly for domgdt to prepare for the next time step (unless the even and odd time steps are correlated with the third index in these arrays).

4.3.3 Calculating and Saving Data

Information about the solution is now gathered in the remainder of the loop over time steps. Instead of printing out values of the variables at mid-depth for each mode number n, one is usually more interested in knowing the maximum amplitudes of these variables, which can occur anywhere inside the box, and their time dependencies if any. One could print the maximum amplitudes for $T_n(z)$, $\omega_n(z)$, and $\psi_n(z)$ and the values of the n and z for which they occur every iprint time steps, where iprint is a user-specified integer.

Data should also be saved that could be used later for postprocessing to analyze the simulation and to make movies. Again the user would need to specify how frequently this data should be constructed and saved. One could save a different file for each movie time step, incorporating the movie step in the name of the file. Alternatively, all the data snapshots could be saved in one file per run.

One also needs to decide whether to save the data as z-dependent Fourier coefficients (z_k, n) or to save the data as fields in purely physical space (x_i, z_k), where the x coordinate is

$$x_i = a \, (i-1)/(N_x - 1) \quad \text{for } i = 1, N_x.$$

The number of grid points in the x-direction, N_x, could be anything but is typically set to something like $a N_z$; it should be set to at least $2N_n$ to "see" the smallest wavelengths, i.e., largest mode numbers n. The latter choice would require constructing and storing $T(x, z)$, $\omega(x, z)$, and $\psi(x, z)$ using Eqs. 2.8, which may be easier to postprocess (see Chapter 5). To avoid calculating all the sines and cosines each time step, which is computationally expensive, these should be precalculated and stored in arrays. The actual velocities, $v_x(x, z)$ and $v_z(x, z)$, could also be constructed, using Eqs. 2.6, and stored in the postprocessing output files. If, on the other hand, the z-dependent Fourier coefficients are stored, the horizontal resolution, N_x, can be decided later. This would also allow a convenient and accurate spectral analysis and the construction of horizontal derivatives. Of course, if the data were stored in x, z-space, the Fourier coefficients could be recovered later with an inverse Fourier transform.

4.3.4 Checking and Changing the Time Step

Besides significantly increasing the computational expense, the nonlinear advection terms add an additional constraint on the size of the computational time step. The *Courant-Friedrichs-Lewy* (CFL) condition (Courant et al., 1928; Durran, 1998) essentially requires the time step to be less than the minimum time a fluid parcel takes to flow between any two grid points for the advection operation to be numerically stable. That is,

$$\Delta t < \Delta z / |v_z|_{max}. \tag{4.7}$$

There is a similar constraint in the x-direction with Δx being approximately a/N_n.

A simple way to understand this constraint is to consider just the one-dimensional advection equation:

$$\frac{\partial f}{\partial t} + v_o \frac{\partial f}{\partial x} = 0; \tag{4.8}$$

that is, the variable $f(x, t)$ is advected by a constant velocity v_o in the x-direction. This velocity, v_o, could be positive or negative and would be the sum of the fluid velocity and the sound speed if we were using the fully compressible equations

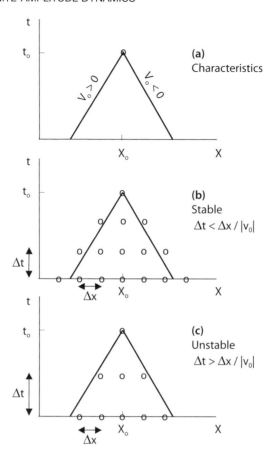

Figure 4.1 (a) A schematic of two characteristics for an advection equation. (b) An example of a stable case for which the CFL condition is satisfied. (c) An example of an unstable case for which the CFL condition is not satisfied; Δx is the same as in (b) but Δt is larger. Circles on (b) and (c) represent points in space-time at which the numerical solution is calculated.

(Chapter 1) or just the fluid velocity when using the Boussinesq (or anelastic) equations. The solution is any pattern that is a function of $(x - v_o t)$, i.e., a pattern that appears constant in time to an observer moving at velocity $dx/dt = v_o$. Consider a point, (x_o, t_o), in the x, t-plane that satisfies this advection equation for a given initial condition. The solution then lies on the "characteristic,"

$$t = t_o + (x - x_o)/v_o, \qquad (4.9)$$

which is illustrated with the pair of solid lines in Fig. 4.1a. The "domain of dependence" of point (x_o, t_o) is the set of space-time points on which the solution existed prior to time t_o, i.e, it is the set of x and t that satisfy Eq. 4.9 for all $t \le t_o$. Now consider a numerical method that uses only nearest-neighbor grid points in x and t to update the solution each time step, like centered spatial and temporal finite differences. The CFL constraint requires the domain of dependence of point (x_o, t_o), i.e., the solid lines in Fig. 4.1, to be inside the *numerical* domain of dependence, i.e., the set of discrete grid points marked as circles in Fig. 4.1, which have influenced the solution at (x_o, t_o). This is satisfied for the numerically stable case illustrated in Fig. 4.1b, where $\Delta t \leq \Delta x / |v_o|$, and is not satisfied for the unstable case illustrated in Fig. 4.1c, where $\Delta t \geq \Delta x / |v_o|$. Note that the CFL condition, Eq. 4.7, is a necessary but not sufficient condition for numerical stability. Also, this condition can be slightly more constraining for higher order numerical methods that use more than just nearest-neighbor grid points in space and time.

If it is likely that the CFL constraint could be violated during a simulation, it would need to be monitored, in both directions, and the time step would need to be reduced whenever the CFL limit becomes less than, say, 120% of the time step. Likewise, the time step should not be too small because many more computational time steps would be computed than needed and because each update of the variables would be less accurate since the changes, added to the previous values, would be much smaller. Therefore, the time step should be increased whenever the CFL limit becomes greater than, say, five times the time step. Near the end of the loop over time steps would be a good place to monitor the CFL constraint and change the time step when needed. Since this involves transformations from spectral to grid space using Eqs. 2.6 and 2.8c and then finding the maximum absolute value of v_z and since the numerical solution does not change significantly in one time step, it is not usually necessary to check the CFL condition every time step; checking it once every 10 steps may be sufficient depending on how time-dependent the solution is.

However, the size of the time step needs to be smaller than both the diffusion limit, Eq. 2.19, and the CFL limit, Eq. 4.7. Since the former is proportional to Δz^2 whereas the latter is proportional to Δz, the former is usually more severe. The exception would be if a relatively large Rayleigh number were prescribed for a given spatial resolution that produces fluid velocities more than four times larger than the grid diffusion velocity ($1/\Delta z$, or in dimensional values, $\nu/\Delta z$ or $\kappa/\Delta z$). This may only happen shortly after the initial phase of a simulation when the fluid velocity increases exponentially in time and peaks before dropping back down to some statistically steady value. In such a case, a CFL check and limit on the time step may only be needed for this initial startup phase. A simple alternative to implementing a CFL check would be to have the user provide an input value for Δt and then check that it at least satisfies the diffusion limit, Eq. 2.19. Then, by trial and error, the user can determine how small it needs to be to avoid a numerical instability for a given set of Ra, Pr, and a. In Chapter 8 we describe an implicit time integration scheme, which avoids the diffusion constraint on the time step; then it is necessary to monitor the CFL constraint and change the time step when needed. As mentioned above, in Chapter 10 we describe the spectral-transform method for computing the nonlinear terms, for which v_x and v_z are calculated every time step; so monitoring the CFL condition then requires little extra work.

Technically, whenever the time step is changed the weight coefficients, 3/2 and $-1/2$, in Eq. 2.18 should be changed for the first step with the new Δt since these values estimate the time derivative at the midway point between t and $t + \Delta t$

assuming a constant Δt. Consider the case when the previous, current, and future times are $(t - \Delta t_{old})$, t and $(t + \Delta t_{new})$, respectively, with $\Delta t_{new} \neq \Delta t_{old}$. In the notation employed in Eq. 2.17, we want

$$\frac{T_{t+\Delta t_{new}} - T_t}{\Delta t_{new}} = c_1 \, G_t - c_2 \, G_{t-\Delta t_{old}} \, .$$

The linear slope of G between $(t - \Delta t_{old})$ and t is $(G_t - G_{t-\Delta t_{old}})/\Delta t_{old}$ and we want to extrapolate to $G_{t+\frac{1}{2}\Delta t_{new}}$. Therefore,

$$G_{t+\frac{1}{2}\Delta t_{new}} = G_t + \frac{(G_t - G_{t-\Delta t_{old}})}{\Delta t_{old}} \left(\frac{\Delta t_{new}}{2} \right)$$

$$= \left(1 + \frac{\Delta t_{new}}{2\Delta t_{old}} \right) G_t - \left(\frac{\Delta t_{new}}{2\Delta t_{old}} \right) G_{t-\Delta t_{old}}. \tag{4.10}$$

When $\Delta t_{new} = \Delta t_{old}$ these coefficients are the same as the 3/2 and $-1/2$ of Eq. 2.18.

4.3.5 Storing the Solution

After the completion of the loop over the time steps, this nonlinear code should provide a way of storing the current solution so later it can be read as *restart* data, allowing the solution to continue to evolve. There are different ways for storing data in files; the two basic ones are formatted (or ASCII) and unformatted (or binary). The data on a formatted file can easily be viewed with an editor. However, unformatted files are usually preferred because the data are written to a disk to the same accuracy as they are known in computer memory and because the data are written in a much more efficient manner in terms of disk space. However, when reading binary data on one computer that was stored on another one needs to know the precision of the numbers (32 or 64 bits per word) and the order the bits were stored in each word (big-endian or little-endian).

The restart data should be written much less frequently than the movie files (Chapter 5) and can be overwritten each time unless several restart files are wanted per run. For the code described here, it would be good to write the values of N_n, N_z, N_x (the number of x grid points, defined in Chapter 5), the current time and time step number. The time step number can be read during a restart and continued for that run. Likewise, the time can continue to increase by adding dt to it each iteration of the time step loop. The tem, omg, psi, dtemdt, and domgdt files should then be written. Actually, only the parts of the time derivative arrays that will represent the previous time step when the solution is restarted need to be stored. A convenient way to name the data files is to begin with one or more letters, indicating the case, followed by the time step number.

4.4 NONLINEAR SIMULATIONS

When the nonlinear code compiles and runs one needs to check if it is correctly solving the model equations, i.e., it needs verification. Choose a Ra several orders of magnitude greater than the critical Ra; note that the larger it is the more energy there will be in the smaller scales and therefore the larger N_n and N_z need to be to resolve these small eddies. Typically, the number of second-order finite-difference grid levels in the vertical direction needed to obtain an accuracy comparable to that of the spectral method employed in the horizontal direction is at least $2/a$ times the number of horizontal spectral modes (e.g., Orszag, 1971b). The "a" here is still the aspect ratio of the box.

If one tries to run a case with too little spatial resolution, the energy that cascades down to the smallest resolved scales (the truncation lengths determined by the choice of N_n and N_z) would artificially build up there because the viscous and thermal diffusivities would be too small to remove the energy fast enough. The result would be a numerical instability, characterized by the values of the variables at these scales growing exponentially with time. Note that as one increases the spatial resolution the temporal resolution also needs to improve, i.e., Δt needs to decrease according to Eqs. 2.19 and 4.7.

4.4.1 A Simple Numerical Test

Since the equations are multivariable, multidimensional, nonlinear, and time-dependent, there will not be a simple analytic solution. However, one can test the numerical method by constructing simple (nonphysical) functions for the temperature, vorticity, and streamfunction that have sinusoidal dependences in space and time and satisfy the streamfunction equation 2.12 and boundary conditions. Then substitute these functions into Eqs. 2.10 and 2.11 and compute the z-dependent and time-dependent "forcing terms" that would need to be added to these two equations for each mode n to make the constructed functions be solutions.

Next add the forcing terms to your nonlinear code and add a check to this modified code that compares the time-dependent numerically produced solution to the time-dependent constructed functions several times during each sinusoidal period. When satisfactory agreement is obtained the code without the forcing terms is ready to be tested with a nonlinear benchmark.

4.4.2 A Nonlinear Benchmark

For example, try the case defined by Ra $= 10^6$, Pr $= 0.5$, and $a = 3$ with $N_n = 50$, $N_z = 101$, and $\Delta t = 3 \times 10^{-6}$. The overall Ra$_{crit}$ for this case is 660.5 and n_{crit} is 2; so this case is more than 1500 times critical. One could simply initialize the $n = 1$ temperature coefficient with $0.01 \sin(\pi z)$ and set all other $n > 0$ coefficients to zero for all z. The resulting nonlinear solution, $T_n(z, t)$, $\omega_n(z, t)$, and $\psi_n(z, t)$, will initially have amplitudes that increase exponentially in time, as they do for the linear problem. However, as the amplitudes grow the nonlinear terms eventually become large enough to stop the exponential growth. Then, for the case suggested here, the nonlinear solution will reach a steady state. That is, the time derivatives of $T_n(z, t)$ and $\omega_n(z, t)$ will slowly vanish, so all variables become constant in time. This will take a several hundred thousand time steps, depending on how many significant figures one wants to see remain constant. The solution evolves to one

Temperature

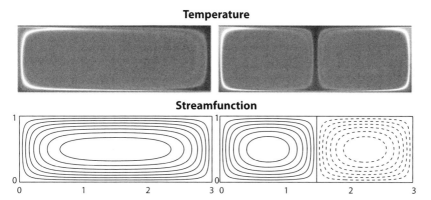

Streamfunction

Figure 4.2 Two steady-state solutions, illustrated with profiles of the temperature and streamfunction plotted on the horizontal (x) and vertical (z) grid for Ra $= 10^6$, Pr $= 0.5$, and $a = 3$ (see *Color Plate* 1a for a color version of this figure). The solution on the left was initialized with an $n = 1$ temperature mode; whereas the one on the right was initialized with both the $n = 1$ and $n = 8$ modes. For temperature, red corresponds to hot buoyant upflow and blue to cold heavy downflow; green is vanishingly small relative to the background temperature. Solid contours of the streamfunction represent positive ψ (i.e., clockwise flow), broken contours are negative ψ (i.e., counterclockwise flow).

circulation cell; that is, temperature and vertical velocity are positive at $x = 0$ (the left side of the box) and negative at $x = a$ (the right side).

This is illustrated on the left side of Fig. 4.2 in terms of the steady-state temperature perturbation, T, and streamfunction, ψ, on the x, z-grid. Typically, reds are used to represent warm temperatures relative to the constant background temperature of 0.5 and blues represent cool temperatures (less than 0.5). Recall that the x-averaged temperature ($n = 0$ mode) is 1 at the bottom boundary and 0 at the top boundary. As described in Section 2.1, contours of (constant values of) the streamfunction represent the streamlines of the flow, i.e., the set of curves that are everywhere tangent to the fluid velocity at a given moment in time. Positive values of ψ (solid contours) represent clockwise circulation and negative values (broken contours) represent counterclockwise circulation. The local density of a set of streamlines (that differ by a constant increment) is proportional to the amplitude of the local fluid velocity. If the flow is in steady state, as it is for this example, streamlines also represent trajectories of fluid parcels over time. The flow is directed clockwise on positive contours of ψ in the x, z-plane and counterclockwise on negative contours.

For this case, the nondimensional values of T_n, ω_n, and ψ_n at $z = 0.32$ (i.e., for $k = 33$, where $k = 1$ is the bottom boundary and $k = 101$ is the top boundary) are given in Table 4.1 after one million numerical time steps (three thermal diffusion times) for horizontal mode numbers, n, up to 20. The solution is fairly well converged at this time. The values in this table can be used to check one's nonlinear code. Note that these values are for the formerly mentioned spatial resolution

Table 4.1 The Steady-State Nondimensional Values of T_n, ω_n, and ψ_n at $z = 0.32$ (for horizontal mode numbers, n, up to 20 corresponding to the case illustrated on the left side in Fig. 4.2; the spatial resolution used was $N_n = 50$ and $N_z = 101$; the nondimensional time step was 3×10^{-6}).

n	T_n	ω_n	ψ_n
0	5.00099E–1	0.00000E + 0	0.00000E + 0
1	2.91441E–2	5.81561E + 3	5.67825E + 2
2	2.52654E–4	−3.58935E + 0	−6.74022E − 1
3	2.90576E–2	1.93069E + 3	1.03898E + 2
4	2.87716E–4	−6.60995E + 0	−5.36102E − 1
5	2.89003E–2	1.14940E + 3	3.23310E + 1
6	3.49647E–4	−1.10564E + 1	−3.95780E − 1
7	2.87013E–2	8.11296E + 2	1.31673E + 1
8	4.78605E–4	−1.47794E + 1	−2.83806E − 1
9	2.82735E–2	6.17174E + 2	6.37517E + 0
10	6.45160E–4	−1.69695E + 1	−2.06439E − 1
11	2.74277E–2	4.88830E + 2	3.46088E + 0
12	8.07511E–4	−2.14400E + 1	−1.53893E − 1
13	2.60559E–2	3.98420E + 2	2.03088E + 0
14	9.77967E–4	−2.61548E + 1	−1.16708E − 1
15	2.41582E–2	3.26806E + 2	1.25792E + 0
16	1.16483E–3	−2.46904E + 1	−8.88572E − 2
17	2.18985E–2	2.64619E + 2	8.09779E − 1
18	1.28586E–3	−2.03670E + 1	−6.77822E − 2
19	1.95530E–2	2.16208E + 2	5.37141E − 1
20	1.25455E–3	−2.19120E + 1	−5.24029E − 2

($N_n = 50$, $N_z = 101$); a comparable or better spatial resolution would produce essentially the same solution but with slightly different distribution of energy among the modes, especially for the higher mode numbers.

At this resolution T_n, ω_n, and ψ_n near mode number 50 are not significantly smaller than those at $n = 20$ for the even number modes; the amplitudes of the odd number modes do drop about two orders of magnitude from mode 20 to 50. This apparent need for higher horizontal resolution (in the even modes) is likely due to the very prominent, shallow, thermal and flow boundary layers that exist on the side boundaries compared to the small perturbations away from the boundaries; Fourier functions are not the best for resolving strong local gradients. In addition, the even modes are not needed for this solution other than in the shallow bottom and top boundary layers.

Note that Eq. 3.9 predicts a critical mode number of 2 for an aspect ratio of 3 *for marginal stability*, i.e., for Ra = Ra$_{crit}$. However, for the chosen supercritical Rayleigh number the initial $n = 1$ temperature perturbation kicks the system into an $n = 1$ steady-state solution, i.e., a single convection cell. If, in addition to initializing the $n = 1$ mode of temperature, one gives a higher mode number a 0.01 $\sin(\pi z)$ initial value, the system converges to the steady-state solution seen on the right side of Fig. 4.2, i.e., two convection cells. Because this double cell solution is symmetric relative to the midpoint in x, all odd n modes become vanishingly small.

Both of the simulations illustrated in Fig. 4.2 were continued at higher resolution ($N_n = 200$, $N_z = 101$) for more than five thermal diffusion times (5 million computational time steps of $\Delta t = 10^{-6}$) with no indication that either steady-state flow pattern was going to change. Therefore, the same set of Ra, Pr, and a appears to have at least two nonlinear solutions; the initial condition determines to which one the system evolves. At this higher resolution the amplitudes at mode number 200 are about four orders of magnitude smaller than their respective values at mode 20. The amplitude of ψ drops eight orders of magnitude between $n = 1$ and $n = 200$.

Note also that the directions of the circulations shown in Fig. 4.2 occur because the initial temperature perturbation is positive on the left side of the box. Solutions also exist with the exact mirror images (in the x-direction) of the temperature profile with the opposite circulation direction. These would be obtained if the odd-n initial temperature coefficients were set to $-0.01 \sin(\pi z)$. The values of all the odd n coefficients in Table 4.1 would then have their signs reversed.

If one does not obtain the steady-state solutions in Fig. 4.2 (or their mirror images), the code needs to be checked for "bugs." Debugging is an art. It can be frustrating, but discovering the bug and fixing it is quite satisfying. The simple, although not usually most efficient, way is to place print statements in the code at various locations to check the values of the variables. Of course, one usually won't know exactly what they should be (except at the boundaries and at the initial time step); the hope is that one might be able to detect which variable becomes too large first and where in the code. Sophisticated debugging tools are available for most programming languages and operating systems. One typically sets "breakpoints" in the code while it is running to be able to decide in real time which variables (including entire arrays) to print out when and where. Debugging can be time-consuming; it motivates one to be very careful when initially writing the code.

4.4.3 Further Studies

When the code is working correctly it is finally ready to be used to study how the chosen values of Ra, Pr, and a affect the pattern and time dependence of the convection. Kinetic energy of the convection is continually being lost by viscous forces (Section 1.1.2) and continually supplied by the gravitational potential energy associated with the net thermal stratification, which is continually maintained (externally) by the isothermal boundary conditions. Increasing the Rayleigh number, Ra, will eventually produce a solution that is continually time-dependent, i.e., does not reach a steady state, although probably will reach a "statistically steady state."

It is sometimes more efficient to start a simulation with a relatively low Ra using relatively coarse spatial and temporal resolutions and then, after reaching steady state, continue the simulation with a somewhat higher Ra. This can be done in steps; but at least at some steps will require increasing the spatial and temporal resolutions, i.e., increasing N_z and N_n and decreasing Δt. This requires a little additional coding that would read in a lower spatial resolution and then fill in the remaining n modes with zeros. Increasing the resolution in the z-direction can be done with a linear interpolation onto the finer grid. One could write a separate code that just reads in the solution at the current spatial resolution and writes out the solution at the desired (higher or lower) resolution. Alternatively, there could be an option in the main code to do this.

It should be obvious that better resolution is needed when increasing the Rayleigh number because this increases the convective driving, which increases the amplitudes of the fluid flow and temperature perturbations while decreasing their length scales. Better spatial resolution is especially needed near the bottom and top boundaries, where shallow thermal boundary layers form. Obtaining better spatial resolution near these boundaries, without increasing the resolution uniformly everywhere, can be achieved by employing a nonuniform vertical grid (Sections 9.1 and 9.2) or by using a Chebyshev polynomial expansion in the vertical direction (Section 9.4).

If the Prandtl number, Pr, is quite different from unity, the fluid flow and temperature perturbations would not need the same amount of spatial resolution. A Prandtl number larger than unity produces temperature structures that in general have smaller length scales than the fluid flow structures, and vice versa. Although it is easier to provide the same spatial resolution for all variables, it may be worth the effort to expand temperature in the different number of modes than the vorticity and streamfunction for cases with Pr very different than unity. In addition, the resolution in one direction may need to be different than what it is in the other due to the aspect ratio, a.

A time-dependent solution is usually more interesting than a steady-state solution. Running at a high enough Rayleigh number to obtain a time-dependent solution becomes more efficient when treating the linear terms implicitly (Chapter 8) and computing the nonlinear terms with a spectral-transform method (Chapter 10). Solutions also tend to be time-dependent, even at relatively low Rayleigh numbers, when the pattern of the convection is free to propagate in the x-direction by using permeable periodic side boundary conditions (Chapter 10) and when magnetic forces (Chapter 11), Coriolis forces (Chapter 13), and buoyancy forces (Chapters 12) produce additional instabilities. However, the easiest way to achieve interesting time dependence at relatively low Rayleigh numbers is by making part of the domain thermally stable and the other part unstable (Chapter 6).

SUPPLEMENTAL READING

Boyd (2001)
Canuto et al. (1988)
Durran (1998)
Peyret (2002)

EXERCISE

1. *Trigonometric identities*
   Derive the trigonometric identities in Eqs. 4.1 using combinations of $\exp(i(a \pm b)) = \cos(a \pm b) + i \ \sin(a \pm b)$.

COMPUTATIONAL PROJECTS

1. *Benchmarking a nonlinear simulation*
   Run the nonlinear case defined in Section 4.4 and compare your steady-state solution to that listed in Table 4.1. Demonstrate, using the suggested spatial resolution of $N_n = 50$ and $N_z = 101$, that a nondimensional time step of 3.6e-6 works; whereas one set to 4.0e-6 does not. Does the latter time step exceed the diffusion limit or the CFL limit?

2. *Stability of steady-state nonlinear solutions*
   Test the stability of the two solutions depicted in Fig. 4.2 by starting with the one-cell steady-state solution and gradually increasing an $n = 2$ temperature perturbation. Likewise, starting with the two-cell steady-state solution add an $n = 1$ perturbation.

3. *Convection with a bolide impact*
   Between chosen times t_1 and t_2 during a simulation of convection in a 2D box alter the boundary conditions at the top boundary on ω_n and ψ_n for $n = 2, 4, 6,$ and 8 in the following way:

   $$\psi_n = -(-1)^{n/2} \left(\frac{a}{n\pi}\right) \left(\frac{V_{max}}{4}\right) g(t),$$

   $$\omega_n = -(-1)^{n/2} \left(\frac{n\pi}{a}\right) \left(\frac{V_{max}}{4}\right) g(t).$$

   For all other n, ψ_n, and ω_n vanish on the top boundary as usual. Set the time dependence during the impact (i.e., for $t_1 \le t \le t_2$) to

   $$g(t) = 1 - \cos(2\pi(t - t_1)/(t_2 - t - 1))$$

   with $g(t) = 0$ at all other times. V_{max} is the chosen maximum velocity of the impact at the top boundary. Note, since ψ_n and ω_n for these four n-modes do not vanish on the top boundary during the impact, check that nowhere in your code have you assumed they do vanish on the top boundary.

4. *Thermal diffusion in a solid above a convecting fluid*
   Simulate thermal convection in the lower half of a box (the fluid part) and solely thermal diffusion in the upper half (the solid part) by forcing $\psi = 0$ everywhere in the upper half at all times. This can be done by modifying the ψ matrix operator (Section 2.5) so the part of the matrix that represents the upper half has only a diagonal of ones. In addition, maintain $\omega = 0$ everywhere in the upper half of the box, i.e., do not update ω for $z \ge 0.5$.

5. *Convection with constant heat flux through the top boundary*

   Replace the constant temperature top boundary condition with a constant (vertical) heat flux boundary condition at the top boundary by forcing the nondimensional $\partial T/\partial z = -1$ using a "ghost point." See the discussion on the use of ghost points in Section 11.2. In that section the vertical gradient of the vector potential, A, is forced to vanish on both the top and bottom boundaries; whereas here the vertical gradient of the temperature, T, needs to be -1 at the top boundary while the value of T is still forced to be 1 at the bottom boundary. Note also, as for A, here T needs to be updated every time step on the top boundary since it no longer is forced to vanish there.

6. *Convection with an x- and t-dependent bottom temperature*

   Replace the constant temperature bottom (and/or top) boundary condition with one that sets all T_n to zero except for $n = 3$. Set T_3 to $\sin \omega_o t$, where t is the time in terms of the numerical time step. Pick a frequency (ω_o) such that $2\pi/\omega_o$ is large relative to a typical convective turnover time (i.e., the depth of the domain divided by the average fluid velocity in the z-direction).

Chapter Five

Postprocessing

The initial check of the nonlinear code can be done by monitoring the values of various T_n, ω_n, and ψ_n at various depths as discussed in Chapter 4. However, to learn from these nonlinear simulations one needs to study them using computer graphics and analysis.

5.1 COMPUTING AND STORING RESULTS

To produce a snapshot like those in Fig. 4.2, one needs to Fourier transform the spectral solution (n-space) to the grid (x-space). (Since the method we have chosen computes the solution in z-space, no transform in that direction is needed.) This is simply done according to Eqs. 2.8. One must, however, choose a set of grid points in the x-direction, $x_i = (i - 1)\Delta x$ for $i = 1 \rightarrow N_x$, where N_x is the number horizontal grid points with uniform grid spacing $\Delta x \equiv a/(N_x - 1)$. As mentioned, considerable computational time can be saved by computing the $\sin(n\pi x_i/a)$ and $\cos(n\pi x_i/a)$ once for all n and i and storing these in arrays instead of computing them every time they are used.

   The resulting snapshots of $T(x, z)$, $\omega(x, z)$, and $\psi(x, z)$ need to be stored in a file during the computer simulation, assuming the Fourier transforms to x-space are done within the main computational code during the simulation. As mentioned in Section 4.3.3, one could alternatively store $T_n(z)$, $\omega_n(z)$, and $\psi_n(z)$ and do the Fourier transforms in the postprocessor.

   If a movie of one or more of these fields is desired, this process needs to be done sequentially during the simulation. An input parameter can be used to set the number of computational time steps that are computed between movie snapshots. This is typically about 100 to 1000 (depending on the degree of time dependence and the Δt); the more computational time steps per movie step the faster the movie will evolve. A different file could be created for each movie snapshot or all the movie snapshots could be stored in one file. The first routine listed in Appendix B illustrates how the former method could be achieved in a Fortran code.

5.2 DISPLAYING RESULTS

A postprocessing code would need to be written that reads these movie files and, using a graphics software package like IDL or Matlab, makes a 2D contoured plot like those of $\psi(x, z)$ in the bottom row of Fig. 4.2 or a 2D color-filled contoured image like those of $T(x, z)$ in Fig. 4.2 for each snapshot. (Note, the "number-crunching" code that solves the nonlinear equations and generates the movie files could also be written using IDL or Matlab; however, typically this code is written in Fortran or C language.) The resulting graphical plots or images can then be combined into an executable movie file using software (like mpeg), which produces additional frames between snapshots via interpolation, that can then be viewed (at typically 30 frames per second) by several common applications.

An example of an IDL program that reads movie files, one per snapshot, is listed in the second routine of Appendix B. This assumes the computational code stores snapshots in individual files. If the resulting movie appears to be in "slow motion," remake it using, for example, every other movie file to speed it up by a factor of 2. On the other hand, if the structure seen in the movie changes too quickly, rerun the simulation saving movie files more frequently.

As mentioned in Section 4.4, contour plots of the streamfunction (as in Fig. 4.2) represent streamlines of the flow, which provide a snapshot of the direction and amplitude of the flow throughout the box. Only if the flow is in steady state do streamlines also represent the trajectories of the fluid parcels. To obtain trajectories for a time-dependent flow one could randomly distribute "tracer particles" in the box and advect them with the flow. A typical numerical algorithm for doing this is a Runge-Kutta scheme, which involves interpolating the fluid velocity between the discrete x, z-grid points and between the discrete time steps (e.g., Press et al., 1992). (A fourth-order Runge-Kutta scheme is described in Section 8.1 as an alternative scheme for integrating a scalar in time.) These particles are called tracers because they do not affect the flow. The trajectories could be visualized in a movie displaying the positions of these tracer particles as they change in time. Sometimes it is useful to make a movie that displays the last 10 (or so) positions of each particle so it is easier to see the particle paths. The longer the line of positions for a given particle the faster it is moving at that time. A simple test would be to check if the trajectories computed this way agree with the contours of $\psi(x, z)$ for a steady-state solution.

Plots and movies of the vorticity $\omega(x, z)$ typically appear similar to those of the streamfunction $\psi(x, z)$. Positive ω tends to represent clockwise circulation in the x, z-plane and negative ω counterclockwise circulation.

One can also compute and plot the fluid velocities, $v_x(x, z)$ and $v_z(x, z)$, using Eqs. 2.6. The x-derivative of ψ (needed for v_z) could be obtained via Eq. 3.2, which would need to be done before the Fourier transform to x-space. Alternatively, this could be approximated via Eq. 2.15 in the x-direction to get v_z as the z-derivative of ψ would be done to get v_x. Note that on all four boundaries the normal components of the velocities vanish and the tangential components can be set equal to their values one grid point inside the boundaries. These velocity components could be combined to plot arrows representing the vector velocity. (See, for example, Figs. 13.9 and 13.11.) The tails of the arrows would be located at the grid points and their heads would point in the direction of the flow at that location and time. The arrow lengths (and sizes of the arrow heads) should be proportional to the amplitude of the velocities. This requires first finding the maximum velocity amplitude for the entire movie. These plots provide the same information as the plots of streaklines and so should be checked that they agree with the contours of ψ.

The temperature perturbation, $T(x, z)$, can be visualized as a contoured plot or as a color-shaded contoured image, as in Fig. 4.2, to show where the fluid is warmer (values greater than 0.5) and cooler (values less than 0.5) relative to the constant background temperature. It is usually a good idea to choose a color table that has a noticeable change in color at 0.5. As mentioned, the color table employed for Fig. 4.2 uses shades of red to represent temperatures higher than the background temperature and shades of blue to represent temperatures lower than the background. Sometimes it is helpful to use a rainbow of colors to more easily see the gradients in the field; however, this can make it difficult to pick out the absolute highs and lows. Alternatively, one could plot a surface over the x, z-plane making the local height of the surface proportional to the local temperature perturbation.

Recall that the density perturbation, $\rho(x, z)$, within the Boussinesq approximation, is proportional to the temperature perturbation (Eq. 1.12). Therefore, comparing plots of the temperature perturbation with those of the fluid flow illustrates how hot (light) fluid tends to rise and cold (heavy) fluid tends to sink. The bottom boundary is held at its constant (hot) value and the top boundary at its constant (cold) value. However, between these boundaries the x-averaged part of the temperature (the $n = 0$ mode) is allowed to evolve. Convection tends to make the temperature profile adiabatic, which for the Boussinesq approximation is isothermal, by advecting hot fluid upward where it is typically cooler and cold fluid downward where it is warmer. Since the boundary values are fixed, shallow thermal boundary layers develop in which the vertical gradient of this mean temperature becomes very steep, i.e., superadiabatic. The mean temperature profile in the bulk of the fluid box approaches a constant value equal to the average of the boundary values. As Ra is increased the boundary layers become thinner and the mean temperature in the bulk of the fluid becomes more nearly isothermal.

The pressure perturbation, $p(x, z)$, can also be obtained. However, since the pressure term was eliminated by taking the curl of the momentum equation, we need to go back to the momentum equation, 1.16. One way to do this is to consider either the x-component or the z-component of this equation on the x, z-grid within the postprocessor, using the known values of the fluid velocity and temperature and using finite-difference methods (Eqs. 2.15 and 2.16) in both directions. The time derivative could be approximated using the Adams-Bashforth time integration scheme, Eq. 2.17, or simply the Euler scheme, which replaces the right side of Eq. 2.17 with just G_t. A more accurate, but more complicated, way would be to consider the Fourier-analyzed version of the x-component of this equation, which requires the pressure to be expanded in $\cos(n\pi x/a)$.

As a final note here, check the images of magnetic field lines in Figs. 13.21 and 13.14, which are produced using a Runge-Kutta scheme similar to that mentioned above, but all at the same time step. That is, instead of following tracer particles advected by the velocity in time, the tracer particles are "advected" by the magnetic field; the resulting "trajectories" form the magnetic field lines. Starting at a prescribed set of locations the three components of the magnetic field are calculated (or interpolated from known values at grid points) to grow each line by a small constant increment. The 3D field is then calculated at the new set of locations and so on. Typically, 1000 equal-length increments are used for each field line. The set of 3D grid points that defines these increments is then input to another graphics routine that plots the lines as narrow tubes with the parts of the tubes in front blocking out those in back.

Several graphics software packages exist that provide additional, more sophisticated visualizations of the scalar and vector data discussed here.

5.3 ANALYZING RESULTS

Besides displaying the various fields discussed in Section 5.2, several additional properties of the solution can be analyzed. As discussed in Section 3.4, the Rayleigh number, Ra, is a rough prediction of what will be the ratio of the convective heat flux in the bulk of the convecting fluid to the conductive heat flux before the onset of convection. It is defined in terms of known boundary conditions and fluid properties before the experiment or simulation is performed. The *Nusselt number*, Nu, on the other hand, is defined as the *measured* ratio of the total heat flow (convective plus conductive) during convection to the conductive heat flow before the onset of convection. "Heat flow" means the x-integrated heat flux. Usually Nu is also defined as the time averaged ratio. Here we denote the horizontally integrated time-averaged value with $\langle\ \rangle$. Therefore,

$$\text{Nu} \equiv \left\langle c_p \rho_o v_z T + \left(-c_p \rho_o \kappa \frac{dT}{dz}\right)\right\rangle_{conv} \bigg/ \left\langle -c_p \rho_o \kappa \frac{dT}{dz}\right\rangle_{no\ conv} . \tag{5.1a}$$

The time average (usually) means that the Nu is independent of z, otherwise there would be convergences and divergences of heat flux that would result in time dependence. That is, Nu should be the same at the bottom and top boundaries and at any z-level within the fluid box. Therefore, since v_z vanishes at the bottom and top boundaries,

$$\text{Nu} = \left[\left\langle \frac{dT}{dz}\right\rangle_{conv} \bigg/ \left\langle \frac{dT}{dz}\right\rangle_{no\ conv}\right]_{z=0\ and\ 1} . \tag{5.1b}$$

Note the dimensional value of $(dT/dz)_{no\ conv}$ is $\Delta T/D$ and the nondimensional value is one.

Recall that when convection is vigorous, $\langle dT/dz\rangle$ approaches zero in the bulk of the convection zone. Therefore, according to Eq. 5.1a, Nu is a measure of what Ra predicts. Experiments and simulations of turbulent convection (e.g., Rogers et al., 2003, and references therein) find that

$$\text{Nu} \propto \text{Ra}^c , \tag{5.2}$$

where the c ranges from about 1/4 to 1/3, with c closer to 1/4 for nonslip boundaries and closer to 1/3 for stress-free boundaries.

Another traditional measurement of nonlinear convection is the *Reynolds number*, Re, which is the ratio of the typical peak or root-mean-square (RMS) fluid velocity, V, to the viscous diffusive velocity for a length scale D:

$$\text{Re} \equiv \frac{V D}{\nu}. \tag{5.3a}$$

When V is scaled by the thermal diffusion velocity, as we have been doing, this is

$$\text{Re} = \frac{V}{\text{Pr}}. \tag{5.3b}$$

Re, which measures the resulting convective vigor, is expected to increase with Ra, which prescribes the convective driving. Experiments and simulations of turbulent convection (e.g., Rogers et al., 2003, and references therein) find that

$$\text{Re} \propto \text{Ra}^{1/2}. \tag{5.4}$$

Another diagnostic test for the nonlinear simulation is the distribution of kinetic energy among the horizontal modes, i.e., the kinetic energy spectrum. Define KE(t) as the total nondimensional kinetic energy in the fluid box per unit length in the y-direction:

$$\text{KE}(t) \equiv \frac{1}{2} \int_0^1 \int_0^a \left[v_x^2 + v_z^2 \right] dx \, dz$$

$$\equiv \sum_{n=1}^{N_n} \int_0^1 \text{KE}_n(z, t) \, dz. \tag{5.5}$$

Using Eqs. 2.6 and 2.8c–2.9b, one sees that

$$\text{KE}_n(z, t) = \frac{a}{4} \left[\left(\frac{\partial \psi_n}{\partial z} \right)^2 + \left(\frac{n\pi}{a} \psi_n \right)^2 \right]. \tag{5.6}$$

The integral over z in Eq. 5.5 can be done numerically using, for example, the Trapezoidal Rule. One could also compute the time-averaged spectrum and the root-mean-square deviation from this average spectrum to provide a measure of the time dependence of the solution as a function of n. It is also sometimes useful know the (nondimensional) root-mean-square velocity, which is related to the kinetic energy:

$$v_{RMS}(t) \equiv \left[\frac{1}{a} \int_0^1 \int_0^a \left[v_x^2 + v_z^2 \right] dx \, dz \right]^{1/2}$$

$$= \left(\frac{2}{a} \text{KE}(t) \right)^{1/2}. \tag{5.7}$$

Note, if the background density were not constant (Chapter 12), it would need to appear inside the vertical integral in Eq. 5.5 but not in Eq. 5.7.

A useful diagnostic, especially during the initial evolution of the simulation, is to plot KE vs. t, using Eqs. 5.5 and 5.6, which would show the initial exponential growth followed by a steady state or statistically steady state. It can also be useful to plot (using contour lines or banded contour colors) the kinetic energy density, $\frac{1}{2}(v_x^2 + v_z^2)$, on a z vs. t graph for a prescribed x location or on an x vs. t graph for a prescribed z location. This nicely shows how kinetic energy density propagates in z (or x) with time. The plots suggested here could be produced with the data in the movie files or, if better temporal resolution is desired, the data could be computed during the simulation and stored.

Justification for truncating the spectral expansions at mode number N_n would be that the kinetic energy per mode at $n = N_n$ is at least a few orders of magnitude less than the peak energy per mode. If the kinetic energy per mode at the truncation scale (i.e., the smallest resolved scale) were not significantly less than that at the larger scales or, worse, if it peaked at the truncation scale, the images of the flow would be dominated by small-scale features and may show periodic oscillations in the x-direction at the highest spatial frequency. This would mean that viscosity is not large enough to remove the energy in the smallest scales. In such cases N_n would need to be increased or Ra would need to be decreased. One could also Fourier analyze the $KE_n(z, t)$ in the vertical z-direction to check if N_z is large enough.

The rate of change of kinetic energy density and the rates that various forces do work to change the local kinetic energy can also be examined in grid space and time. The kinetic energy density equation (1.7) for the nondimensional Boussinesq problem is

$$\frac{\partial}{\partial t}\left(\tfrac{1}{2}v^2\right) = -(\mathbf{v}\cdot\nabla)\left(\tfrac{1}{2}v^2\right) - \mathbf{v}\cdot\nabla p + \mathrm{RaPr}\, v_z T + \mathrm{Pr}\, \mathbf{v}\cdot\nabla^2\mathbf{v}. \qquad (5.8)$$

For our 2D model, v^2 is $v_x^2 + v_z^2$. The first term on the right is the advection of kinetic energy density, which comes from the velocity dotted with the velocity advection term in Eq. 1.16, using Eq. 2.2. The next three terms on the right of Eq. 5.8 are the rates pressure gradient, buoyancy, and viscous forces do work per volume on the fluid, respectively. The sum of the advection and pressure terms could be written as the convergence of the flux of kinetic energy density and pressure, $-\nabla\cdot(\tfrac{1}{2}v^2 + p)\mathbf{v}$. This is a convenient form when integrating over the entire fluid box to get the rate of change of the total kinetic energy because the integrated advection and pressure terms vanish, as can be seen when applying the divergence theorem and the impermeable (and fixed) boundary conditions. Therefore, the time rate of change of total kinetic energy in the box equals the net rate that buoyancy and viscosity do work; the former being positive and the latter negative; in steady state, of course, the net rate vanishes.

The terms in Eq. 5.8 could be computed using finite differences in the x- and z-directions for each movie time step or averaged in time. Plots of these terms in the x, z-plane show that buoyancy tends to drive the upflows and downflows doing positive work, approximately balanced by the hindering pressure gradient and viscous forces which do negative work in these regions. The horizontal flows are driven by pressure gradients and approximately balanced by the viscous forces. The unbalanced work rates change the local kinetic energy density; in steady state the sum of the terms on the right of Eq. 5.8 vanish everywhere. It can also be useful to (numerically) integrate in the x-direction all the terms in Eq. 5.8 and plot the resulting profiles in z.

The internal energy density is proportional to the temperature, which when integrated over x and divided by a is just the mean profile $T_0(z, t)$. However, one can examine the spectrum of the temperature variance. Integrating the square of the temperature over x gives an expression similar to Eq. 5.6:

$$a \left[(T_0)^2 + \tfrac{1}{2} \sum_{n=1}^{N_n} (T_n)^2 \right] . \tag{5.9}$$

This can be analyzed in the same manner as the kinetic energy spectrum.

The nondimensional Boussinesq internal energy density equation is Eq. 1.17. The same type of time and spatial analyses can be performed on this equation as was discussed for the kinetic energy equation 5.8. In particular, such analyses show how advection and diffusion of temperature tend to balance locally. Also, since, because of Eq. 1.15, $-(\mathbf{v}\cdot\nabla)T = -\nabla\cdot(T\mathbf{v})$ and $\nabla^2 T = -\nabla\cdot(-\nabla T)$, the right side of Eq. 1.17 is $-\nabla\cdot(T\mathbf{v} - \nabla T)$, i.e., the convergence of the sum of the (nondimensional) convective and diffusive heat fluxes. That is, the local temperature increases with time when and where there is a convergence of total heat flux and decreases when and where there is a divergence. Plots of the z-components of these two fluxes show how they tend to add to a constant vertical heat flux.

SUPPLEMENTAL READING

Press et al. (1992)

EXERCISES

1. *Kinetic energy spectrum*
   Derive the expression for the Fourier coefficients of kinetic energy (Eq. 5.6).
2. *Temperature variance spectrum*
   Derive the expression for the Fourier coefficients of the variance of temperature (Eq. 5.9).

COMPUTATIONAL PROJECTS

1. *Conductive and convective heat flows*
   Show for the *Thermal diffusion in a solid above a convecting fluid* project in Chapter 4 that, when averaged in time, the total conductive heat flow through any constant z-level in the upper half of the box equals the sum of the conductive and convective heat flows through any constant z-level in the lower half of the box.
2. *Mean temperature profile*
   Run a series of nonlinear convection simulations and, for each, plot the mean ($n = 0$) temperature perturbation as a function of z to illustrate how the thickness of the thermal boundary layer depends on Ra for a given Pr and aspect ratio a, on Pr for a given Ra and a and on a for a given Ra and Pr.

3. *Nusselt and Reynolds numbers vs. Rayleigh number*

   Run a series of convection simulations with different Rayleigh numbers for a given Prandtl number and aspect ratio and measure how the Nusselt number varies with Rayleigh number and compare your results to Eq. 5.2. For the same series of cases, measure how the Reynolds number varies with Rayleigh number and compare your results to Eq. 5.4.

4. *Kinetic energy diagnostics*

   Run a simulation of convection for a given aspect ratio and Prandtl number and for a Rayleigh high enough that the solution remains time-dependent. Using Eqs. 5.5 and 5.6, plot the time evolution of the total kinetic energy, KE vs. t. After reaching a statistically steady state, compute the spectrum of the time-averaged kinetic energy, i.e., KE_n vs. n. Also, plot the terms in the kinetic energy equation as a function of z, averaged in x and t.

5. *Temperature variance and heat flow diagnostics*

   For the case in the *Kinetic energy diagnostics* project plot the spectrum of the time-averaged and volume-averaged temperature variance (Eq. 5.9). Also, plot the time-averaged and x-integrated convective and conductive heat flows as functions of z.

Chapter Six

Internal Gravity Waves

In Section 1.1.1 we discuss how a mean (x-averaged) superadiabatic temperature gradient supports thermal convection (assuming a supercritical Rayleigh number) and how a mean subadiabatic temperature gradient supports internal gravity waves. However, so far we have focused on how to develop a numerical model for the former, prescribing a higher fixed temperature at the bottom boundary than at the top boundary. Recall that within the Boussinesq approximation the background is considered both isothermal and adiabatic, so where the mean temperature gradient, $\partial T_0/\partial z$, is negative the thermal stratification is superadiabatic and convectively unstable and where it is positive the thermal stratification is subadiabatic and convectively stable. In this chapter we focus on internal gravity waves in a stable thermal stratification.

6.1 LINEAR DISPERSION RELATION

When the amplitude of the fluid velocity is small relative to the amplitude of the phase velocity, i.e., when the fluid moves much more slowly than the pattern of the wave propagates, a linear analysis, which neglects advection, provides insight to the relation between the wavelength and frequency of internal gravity waves. In addition, when thermal and viscous diffusion play relatively minor roles the system can be further simplified by neglecting diffusion (e.g., Kundu & Cohen, 2008).

Consider the *dimensional* versions of the temperature equation 1.17, the vorticity equation 2.4, and the streamfunction equation 2.7 and drop the diffusion and nonlinear advection terms. As we did in Section 3.1, employ a linear approximation for the advection of temperature by prescribing a constant (in space and time) mean temperature gradient, $d\overline{T}/dz = \Delta T/D$, which here is positive. Using Eqs. 2.6 to write v_z in terms of ψ, we then have the following coupled set of linear differential equations describing T, ω, and ψ as functions of x, z, and t:

$$\frac{\partial T}{\partial t} = -\frac{\partial \psi}{\partial x}\frac{d\overline{T}}{dz}, \tag{6.1}$$

$$\frac{\partial \omega}{\partial t} = -g_o\alpha\frac{\partial T}{\partial x}, \tag{6.2}$$

$$\omega = -\left(\frac{\partial^2}{\partial x^2} + \frac{\partial^2}{\partial z^2}\right)\psi. \tag{6.3}$$

This set of three equations and three unknown functions can easily be reduced to one equation by taking the time derivative of Eq. 6.2 and substituting in Eq. 6.1 for the resulting time derivative of T and substituting in Eq. 6.3 for ω. The result is the internal gravity wave equation:

$$\left(\frac{\partial^2}{\partial x^2} + \frac{\partial^2}{\partial z^2}\right)\frac{\partial^2 \psi}{\partial t^2} = -N^2 \frac{\partial^2 \psi}{\partial x^2} . \tag{6.4}$$

The square of the dimensional Brunt-Väisälä frequency is

$$N^2 \equiv g_o \alpha \frac{d\overline{T}}{dz} ; \tag{6.5}$$

N has units of time^{-1}. The square of its *nondimensional* value is

$$N_{nondim}^2 = \text{Ra Pr} .$$

Consider a propagating plane wave solution to Eq. 6.4 in the bulk of the fluid (far from the boundaries) with a frequency $\bar{\omega}$ and a vector wavenumber $\mathbf{k} = k_x \hat{\mathbf{x}} + k_z \hat{\mathbf{z}}$:

$$\psi(x, z, t) = \psi_o \, e^{i(k_x x + k_z z - \bar{\omega} t)} \tag{6.6}$$

and similarly for T, ω. Here, $i = (-1)^{1/2}$. Substituting Eq. 6.6 into Eq. 6.4 gives the wave frequency as a function of the wavenumbers, i.e., the *dispersion relation*:

$$\bar{\omega} = \pm \frac{N\,k_x}{|\mathbf{k}|}$$

$$= \pm N \cos\theta , \tag{6.7}$$

where $|\mathbf{k}| = |(k_x^2 + k_z^2)^{1/2}|$. The angle θ is the angle between the direction of the vector wavenumber, $\mathbf{k}$ (i.e., the direction of the phase propagation), and the x-direction. The $\pm$ sign means that the internal gravity wave propagates in the positive x-direction if $\bar{\omega}$ and k_x are both positive or both negative and in the negative x-direction if not.

Now, using the plane wave expression for ψ (Eq. 6.6) and Eqs. 2.6 for v_x and v_z, mass conservation (Eq. 1.11) reduces to

$$\nabla \cdot \mathbf{v} = i(k_x v_x + k_z v_z) = i(\mathbf{k} \cdot \mathbf{v}) = 0 . \tag{6.8}$$

That is, the fluid motion is always perpendicular to the direction of the phase propagation, i.e, this is a transverse wave. This explains why, according to Eq. 6.7, the maximum internal gravity wave frequency, N, occurs when the phase of the wave propagates only in the x-direction since then the fluid motions are only in the z-direction, parallel to gravity, and therefore providing the maximum buoyancy-restoring forces.

Consider a mode for which k_x, k_z, and $\bar{\omega}$ are all positive. A snapshot of such a wave is illustrated in Fig. 6.1. The lines represent the current locations of the maximum amplitudes of the fluid velocity, with arrows indicating the current directions. That is, these are lines of constant phase. In this example, Eq. 6.8 says that when and where v_x is positive v_z is negative, and vice versa. The velocity of this pattern, i.e., the phase velocity, is

$$\mathbf{c} = \frac{\bar{\omega}}{|\mathbf{k}|}\hat{\mathbf{k}}$$

$$= \frac{k_x N}{|\mathbf{k}|^3}(k_x \hat{\mathbf{x}} + k_z \hat{\mathbf{z}}) , \tag{6.9}$$

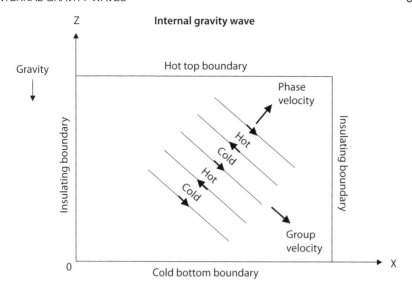

Figure 6.1 A schematic of an internal gravity plane wave in a stably stratified fluid. The snapshot shows the fluid velocity (arrows) constant along lines of constant phase, with the pattern of the phase propagating perpendicular to the fluid velocity and with the group velocity parallel (or antiparallel) to the fluid velocity.

where $\hat{k} = \mathbf{k}/|\mathbf{k}|$, the unit vector in the direction of the phase propagation. However, the velocity at which the wave energy is transported by a superposition of many waves with different frequencies and wavenumbers, i.e., the group velocity, depends on how the wave frequency varies with wavenumber:

$$
\begin{aligned}
\mathbf{c}_g &= \frac{\partial \bar{\omega}}{\partial k_x}\hat{x} + \frac{\partial \bar{\omega}}{\partial k_z}\hat{z} \\
&= \frac{k_z \mathrm{N}}{|\mathbf{k}|^3}(k_z\hat{x} - k_x\hat{z}) .
\end{aligned}
\tag{6.10}
$$

Comparing the vector components of Eqs. 6.9 and 6.10, one can see that the group velocity is perpendicular to the phase velocity with the horizontal components of these two velocities in the same direction and the vertical components in opposite directions. For our example in Fig. 6.1 $\mathbf{k}$ and $\mathbf{c}$ are directed to the upper right and $\mathbf{c}_g$ is directed to the lower right.

A way to picture this is to imagine a localized wave packet made up of a superposition of many modes, each with different frequencies and wavenumbers (although similar phase velocities). The packet moves with the group velocity while the pattern of the fluid velocity within the packet propagates with the average phase velocity perpendicular to the direction the packet travels.

Information about the temperature (and therefore density) perturbations can be obtained by examining Eq. 6.1, which shows that the amplitudes of v_z and $\partial T/\partial t$ are in phase. Since T is 90° out of phase with $\partial T/\partial t$, it is also 90° out of phase with v_z. Therefore, the peak temperature perturbations occur midway between velocity phase lines in Fig. 6.1. On the lines representing negative v_z, T is increasing with time; therefore, since this pattern propagates to the upper right, high temperatures exist below these phase lines and low temperatures above. The physical reason for this (in this stable thermal stratification for which $dT/dz > 0$) is that warm fluid from above is flowing down, causing the local temperature to increase with time. Likewise, cold fluid from below flows upward along lines of positive v_z, reducing the local temperature.

It can also be shown (e.g., Kundu & Cohen, 2008) that the rate of change of wave energy density (i.e., the sum of kinetic and gravitational potential energy densities averaged over a horizontal wavelength) equals the convergence of wave energy flux. Wave energy flux is wave energy density times group velocity. That is, the local energy in a wave propagates parallel to the wave fronts; when and where it converges (diverges) the local energy density increases (decreases).

6.2 CODE MODIFICATIONS AND SIMULATIONS

Now let's consider what modifications would be needed to convert one's thermal convection code to a code that simulates internal gravity waves, including the nonlinear and diffusive terms. First change the boundary conditions on the mean temperature to $T_0(z = 0) = 0$ and $T_0(z = 1) = 1$ to force a stable stratification. In addition, change the initial condition for the mean temperature to $T_0(z) = z$. Now the temperature scale, ΔT, represents the top minus the bottom boundary temperature and the Rayleigh number, Ra, is still defined as positive.

Another consideration is the amplitude of the numerical time step, Δt. As was shown in Section 6.1, the maximum frequency of internal gravity waves is the Brunt-Väisälä frequency, N, which for our nondimensional formulation is $(\text{RaPr})^{1/2}$. This is in radians per time. Therefore, if one wishes to have at least, say, 50 time steps per oscillation, the nondimensional Δt should be set to be no larger than $2\pi/(50\,(\text{RaPr})^{1/2})$. A CFL-like constraint on Δt also exists in terms of the grid spacing and the group velocity. For a wave propagating at $45°$, for which $k_z = k_x = n\pi/a$, the constraint is roughly $\Delta t < 2\pi/(a N_z(\text{RaPr})^{1/2})$, where here n has been taken to be 1, the mode with the greatest vertical group velocity. Of course, Δt also needs to satisfy the usual constraints due to diffusion (Eq. 2.19) and advection (Eq. 4.7).

6.2.1 Internal Gravity Wave Simulations

With these code modifications a simple internal gravity wave could be simulated by initializing $T_n(z)$ or $\omega_n(z)$ with a $\sin(m\pi z)$ vertical profile for some chosen values of n and m. If one chooses to initiate $T_n(z)$ the amplitude can be of order unity, i.e., similar to the total difference in the mean temperatures at the bottom and top boundaries. This produces an oscillation with the pattern dominated by the chosen initial condition with the amplitude decaying in time. If only one mode n is excited, a standing wave usually results.

A more interesting simulation can be produced by exciting a more localized wave packet. For example, set Ra to 10^8 and Pr and a both to 1. Set the initial $T_n(z)$ and $\psi_n(z)$ to zero for all $n > 0$. Initialize the vorticity to be zero everywhere except at the mid-depth k-level where, for all even n, set $\omega_n(k_{mid}) = -(-1)^{n/2} n/N_n$; set $\omega_n(k_{mid}) = 0$ for all odd n. This makes a peak in ω at the center point in the box at $t = 0$ with smaller amplitude ripples at all other values of x at mid-depth.

The resulting internal gravity wave simulation is quite interesting. The temperature and streamfunction evolve through what appears to be a nonrepeating series of patterns, symmetric in x and z about the central point. A snapshot is illustrated in the left-hand panels of Fig. 6.2. At this point in time the two hot (red) patterns on opposite sides of the central point are propagating horizontally away from the central point. These are hot because fluid was advected downward in these regions from above, where the mean temperature is higher because to the stable thermal stratification. Note the downward streamlines just in front of these regions indicating downward flow, which is now heating up the regions ahead of the currently hot regions. Cold fluid is being advected upward in the central region of the box, causing the temperature there to decrease (i.e., become more blue).

As the simulation evolves the shapes and propagation directions of the temperature and flow patterns continue to change while maintaining the symmetry about the central point. To break this symmetry one could choose a different aspect ratio or excite the wave at some other location in the box or add an $n = 1$ mode, for example, to the initial vorticity instead of initializing only even modes. However, the flow and temperature boundary conditions on the four sides of the box limit the types of patterns that can develop.

The amplitudes of the waves in this simulation decay because the wave is excited only at $t = 0$; however, they decay very slowly relative to the evolution of the pattern because of the high Ra. To produce a nondecaying simulation one could apply a continuous excitation with a prescribed frequency or set of frequencies.

6.2.2 Internal Gravity Waves Excited by Adjacent Convection

Alternatively, one could maintain part of the box convectively unstable $(dT_0/dz < 0)$ and the other stable $(dT_0/dz > 0)$. This could serve as a crude model of, for example, the interior of the sun, which has a stable (radiative) region in its deep interior and an unstable (convective) zone in roughly the outer 30% of its radius. In the sun's stable interior radiative transfer is efficient enough to carry the entire solar luminosity. However, at about 70% of the solar radius the temperature is low enough that hydrogen is no longer completely ionized, which increases the opacity and therefore decreases the efficiency of radiative transport; that is, the thermal diffusivity is less there than in the deep interior. The resulting convergence of heat flux there makes the mean vertical temperature gradient above this radius steeper than an adiabatic gradient. The resulting thermal convection in this unstable region carries most of the luminosity. Downwelling convective plumes in this outer region continually excite internal gravity waves in the inner stable region by overshooting. The steepness of the mean temperature gradient, dT_0/dz, in each of these

Temperature

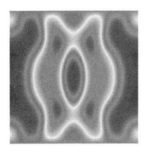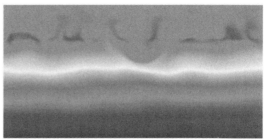

Streamfunction

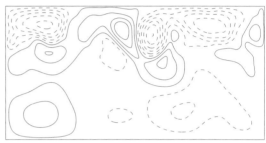

Figure 6.2 Snapshots of two time-dependent solutions, illustrated with profiles of temperature and streamfunction plotted on the horizontal (x) and vertical (z) grid (see
Color Plate 1b for a color version of this figure). The snapshot illustrated in the
left-hand panels is for Ra = 10^8, Pr = 1, and a =1; it has a stable mean temperature profile and was initialized with a localized perturbation in ω at the center of
the box, as described in the text. The snapshot illustrated in the right-hand panels
is for Ra = 10^8, Pr = 1, and a = 2; it has a stable mean temperature profile
in the lower three-quarters of the box maintained with a thermal forcing time of
10^{-4} and an unstable profile in the upper quarter maintained with a top boundary
temperature of 0.9 (see description in the text). For temperature, red represents
hot buoyant fluid and blue cold heavy fluid. The horizontal-mean temperature,
$T_0(z)$, is not included in the image on the left so the small temperature perturbations can be seen. The mean temperature profile is included in the image on
the right, displaying the significant increase in mean temperature with height in
the stable region and the nearly adiabatic (i.e., constant) mean temperature in the
upper convection zone. Solid contours of the streamfunction represent positive
ψ (i.e., clockwise flow), broken contours are negative ψ (i.e., counterclockwise
flow). Both simulations have a resolution of N_z = 201 and N_n = 100; $\Delta t = 10^{-7}$
for the left and 10^{-6} for the right simulation.

regions controls the vigor of the convection, the amount of convective overshoot,
and the frequency of the waves.

To establish and maintain part of the box superadiabatic and part subadiabatic,
a time-independent reference temperature could be prescribed as a function of z.
The (total) horizontal-mean temperature would than be the sum of this reference

temperature and the $n = 0$ temperature perturbation relative to it, which would be allowed to evolve in time.

Alternatively, one could continue to have a constant reference state temperature but prescribe a time-independent reference state heating/cooling term, $\overline{Q}$, as a function of z, which would be added to the right side of Eq. 2.10 for (only) $n = 0$. If, as in the sun, one wishes to simulate a superadiabatic convection zone overlying a subadiabatic stable zone, $\overline{Q}$ would be defined as zero everywhere except in a local region spanning the desired interface of these two zones where it would be positive (i.e., a heating rate), possibly with a gaussian shape in z. This could model the convergence of the upward radiative heat transfer due to a larger opacity in this region and would cause the mean ($n = 0$) temperature perturbation to increase with z below the interface (i.e., be subadiabatic) and to decrease with z above it (i.e., be superadiabatic). If, on the other hand, one wishes to simulate convection below a stable region, a similar $\overline{Q}$ could be prescribed, but now it would be negative (i.e., a cooling rate) to model the divergence of radiative heat flux.

Yet another method for maintaining part of the box superadiabatic and part subadiabatic would be to add a thermal forcing term to the right side of the temperature equation (2.10) for $n = 0$ that continually nudges $T_0(z)$ toward the desired profile. This would mimic some underlying heating and cooling sources like those resulting from the convergence and divergence of radiative heat flux.

As an example of this last method, choose the interface z-level between the upper convection zone and the lower stable region to be at the nondimensional height of, say, $z_s = 0.75$, making the corresponding k-index be $k_s = [0.7 (N_z - 1) + 1]$. Then define the desired mean temperature in the lower stable region to be $T_s(k) = z(k)/z(k_s)$. In addition, choose the (nondimensional) mean temperature at the top boundary to be a little less than the value at the interface; let's say $T_0(N_z) = 0.9$. Set the initial $T_0(k)$ to $T_s(k)$ in the lower stable region and set it to $[(1 - z') + z' T_0(N_z)]$ in the upper convection zone, where $z'(k) \equiv 1 - [(1 - z(k))/(1 - z(k_s))]$, i.e., a linear profile that satisfies the interface and top boundary conditions. Initialize $T_1(k)$ and any other desired $T_{n>0}$ with an amplitude proportional to $\sin(\pi z')$; set all $T_{n>0} = 0$ below the interface. Then, in the time integration loop, where the linear term in the temperature equation (Eq. 2.10) is added in, add the following forcing term to the $n = 0$ temperature time derivative:

$$-(T_0(k) - T_s(k))/\tau .$$

Here τ is the prescribed time scale for the thermal forcing. The smaller the value of τ the larger the forcing and so the closer the mean temperature is maintained to the desired temperature, $T_s(z)$. For this example let's set $\tau = 10^{-4}$. Note, no forcing is required in the convection zone, other than the prescribed value of T_0 at the top boundary.

A snapshot of such a scenario with parameters Ra $= 10^8$, Pr $= 1$, and $a = 2$ is displayed in the right-hand panels of Fig. 6.2. Convection is dominated by strong downwelling plumes and is efficient enough to maintain a nearly z-independent mean temperature in the bulk of the upper region. Internal gravity waves are excited whenever the convective plumes slightly penetrate into the stable region. The wave amplitude is too small to significantly modify the mean temperature profile in the stable region, as is seen by the large positive gradient of the mean temperature (from blue at the bottom to red at the interface).

The waves are much more discernible when temperature is plotted without the mean temperature, $T_0(z)$. A movie of this simulation, without the mean temperature, shows that the waves are approximately plane waves; although considerable interference occurs with waves excited by different plumes and with those reflected off the boundaries of the box. Since the direction of a plane wave is governed by its frequency according to Eq. 6.7, a given frequency can produce waves traveling in one of four different directions (up or down, left or right), all making an angle θ with the x-axis, i.e., $\pi/2 \pm \theta$ with the direction of gravity. Also, for our rectangular domain, the four (impermeable) boundaries are all either parallel or perpendicular to the x-axis and z-axis. Therefore, when a plane wave reflects off one of these boundaries, the angle of reflection equals the angle of incidence. This would also be true in a spherical Boussinesq domain (Chapter 10) for which the bottom and top curved boundaries are locally perpendicular to gravity. However, for an impermeable boundary not parallel or perpendicular to the local gravity the angles of incidence and reflection would not be equal. To reduce wave reflection, and therefore interference, in our box simulation, see Section 10.2 for the side boundaries and Section 10.1 for the top and bottom boundaries.

The resulting transport of energy can be inferred from such a movie. The phase velocities of the waves in the stable region are mainly toward the upper right or toward the upper left; the group velocities have a downward component, advecting kinetic energy away from the source at the interface. Also clearly seen is how cold downwelling plumes quickly heat up relative to the subadiabatic mean temperature of the stable region as they penetrate, nearly adiabatically, through the interface. This dynamic process tends to push the interface slightly deeper and, by adiabatic heating, warms up the region just below the interface. A similar process may occur in the solar interior.

6.3 WAVE ENERGY ANALYSIS

The wave properties predicted by the linear diffusionless analysis presented in Section 6.1 can be compared to those of the waves generated in nonlinear simulations of small amplitude waves when running with a high Ra and significant spatial and temporal resolution. A dispersion relation, for example, can be produced for a 2D numerical simulation of internal gravity waves by plotting the streamfunction for each Fourier mode, ψ_n, as a function of the frequency, $\bar{\omega}$ (e.g., Rogers & Glatzmaier, 2005b). This can then be compared to that predicted by Eq. 6.7. In addition, the analyses discussed in Section 5.3 for simulations of thermal convection can be performed on simulations of internal gravity waves.

SUPPLEMENTAL READING

Kundu & Cohen (2008)

EXERCISES

1. *Dispersion relation with diffusion*
   Derive the dispersion relation for internal gravity waves in a Boussinesq fluid, like that of Eq. 6.7, but with viscous and thermal diffusion included.
2. *Group velocity*
   Derive the expression for the group velocity given in Eq. 6.10 for a linear nondiffusive internal gravity wave.
3. *Internal gravity wave energy flux*
   Show that the rate of change of wave energy density (i.e., the sum of kinetic and gravitational potential energy densities averaged over a horizontal wavelength) equals the convergence of wave energy flux (see, e.g., Kundu & Cohen, 2008).

COMPUTATIONAL PROJECTS

1. *Internal gravity waves excited by a bolide impact*
   Do the *Convection with a bolide impact* project of Chapter 4 but now with the object impacting a stably stratified fluid.
2. *Internal gravity waves excited by a continuous central source*
   Produce simulations of internal gravity waves in a stably stratified fluid box with a given Brunt-Väisälä frequency (Eq. 6.5). Excite these waves with a forced, continuous oscillation of ψ in a small region in the center of the box. The localization in z is easily done by choosing a small number of z-levels. The localization in x requires a Fourier expansion of modes that produces a fully constructive interference only at the center of the box. Demonstrate that the chosen frequency of the forced oscillation determines the angle of the resulting phase propagation according to Eq. 6.7.
3. *Plot of simulated dispersion relation*
   In our nondimensional spectral formulation, the horizontal wavenumber for ψ is $k_x = n\pi/a$ and the vertical wavenumber is $k_z = m\pi$, where m is the vertical mode number. Using the nonlinear simulation from *Internal gravity waves excited by a continuous central source* of internal gravity waves excited by an internal source, compute the amplitude of $\psi_{nm}(t)$ from $\psi_n(z, t)$ using the inverse of the sine transform in Eq. 3.7c. Then Fourier transform the time series of $\psi_{nm}(t)$, for each set of integers n and m, to frequency space, $\psi_{nm}(\bar{\omega})$. Plot the resulting $\psi_{nm}(\bar{\omega})$ on an $\bar{\omega}$ vs. k plot and on an $\bar{\omega}$ vs. k_x plot, where k is the total wavenumber equal to $(k_x^2 + k_z^2)^{1/2}$. Compare your plot to the linear nondiffusive dispersion relation (6.7) using the z-averaged Brunt-Väisälä frequency.
4. *Vertical dependence of dispersion relation*
   Continue the analysis of the *Plot of simulated dispersion relation* project but, instead of Fourier transforming ψ_n to k_z-space, select three different z-levels and Fourier transform the time series of ψ_n on each of these three levels to frequency space. Then make $\bar{\omega}$ vs. k_x plots for these three z-levels to see how the dispersion relation depends on z.

Chapter Seven

Double-Diffusive Convection

So far we have considered the diffusion of just one scalar quantity, temperature, and just one source of buoyancy, also temperature. Consider now a fluid composed of two constituents, a primary constituent and a small concentration of a secondary constituent with a different density, for example, salt in water (an ocean), MgO in a magmatic melt (a magma chamber), sulfur in liquid iron (a planetary core), or heavy elements in hydrogen (a giant planet or star). In these cases, buoyancy is partly thermal buoyancy, which is what we have considered so far, and partly compositional buoyancy due to variations in the concentration of the secondary constituent, which diffuses much more slowly than the fluid temperature. In fact, the diffusivity of the secondary constituent within the fluid can be many orders of magnitude smaller than the thermal diffusivity. Double-diffusive instabilities can occur in such a two-constituent fluid when it is in hydrostatic equilibrium with a stable density stratification (i.e., density decreases with height faster than an adiabatic density stratification) if the thermal stratification is stable and the compositional stratification is unstable or vice versa. After the instability grows for a sufficient amount of time under the right conditions the nonlinear processes can produce "staircase" profiles in the vertical direction of the horizontal-mean temperature and composition. These profiles have layers of convection separated by layers of internal gravity waves. Such processes can significantly increase the vertical transport of heat and composition over what would otherwise be due to just diffusion. Considerable research has been focused on double-diffusive convection (e.g., Stern, 1960; Veronis, 1965, 1968; Schubert, 1968; Baines & Gill, 1969; Stevenson & Salpeter, 1977; Piacsek & Toomre, 1980; Spiegel & Weiss, 1982; Schmitt, 1994; Hughes & Weiss, 1995; Merryfield, 1995; Hansen & Yuen, 1995; Stern et al., 2001; Charbonnel & Zahn, 2007; Radko, 2008; Stellmach et al., 2011; Rosenblum et al., 2011). The goals are to better understand the details of the small-scale dynamics, which can have a scale of only a few centimeters in the ocean, and to parameterize the enhanced transports as turbulent "eddy" diffusivities that could be used in large-scale global models. Here we provide a brief introduction to the study of the initial instability and eventual nonlinear evolution.

The thermal and flow structures that develop at the onset of these instabilities depend on the initial horizontally averaged temperature and compositional gradients. Here we first consider the "salt-fingering" instability and then the "semiconvection" instability. To avoid the influences of boundary layers, many numerical modeling studies of double-diffusive convection prescribe all boundaries of the domain to be permeable and periodic (Section 10.2) and apply horizontal-mean (background) vertical gradients of temperature and composition that are constant in space and time. These boundary conditions allow "elevator modes," which are vertical flows alternating in the horizontal direction that are not inhibited by the bottom and top boundaries and therefore are excited at lower Rayleigh numbers than cellular flows within impermeable boundaries. However, to minimize the number of modifications that would need to be made to the model developed in this Part 1, here we continue to prescribe a fluid domain in a 2D box with impermeable boundaries that are isothermal on the bottom and top and insulating on the sides. In this respect, our linear instability analyses are similar to the original studies by Stern (1960) for salt-fingering and by Veronis (1965) for semiconvection.

7.1 SALT-FINGERING INSTABILITY

Consider salt in water with the secondary constituent being the salt (Stern, 1960). Salt water has a greater mass density than that of fresh water and salt diffuses much more slowly in water than does temperature; the thermal diffusivity, κ, is roughly a hundred times greater than the salt diffusivity, κ_ξ. The ratio of these has traditionally been given the symbol τ:

$$\tau \equiv \frac{\kappa_\xi}{\kappa}, \tag{7.1}$$

Note that $1/\tau$ is usually called the *Lewis number*.

A typical situation within the near surface of the equatorial ocean is warm salty water above cold fresh water because of solar heating and evaporation at the surface. That is, using the Boussinesq approximation, the horizontal-mean thermal stratification is stable,

$$\frac{\partial T}{\partial z} > 0 \,, \tag{7.2}$$

and, since salt water is heavier than fresh water, the horizontal-mean compositional stratification is unstable,

$$\frac{\partial \xi}{\partial z} > 0 \,. \tag{7.3}$$

Here, ξ is the perturbation in the local salt concentration relative to a constant (well-mixed) background concentration, i.e., the local density of salt over the local density of the salt water mixture, which is sometimes called the "mixing ratio."

The density perturbation is now determined by an equation of state involving both the temperature perturbation, T (Eq. 1.12), and the perturbation salt concentration, ξ:

$$\rho = \rho_o(\beta\xi - \alpha T) \,, \tag{7.4}$$

where ρ_o is the usual constant reference state density, α is the thermal expansion coefficient,

$$\alpha \equiv -\frac{1}{\rho}\left(\frac{\partial \rho}{\partial T}\right)_{p,\xi} > 0 \,, \tag{7.5}$$

and the density dependence on the salt concentration (the compositional contraction coefficient) is

$$\beta \equiv \frac{1}{\rho}\left(\frac{\partial \rho}{\partial \xi}\right)_{p,T} > 0 . \qquad (7.6)$$

Here we assume both α and β are positive constants (Eqs. 7.5 and 7.6), which is not always the case, and we assume the horizontal-mean density stratification is stable across the depth of the domain,

$$\Delta\rho \equiv \rho_{top} - \rho_{bot} = \rho_o(\beta\Delta\xi - \alpha\Delta T) < 0 , \qquad (7.7)$$

where we define

$$\Delta T \equiv |T_{top} - T_{bot}| \quad \text{and} \quad \Delta\xi \equiv |\xi_{top} - \xi_{bot}| , \qquad (7.8)$$

i.e., also both as positive constants.

Consider what happens as a small parcel of fluid, originally in pressure, thermal, and compositional equilibrium with its surroundings, is displaced upward into warmer surroundings. As in our discussion of the purely thermal convective instability (Section 1.1.1), here we assume the parcel is moved slowly enough that it remains in pressure equilibrium. However, instead of assuming it moves fast enough to ignore heat exchange with the surroundings, here, since $\kappa \gg \kappa_\xi$, we account for the diffusive transfer of heat from the surroundings to the parcel, but we neglect the diffusion of salt. Therefore, as the parcel is displaced upward it heats up and comes into thermal equilibrium with its new surroundings while retaining the smaller salt concentration it had at its original position. This makes the parcel buoyant; therefore it continues to rise. Likewise, a parcel displaced downward continues to sink. This produces long thin salty downflows ("salt-fingers") separated by long thin fresh-water upflows. An estimate of the fingering length scale is given by Stern (1960):

$$\left(\frac{\kappa \nu}{g_o \alpha \, dT/dz}\right)^{1/4} . \qquad (7.9)$$

Like thermal convection, the marginal stability for this double-diffusive instability occurs when a Rayleigh-like number just exceeds some critical value. To study this instability we need to add to the Boussinesq equations (1.11 and 1.14) the following *dimensional* advection-diffusion equation that describes the evolution of the salt concentration,

$$\frac{\partial \xi}{\partial t} = -(\mathbf{v} \cdot \nabla)\xi + \kappa_\xi \nabla^2 \xi ;$$

and we need to add a term to the momentum equation (1.13) that accounts for the full density perturbation, Eq. 7.4, in the buoyancy term. Scaling ξ with $\Delta\xi$ gives the *nondimensional* set of equations for double-diffusive convection:

$$\nabla \cdot \mathbf{v} = 0, \tag{7.10}$$

$$\frac{\partial \mathbf{v}}{\partial t} = -(\mathbf{v} \cdot \nabla)\mathbf{v} - \nabla p + (\mathrm{Ra}\, T - \mathrm{R}_\xi\, \tau\, \xi)\, \mathrm{Pr}\, \hat{z} + \mathrm{Pr}\, \nabla^2 \mathbf{v}, \tag{7.11}$$

$$\frac{\partial T}{\partial t} = -(\mathbf{v} \cdot \nabla)T + \nabla^2 T, \tag{7.12}$$

$$\frac{\partial \xi}{\partial t} = -(\mathbf{v} \cdot \nabla)\xi + \tau\, \nabla^2 \xi. \tag{7.13}$$

We define the *compositional* Rayleigh number as

$$\mathrm{R}_\xi \equiv \frac{g_o \beta \Delta \xi D^3}{\nu \kappa_\xi}. \tag{7.14}$$

The boundary and initial conditions on ξ are prescribed the same way they are prescribed for the temperature perturbation.

Note, sometimes the compositional Rayleigh number, R_ξ, is defined with κ_ξ in Eq. 7.14 replaced by κ. With this alternate choice, R_ξ in Eq. 7.11 and in everything that follows would be replaced by R_ξ / τ.

We proceed from here as we have for thermal convection; that is, we formulate a vorticity equation by taking the curl of the momentum equation and do a Fourier expansion in the horizontal direction and a finite-difference method in the vertical direction. Like T (Eq. 2.8a), ξ is expanded in cosines to satisfy its side boundary conditions. This adds the term

$$- \mathrm{R}_\xi\, \tau\, \mathrm{Pr} \left(\frac{n\pi}{a} \right) \xi_n \tag{7.15}$$

to the right side of Eq. 2.11 and adds the spectral equation

$$\frac{\partial \xi_n}{\partial t} = -[(\mathbf{v} \cdot \nabla)\xi]_n + \tau \left(\frac{\partial^2 \xi_n}{\partial z^2} - \left(\frac{n\pi}{a} \right)^2 \xi_n \right) \tag{7.16}$$

to the system of coupled equations. The constraint on the numerical time step now technically also involves the diffusion of ξ; however, since τ is assumed to be (much) smaller than unity, in practice the usual limit, Eq. 2.19, is a stronger constraint.

We first consider an instability that grows without an oscillation. The critical "thermohaline" Rayleigh number for the condition of "marginal stability," i.e., the condition for which a perturbation neither grows nor decays, is then obtained by replacing the nonlinear advections of T and ξ with their linear approximations (here setting their prescribed nondimensional background gradients to +1), expanding all coefficients in $\sin(m\pi z)$ as in Eqs. 3.7a–c, and setting the time derivatives to zero for this *nonoscillating* instability. The resulting system of algebraic equations is then solved as in Section 3.4. This gives the critical thermohaline Rayleigh number (i.e., the difference of the two Rayleigh numbers) for vertical mode number m and horizontal mode number n:

$$(R_\xi - Ra)_{crit} = \left[\frac{g_o D^3}{\nu} \left(\frac{\beta \Delta \xi}{\kappa_\xi} - \frac{\alpha \Delta T}{\kappa} \right) \right]_{crit}$$

$$= \left(\frac{\pi}{a} \right)^4 \frac{\left(n^2 + (am)^2 \right)^3}{n^2} \equiv R_{mn}. \tag{7.17}$$

For the onset of a nonoscillating salt-fingering instability, $(R_\xi - Ra)$ needs to be greater than the right side of Eq. 7.17. That is, the unstable compositional Rayleigh number needs to be sufficiently larger than the stable thermal Rayleigh number for this *nonoscillating* salt-fingering instability to grow. As for purely thermal convection (Section 3.4), this critical condition is independent of the Prandtl number and the most unstable mode occurs as a single cell ($m = n = 1$) in a box with an aspect ratio of $a = \sqrt{2}$, i.e., the minimum R_{mn} is $27\pi^4/4$. Note, this large horizontal cell size occurs because the thermohaline Rayleigh number considered here is just barely critical; for more supercritical cases the horizontal scale of the dominant mode is significantly smaller than the depth of the box because the higher vertical velocities require thinner cells to be able to diffuse heat fast enough (Stern, 1960).

The choice of R_ξ and Ra is also limited by the stable density stratification constraint for salt-fingering, Eq. 7.7, which is satisfied by requiring

$$R_\xi \tau < Ra. \tag{7.18}$$

Combining Eqs. 7.17 and 7.18 gives the range in which the thermal Rayleigh number needs to be for a nonoscillating salt-fingering instability:

$$R_\xi \tau < Ra < R_\xi - R_{mn}. \tag{7.19}$$

According to this constraint, if Ra were less than $R_\xi \tau$, there would be full-scale compositional convection; and if Ra were greater than $R_\xi - R_{mn}$, the system would be dynamically stable.

7.2 SEMICONVECTION INSTABILITY

Double diffusion can also drive an instability when the vertical gradients of temperature and of the more-dense less-diffusive constituent are both negative (Veronis, 1965). This process is called "semiconvection" in the astrophysical community and "diffusive-convection" in the geophysical fluid dynamics community. In this case, again using the Boussinesq approximation, the horizontal-mean thermal stratification is unstable,

$$\frac{\partial T}{\partial z} < 0, \tag{7.20}$$

and the horizontal-mean compositional stratification is stable,

$$\frac{\partial \xi}{\partial z} < 0. \tag{7.21}$$

For example, consider the upper layer of the Arctic Ocean where cold fresh water from the melting of sea ice can exist above a relatively thin layer of warm salty water that has circulated in from the Atlantic Ocean. The overall (perturbation) density stratification within this region is again stable:

$$\Delta\rho \equiv \rho_{top} - \rho_{bot} = \rho_o(\alpha\Delta T - \beta\Delta\xi) < 0 . \tag{7.22}$$

Since both ΔT and $\Delta\xi$ are still defined as positive, their contributions to the drop in the perturbation density across the domain in Eq. 7.22 for this scenario are reversed relative to those in Eq. 7.7.

Another example might be the region just above a nuclear-burning convective core of a massive star (more massive than the sun) where the mean temperature gradient is superadiabatic but the heavy element concentration decreases sufficiently fast with radius (because of hydrogen burning) that the overall (perturbation) density stratification is stable (Eq. 7.22), which precludes full-scale convection according to the *Ledoux criterion* (Ledoux, 1947). (The *Schwarzschild criterion* says that where there is no compositional gradient the mean temperature gradient needs to be subadiabatic to preclude full-scale convection.) However, since τ could be as small as 10^{-7} in stellar interiors, semiconvection might exist in such a Ledoux-stable region, which would enhance the fluxes of heat and composition above their diffusive values.

To see how semiconvection can occur, again consider an upward displacement of a small parcel of fluid during which heat is now transferred from the parcel to the surroundings while maintaining a heavy-element concentration higher than its new surroundings. Therefore, when released, the parcel sinks, quickly heats up as heat is now transferred into the parcel, and, after passing its original position, becomes compositionally buoyant. That is, now there is an internal gravity wave oscillation. The oscillation is "overstable," i.e., for a slightly supercritical situation, the amplitude of the oscillation grows with time because the parcel's temperature lags behind that of its surroundings as it moves away from its original position. Consequently, this temperature perturbation, via thermal buoyancy, tries to accelerate the parcel a little farther from its original position during each oscillation while the dominant compositional buoyancy always accelerates it toward its original position.

Here again we first consider a nonoscillating instability. The system of equations for a nonoscillating onset of semiconvection is identical to that for salt-fingering (Eqs. 7.10–7.16). We also continue to define Ra, R_ξ, ΔT, and $\Delta\xi$ (Eqs. 7.8) as positive constants but with the understanding that the *nondimensional* boundary conditions for semiconvection are

$$T_{bot} = 1, \qquad T_{top} = 0, \qquad \xi_{bot} = 1, \qquad \xi_{top} = 0,$$

the opposite of what they are for salt-fingering.

We can therefore find the critical thermohaline Rayleigh number for *nonoscillating* semiconvection using the same method we used in Section 7.1 to arrive at Eq. 7.17, except that now, in the linear approximation for advection of T and ξ, the nondimensional vertical gradients of the background temperature and composition are both set to -1. This reverses the order (relative to that for salt-fingering) of R_ξ and Ra in the semiconvection expression for the critical thermohaline Rayleigh number:

$$(\text{Ra} - \text{R}_\xi)_{crit} = \left[\frac{g_o D^3}{\nu}\left(\frac{\beta\Delta\xi}{\kappa_\xi} - \frac{\alpha\Delta T}{\kappa}\right)\right]_{crit}$$

$$= \text{R}_{mn} \, . \tag{7.23}$$

(Recall that Ra and R_ξ are still both defined as positive.) For the onset of a nonoscillating semiconvection instability, $(\text{Ra} - \text{R}_\xi)$ needs to be greater than the right side of Eq. 7.23. That is, the unstable thermal Rayleigh number now needs to be sufficiently larger than the stable compositional Rayleigh number for a nonoscillating semiconvection instability to grow.

The choice of R_ξ and Ra is also limited by the stable (perturbation) density stratification constraint for semiconvection, Eq. 7.22, which is satisfied by requiring

$$\text{R}_\xi \tau > \text{Ra} \, . \tag{7.24}$$

However, since τ is assumed to be less then unity, Eqs. 7.23 and 7.24 cannot both be satisfied, which suggests that semiconvection begins as an oscillating instability.

7.3 OSCILLATING INSTABILITIES

Consider now the possibility that the onsets of these instabilities at marginal stability have an amplitude that oscillates in time. To find the critical thermohaline Rayleigh number for such an onset and the dispersion relation for the oscillation we revisit the set of linear equations; but now, instead of setting the time derivatives to zero, we assume the spectral coefficients for the temperature, vorticity, and streamfunction have an $\exp(-i\bar{\omega}t)$ time dependence, where $\bar{\omega}$ is strictly a real constant. (The negative sign in the exponent is an arbitrary choice.) We want to find the conditions that would result in a zero growth rate of the oscillation amplitude (i.e., a zero imaginary part of $\bar{\omega}$) to find the marginal stability constraint on the Rayleigh numbers for the onset of an oscillating instability.

Solving the same set of linear equations as above, but now with the sinusoidal time dependence, we get a cubic equation for the frequency (Stern, 1960; Veronis, 1965):

$$\bar{\omega}^3 + C_2\bar{\omega}^2 + C_1\bar{\omega} + C_0 = 0 \, , \tag{7.25}$$

where the constant coefficients

$$C_0 \equiv i\tau\text{Pr}\left[-k^6 \pm \left(\text{R}_\xi - \text{Ra}\right)\left(\frac{n\pi}{a}\right)^2\right] \, ,$$

$$C_1 \equiv -\text{Pr}\left[\left(1 + \tau + \frac{\tau}{\text{Pr}}\right)k^4 \pm \left(\text{R}_\xi\tau - \text{Ra}\right)\left(\frac{n\pi}{a}\right)^2\frac{1}{k^2}\right] \, ,$$

$$C_2 \equiv ik^2(1 + \text{Pr} + \tau) \, ,$$

with a 2D wavenumber squared defined as

$$k^2 \equiv \left[(m\pi)^2 + \left(\frac{n\pi}{a}\right)^2\right] \, .$$

Here, the $\pm$ means use the $+$ for salt-fingering and use the $-$ for semiconvection.

Equation 7.25 has three roots. One is $\bar{\omega} = 0$, the nonoscillating case, which we have already considered. That is, $C_0 = 0$ is Eq. 7.17 for salt-fingering and Eq. 7.23 for semiconvection. As mentioned, Eq. 7.23 is incompatible with our double-diffusive condition that the overall perturbation density decreases with height (Eq. 7.24).

The other two nonzero roots are obtained by separately setting the real and imaginary parts of Eq. 7.25 to zero. Setting the real part to zero gives the (nondimensional) dispersion relation for the oscillation:

$$\bar{\omega}^2 = k^4 (\text{Pr} + \tau \text{Pr} + \tau) + \frac{N_{nondim}^2}{k^2} \left(\frac{n\pi}{a} \right)^2 , \tag{7.26}$$

where the nondimensional Brunt-Väisälä frequency is

$$N_{nondim}^2 \equiv (\text{R}_\xi \tau - \text{Ra})\text{Pr} . \tag{7.27}$$

Note, the dimensional Brunt-Väisälä frequency is

$$N^2 \equiv g_o \left(\alpha \frac{d\overline{T}}{dz} - \beta \frac{d\overline{\xi}}{dz} \right) .$$

For semiconvection, N_{nondim} is real because of Eq. 7.24; the onset of semiconvection is like an internal gravity wave driven by the imbalance resulting from double diffusion.

Substituting Eq. 7.26 into the imaginary part of Eq. 7.25 gives the stability constraint:

$$\left(\text{Ra} - \text{R}_\xi \tau \frac{(\text{Pr} + \tau)}{(\text{Pr} + 1)} \right)_{crit} = \frac{(\text{Pr} + 1 + \tau)(1 + \tau + \tau/\text{Pr}) \pm \tau}{(\text{Pr} + 1)} \text{R}_{mn} . \tag{7.28}$$

Again, the $\pm$ means use the $+$ for salt-fingering and use the $-$ for semiconvection. For a growing salt-fingering instability the corresponding supercritical value of the left side of Eq. 7.28 needs to be less than the right side; that is, R_ξ needs to be larger than its critical value (all other parameters being the same). On the other hand, for a growing semiconvection instability the corresponding supercritical value of the left side needs to be greater than the right; that is, Ra needs to be larger than its critical value (all other parameters being the same). In the limit of $\tau \to 0$,

$$\left(\text{Ra} - \text{R}_\xi \tau \frac{\text{Pr}}{(\text{Pr} + 1)} \right)_{crit} \to \text{R}_{mn} . \tag{7.29}$$

Note, $\text{R}_\xi \tau$ is finite.

Overstable semiconvection requires both Eqs. 7.24 and 7.28 to be satisfied and therefore occurs when $\text{R}_\xi \tau$ is in the range

$$\text{Ra} < \text{R}_\xi \tau < \frac{(\text{Pr} + 1)}{(\text{Pr} + \tau)} \left[\text{Ra} - \frac{(\text{Pr} + \tau)(1 + \tau)}{\text{Pr}} \text{R}_{mn} \right] , \tag{7.30}$$

assuming Ra is less than the far right side of Eq. 7.30 for the chosen set of parameters. Recall that for semiconvection, $\text{R}_\xi \tau$ is the stabilizing effect. If $\text{R}_\xi \tau$ were greater than the far right side of Eq. 7.30 (and greater than Ra), the system would be stable; if $R_\xi \tau$ were less than Ra, there would be full-scale thermal convection. In the limit of $\tau \to 0$ (Eq. 7.29) and Pr $\to 0$, this constraint approaches the purely thermal convection constraint (Eq. 3.8) even though the overall perturbation density stratification is stable.

Overstable salt-fingering requires both Eqs. 7.18 and 7.28 to be satisfied and therefore occurs when $R_\xi \tau$ is in the range

$$\text{Ra} > R_\xi \tau > \frac{(\text{Pr}+1)}{(\text{Pr}+\tau)} \left[\text{Ra} - \frac{(\text{Pr}+1+\tau)(1+\tau+\tau/\text{Pr})+\tau}{(\text{Pr}+1)} R_{mn} \right] , \quad (7.31)$$

assuming Ra is greater than the far right side of Eq. 7.31 for the chosen set of parameters. Recall though that a nonoscillating salt-fingering instability occurs when Ra is less than $R_\xi - R_{mn}$ (Eq. 7.19); so this also needs to be checked to determine the type of salt-fingering instability that occurs for a chosen set of parameters and the mode mn that is the most unstable.

7.4 STAIRCASE PROFILES

Salt-fingering and semiconvection are both driven mainly by the stratification of the more-dense less-diffusive constituent. However, in both scenarios, after evolving beyond the onset of the instability, thermal diffusion between the moving parcel and the surroundings can alter the initial linear vertical profile of the horizontal-mean temperature into a "staircase" profile. This evolution of the temperature profile can be studied via nonlinear simulations. The first salt-fingering simulation was produced by Piacsek & Toomre (1980) and the first semiconvection simulation for stars by Merryfield (1995).

7.4.1 Salt-Fingering Staircase

Consider salt-fingering. The mean temperature profile (relative to the adiabat) over which fluid parcels rise (sink) as they transfer heat from (to) the surroundings becomes nearly isothermal (i.e., adiabatic). That is, the convection, driven by compositional buoyancy, tends to reduce the horizontal-mean vertical temperature gradient. However, this process has to be confined to local regions if the temperature (relative to the adiabat) is forced to be greater at the top boundary than at the bottom boundary. That is, within thin interfaces between these local well-mixed nearly adiabatic convective layers the mean temperature needs to increase significantly with height in order to have a net increase in temperature from the bottom boundary to the top boundary. Thermally driven internal gravity waves are continually excited within these thin, strongly subadiabatic interfaces by the convection that occurs adjacent to the interfaces. A schematic of the resulting "staircase" profile of the horizontal-mean temperature is displayed in Fig. 7.1.

The vertical fluxes of heat and composition through an evolved staircase profile are significantly greater than what they would be if they were only due to diffusion down the initial linear gradients. This occurs because compositional convection in

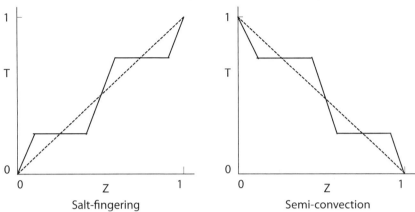

Figure 7.1 Schematic plots of evolved (solid lines) horizontal-mean temperature T (relative
to the adiabat) vs. height z. The initial profiles of T are indicated as dashed lines.
The nearly constant (adiabatic) T levels represent regions of compositionally
driven convection for salt-fingering and compositionally driven internal gravity
waves for semiconvection. The interfaces defined by the nearly step changes in T
are strongly subadiabatic for salt-fingering and strongly superadiabatic for semiconvection.

the well-mixed layers is so much more efficient than diffusion; and, within the thin,
strongly stable interfaces that develop, internal gravity waves are very efficient in
transporting heat and composition. That is, the enhanced eddy fluxes are due to
advective processes. For salt-fingering, the effective eddy compositional diffusivity
tends to be greater than the eddy thermal diffusivity. The kinetic energy of this flow
structure is obtained from the continual conversion of the gravitational potential energy into kinetic energy within the net unstable compositional stratification, which
is (externally) maintained by the boundary conditions.

Salt-fingering staircase profiles have been observed in the oceans, mainly in the
tropics where the greater solar heating and resulting evaporation rate at the surface
produce warm more-salty surface water above cold fresher deep water. This process
has been studied via laboratory experiments and in numerical simulations (e.g.,
Stellmach et al., 2011). The process usually begins with a large number of relatively
thin convective layers; but, as it evolves, convective layers merge and the change
in temperature within the stable interfaces increases until the system settles into a
preferred configuration.

7.4.2 Semiconvection Staircase

A staircase temperature profile can also develop with semiconvection. However,
in this process internal gravity waves, driven by compositional buoyancy, tend to
increase the horizontal-mean vertical temperature gradient, which in this case also
makes it nearly adiabatic because now rising (sinking) parcels heat up (cool down)

the surroundings via thermal diffusion. Again, this cannot extend over the entire vertical domain if the temperature (relative to the adiabat) at the bottom boundary is forced to be greater than that at the top boundary. Therefore, local well-mixed nearly adiabatic regions of internal gravity waves become separated by thin interfaces in which the temperature decreases significantly with height. Thermal convection is driven within these thin strongly superadiabatic interfaces. A schematic of the resulting "staircase" profile of the horizontal-mean temperature is displayed in Fig. 7.1.

The vertical fluxes of heat and composition through the evolved staircase profile in semiconvection are also significantly greater than what they would be if they were only due to diffusion down the initial linear profiles. Here it occurs because of advective transport in the well-mixed layers by internal gravity waves and, within the thin interfaces, by the thermal convection. For semiconvection, the effective eddy thermal diffusivity tends to be greater than the eddy compositional diffusivity. The kinetic energy is again obtained from the continual conversion of the gravitational potential energy, now due to the net unstable thermal stratification, which is (externally) maintained by the boundary conditions.

7.4.3 Other Double-Diffusive Processes

We have been assuming that the externally maintained gradients of temperature and composition are only in the vertical direction. In reality they can also have a small horizontal component, which causes "lateral intrusions" of heat and composition. These relatively thin and slightly tilted layers can transport heat and composition over very large horizontal distances.

We have also been assuming that the secondary constituent is more dense than the primary. However, this is not required for a double-diffusive instability. Consider the Earth's core, for example, which has a solid inner region and a liquid outer region; both are mainly iron with small concentrations of *less dense* elements. As the core slowly cools, liquid iron plates onto the solid inner core more easily than the lighter elements, which produces a source of compositional buoyancy at the bottom boundary of the fluid outer core (the "inner-core boundary," ICB). Latent heat is also released in this process and therefore provides a source of thermal buoyancy. Heat is convectively transported to the top boundary of the fluid core (the "core-mantle boundary," CMB) where it diffuses into the mantle. However, the lighter constituent, which is also convectively transported to the CMB, accumulates there and possibly develops a compositionally stable stratification within the unstable thermal stratification. Semiconvection might therefore occur in this upper region of the Earth's outer fluid core. In such a scenario the β (Eq. 7.6) would be negative and horizontal-mean compositional gradient, $d\xi/dz$, would be positive.

Double-diffusive processes and their evolved structures can be affected by many other instabilities caused by, for example, shear flows, Coriolis forces, turbulence, chemical reactions, radiation pressure (which depends on the particular secondary element and its excitation and ionization state), and magnetic pressure. For example, several studies (e.g., Schubert, 1968; Spiegel & Weiss, 1982; Hughes & Weiss, 1995) have been made of double-diffusive instabilities due to buoyancy being partly thermal and partly magnetic in 2.5D models (Sections 10.5 and 10.6) for which the background magnetic field is horizontal but in the direction in which no variables depend (the y-direction for a box, the longitudinal direction for a spherical shell). In these studies magnetic buoyancy is due to magnetic pressure (Chapter 11) and magnetic diffusivity is assumed to be much less than thermal diffusivity. It has been suggested that this double-diffusive magnetic instability initiates disruptions of toroidal magnetic field stored just below the solar convection zone, which then buoyantly rise to the solar surface, forming sunspots.

7.5 DOUBLE-DIFFUSIVE NONLINEAR SIMULATIONS

Our nonlinear 2D model, Eqs. 2.10 – 2.12, can easily be modified to simulate the evolution of these two double-diffusive processes by adding the expression 7.15 to the right side of Eq. 2.11 and updating ξ using Eq. 7.16. As mentioned above, ξ is expanded in $\cos(n\pi x/a)$ and its nondimensional boundary and initial conditions are the same as those for T. The nonlinear advection in Eq. 7.16 can be calculated exactly as it is in Eq. 2.10 (Chapter 4). However, one needs to choose R_ξ, Ra, τ, and Pr to satisfy Eq. 7.19 if salt-fingering is desired. An effective *density ratio*,

$$\frac{\alpha \Delta T}{\beta \Delta \xi} = \frac{\text{Ra}}{R_\xi \tau},$$

for salt-fingering is about 1.1 (Stellmach et al., 2011). If, on the other hand, semi-convection is desired, Eq. 7.30 should to be satisfied, which would make the density ratio less than one. Besides considering the parameter regimes for double-diffusive instabilities when choosing values for Ra, R_ξ, and τ, also estimate the horizontal length scale for individual flow structures, Eq. 7.9. Also recall that for salt-fingering both $T_{n=0}$ and $\xi_{n=0}$ are set to 0 at the bottom boundary and to 1 at the top boundary when using a nondimensional code; the opposite boundary conditions are set for semiconvection. A small initial $\sin \pi z$ perturbation in, say, the $n = 1$ and $n = N_n/2$ modes of T or ξ is needed to trigger the instability.

A major computational challenge for double-diffusive convection simulations is to use a realistically small τ because, roughly speaking, the smaller κ_ξ is relative to κ the smaller the typical compositional length scale will be relative to the temperature length scale. When representing both T and ξ with the same spatial resolution, as we have assumed here, that spatial resolution needs to be adequate for simulating the fine-scale structures of ξ and therefore will likely be much greater than is needed for structures of T. A related challenge is to simultaneously resolve the small-spatial-scale short-time-scale structures (fingers) and the large-spatial-scale long-time-scale structures (staircases). The Galerkin method outlined in this Part 1 is not a particularly efficient method to deal with these computational challenges.

A more efficient way to deal with the need for an extremely small τ could be to treat the evolution of ξ using a Lagrangian method, instead of an Eulerian method (Section 1.1.2), which would approximate the extreme limit of a nondiffusive secondary constituent. For example, a "particle-in-cell" method uses millions of *tracer particles* to represent the distribution of ξ. These tracer particles are advected each time step by the fluid velocity using, for example, a Runge-Kutta scheme. The accuracy of the advection is determined by the order of the Runge-Kutta scheme, the order of the interpolation of the fluid velocity to the particle positions each time step, and the size of the time step. The degree that these are inaccurate contributes to "artificial" (i.e., numerical) diffusion. The fluid domain is divided into cells, each containing a central grid point; and the Eulerian value of ξ at a each grid point is proportional to the number of tracer particles in the cell containing the grid point. The concentration, ξ, in grid space could then be transformed to spectral space each time step if the Fourier spectral method is used in the horizontal direction. Alternatively, a fully finite-difference method (Section 9.3) could be employed.

The numerical methods introduced in Part 2 could also help with the computational challenges of simulating double-diffusive convection. For example, if a spectral method is desired, the spectral-transform method of computing the nonlinear terms would allow significantly greater numerical resolution compared to the Galerkin method. Also, to avoid the influence of impermeable boundaries and non-conducting side boundaries (which inhibit staircase structures), one could employ periodic boundary conditions on all the boundaries and maintain the horizontal-mean vertical gradients of temperature and composition by imposing constant (in both space and time) background gradients, which would provide linear parts to the advection of temperature and composition in addition to the nonlinear advection terms. With such a model a fully spectral method using Fourier expansions in all directions could be a reasonable choice (Stellmach et al., 2011).

SUPPLEMENTAL READING

Brandt & Fernando (1995)

EXERCISES

1. *Onset of a nonoscillating salt-fingering instability*
   Derive the analytic expression for the critical thermohaline Rayleigh number for the onset of nonoscillating salt-fingering (Eq. 7.17).
2. *Onset of an oscillating double-diffusive instability*
   Derive the cubic equation (Eq. 7.25) for the frequency of an oscillating double-diffusive instability and solve for the dispersion relation (Eq. 7.26) and stability constraint (Eq. 7.28).

COMPUTATIONAL PROJECTS

1. *Double-diffusive linear stability analyses*
   Convert a linear convection code to a linear double-diffusive code and check, using the procedure described in Section 3.2, that it predicts the critical

thermohaline Rayleigh number for salt-fingering, Eq. 7.17, and for semiconvection, Eq. 7.30. Recall that for the salt-fingering instability the linear approximations to the advection of T and ξ are constructed by setting their prescribed nondimensional background gradients to $+1$; for the semiconvection instability, these are set to -1.

2. *Double-diffusive nonlinear simulations: initial linear profiles*
Convert a nonlinear 2D Boussinesq convection code to a nonlinear double-diffusive code. For a τ somewhat less than one, a $\text{Pr} = 1$, and $a = \sqrt{2}$, run a salt-fingering simulation and a semiconvection simulation, choosing appropriate values for Ra and R_ξ for each case, considering the horizontal length scale estimate of fingers, Eq. 7.9. Use linear profiles of the horizontal mean $(n = 0)$ T and ξ for initial conditions, i.e., the dashed lines in Fig. 7.1. Plot the total kinetic energy and vertical heat flow as functions of time. Analyze how the pattern and evolution of the flow and the horizontal-mean profiles of temperature and composition depend on Ra and R_ξ.

3. *Double-diffusive nonlinear simulations: initial step-function profiles*
Run double-diffusive simulations and analyses as outlined in the *Double-diffusive nonlinear simulations: initial linear profiles* exercise but now with initial step-function profiles for the horizontal-mean T and ξ. That is, for salt-fingering set the initial $n = 0$ parts of the nondimensional temperature and composition to 1 for $z > 0.5$ and to 0 for $z < 0.5$; reverse this initial condition for semiconvection. Set the $n = 1$ and $n = N_n/2$ modes of ξ to 0.1 at $z = 0.5$ and zero at all other z-levels. Monitor how the thicknesses of the temperature and compositional interfaces grow with time. For comparison, simulate a "Rayleigh-Taylor" instability by setting the initial $n = 0$ part of T to 0 and ξ to 1 for $z > 0.5$ and vice versa for $z < 0.5$, including the corresponding bottom and top boundaries; that is, make both the temperature and composition unstable. This Rayleigh-Taylor case could also be done with the original non-double-diffusive code using just T.

Temperature

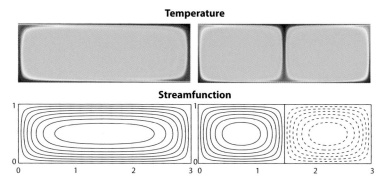

Streamfunction

Plate 1a Two steady-state solutions, illustrated with profiles of the temperature and streamfunction plotted on the horizontal (x) and vertical (z) grid for Ra $= 10^6$, Pr $= 0.5$, and $a = 3$. The solution on the left was initialized with an $n = 1$ temperature mode; whereas the one on the right was initialized with both the $n = 1$ and $n = 8$ modes. For temperature, red corresponds to hot buoyant upflow and blue to cold heavy downflow; green is vanishingly small relative to the background temperature. Solid contours of the streamfunction represent positive ψ (i.e., clockwise flow), broken contours are negative ψ (i.e., counterclockwise flow).

Temperature

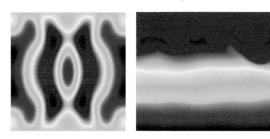

Streamfunction

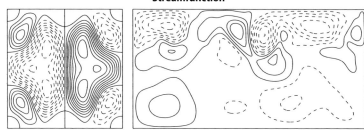

Plate 1b Snapshots of two time-dependent solutions, illustrated with profiles of temperature and streamfunction plotted on the horizontal (x) and vertical (z) grid. The snapshot illustrated in the left-hand panels is for Ra $= 10^8$, Pr $= 1$, and $a = 1$; it has a stable mean temperature profile and was initialized with a localized perturbation in ω at the center of the box, as described in the text. The snapshot illustrated in the right-hand panels is for Ra $= 10^8$, Pr $= 1$, and $a = 2$; it has a stable mean temperature profile in the lower three-quarters of the box maintained with a thermal forcing time of 10^{-4} and an unstable profile in the upper quarter maintained with a top boundary temperature of 0.9 (see description in the text). For temperature, red represents hot buoyant fluid and blue cold heavy fluid. The horizontal-mean temperature, $T_0(z)$, is not included in the image on the left so the small temperature perturbations can be seen. The mean temperature profile is included in the image on the right, displaying the significant increase in mean temperature with height in the stable region and the nearly adiabatic (i.e., constant) mean temperature in the upper convection zone. Solid contours of the streamfunction represent positive ψ (i.e., clockwise flow), broken contours are negative ψ (i.e., counterclockwise flow). Both simulations have a resolution of $N_z = 201$ and $N_n = 100$; $\Delta t = 10^{-7}$ for the left and 10^{-6} for the right simulation.

	Temperature	**Streamfunction**

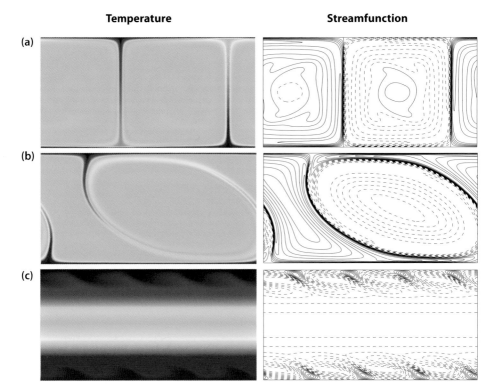

Plate 2 Three snapshots of a Boussinesq simulation with permeable periodic side boundaries. The parameters are Ra $= 10^7$, Pr $= 1$, and $a = 2$. The snapshots are at (a) 0.01, (b) 0.017, and (c) 0.1 thermal diffusion times. The temperature perturbation is represented in the left column with maximum value (1) being dark red to minimum value (0) in dark blue. The corresponding streamlines are plotted on the right with broken lines representing counterclockwise flow. A second-order finite-difference method on a uniform vertical grid ($N_z = 256$) is employed with a full Fourier spectral method in the horizontal direction ($N_x = 512$, $N_n = 170$). A spectral-transform method is used to compute the nonlinear terms. The solution is integrated in time with a semi-implicit time integration scheme using a (nondimensional) numerical time step of $\Delta t = 10^{-7}$.

Plate 3 A snapshot of temperature for a Boussinesq simulation of thermal convection in a 2D annulus. The parameters for this case are Ra $= 10^{11}$, Pr $= 0.2$, and $r_{bot}/r_{top} = 0.2$. Yellows represent hot fluid and blues represent cold fluid relative to the background temperature.

Temperature   **Magnetic field lines**

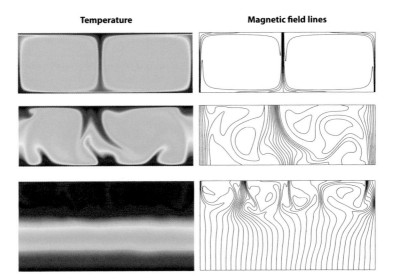

Plate 4a   Snapshots of three magnetoconvection solutions, illustrated with profiles of temperature and vector potential (i.e., magnetic field lines) plotted on the horizontal (x) and vertical (z) grid. The
snapshot illustrated in the top row is the case on the right in Fig. 4.2 with the addition of a
relatively weak vertical background magnetic field ($Q = 10^2$ and $q = 1$). Initially it appears to
be in steady state, like the nonmagnetic version; but after about one thermal diffusion time the
pattern switches to a single cell like that on the left side of Fig. 4.2. The snapshot illustrated
in the middle row is the same case but with a much more intense background field ($Q = 10^4$
and $q = 1$). This case is quite time-dependent. The snapshot illustrated in the bottom row is
the case on the right in Fig. 6.2 with a relatively intense background field ($Q = 10^4$, $q = 1$).
It too is time-dependent as it was without the field. For temperature, red represents hot buoyant fluid and blue cold heavy fluid. The horizontal-mean temperature is included in these images.
Solid contours of the vector potential are magnetic field lines, entering the bottom boundary and
exiting the top.

Temperature   **Magnetic field lines**

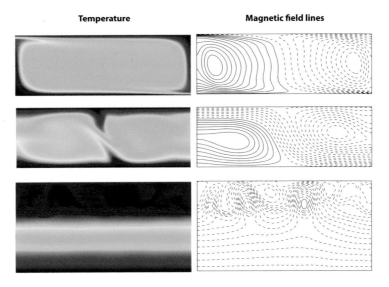

Plate 4b   Snapshots of three magnetoconvection solutions, illustrated with profiles of temperature and vector
potential (i.e., magnetic field lines) plotted on the horizontal (x) and vertical (z) grid. The cases
represented in the top, middle, and bottom rows are the same as those in Fig. 11.1 except that here
there is a horizontal background field. Note the dashed lines represent negative vector potentials,
because of the arbitrary choice we made for the top and bottom boundary values of **A**.

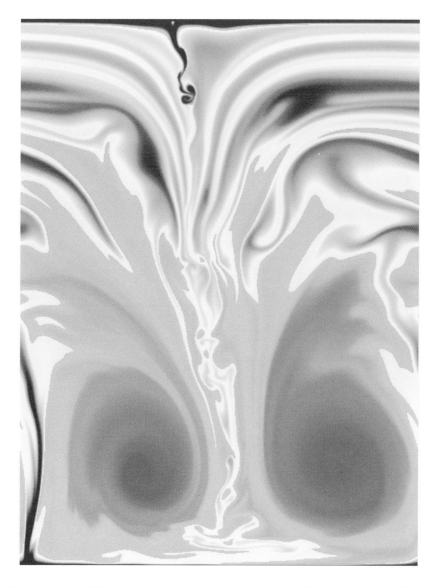

Plate 5 A snapshot of the entropy perturbation in an anelastic simulation of thermal convection in a 2D box with Ra $= 10^{10}$, Pr $= 1$, $N_\rho = 5$, and an aspect ratio of 0.75 (see *Color Plate* 5 for a color version of this figure). Reds represent hot fluid, yellows warm, and blues cold fluid relative to the volume-averaged mean entropy.

Plate 6 A snapshot of the entropy perturbation in an anelastic simulation of thermal convection within a 2D annulus after a million computational time steps with $N_\rho = 3$, Ra $= 1.3 \times 10^{10}$, and Pr $= 1$. Reds represent hot fluid and blues represent cold fluid relative to the constant reference state entropy, with the crossover from blue to yellow at $S = S_{bot} + \Delta S/2$.

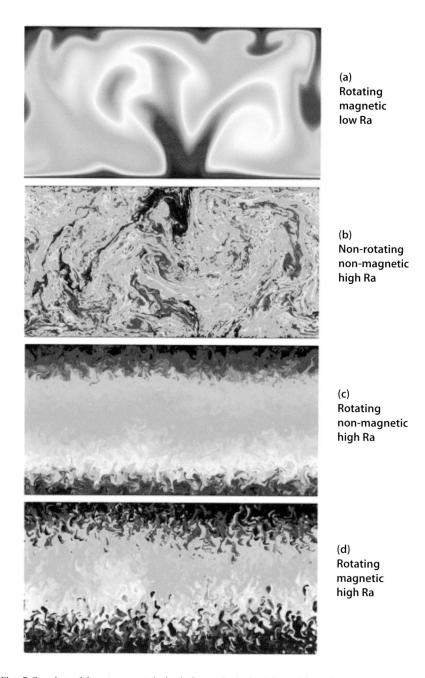

(a)
Rotating
magnetic
low Ra

(b)
Non-rotating
non-magnetic
high Ra

(c)
Rotating
non-magnetic
high Ra

(d)
Rotating
magnetic
high Ra

Plate 7 Snapshots of the entropy perturbation in four anelastic simulations of thermal convection in a 2D box
with an adiabatic polytropic reference state defined by $n = 1$, $N_\rho = 0.2$, and Pr $=$ q $= 1$. Case (a)
has Ra $= 3 \times 10^6$, Ek $= 10^{-4}$, and Q $= 10^4$. The other three cases have Ra $= 3 \times 10^{12}$. Case (b) is
nonrotating and nonmagnetic. Cases (c) and (d) are rotating with Ek $= 10^{-9}$. Case (c) is nonmagnetic
and case (d) has a vertical background magnetic field with Q $= 10^6$. Reds represent hot fluid, yellows
warm fluid, and blues cold fluid. (This material is reproduced from Glatzmaier (2005a) with permission
from *Taylor and Francis Group*, LLC, a division of *Informa plc*.)

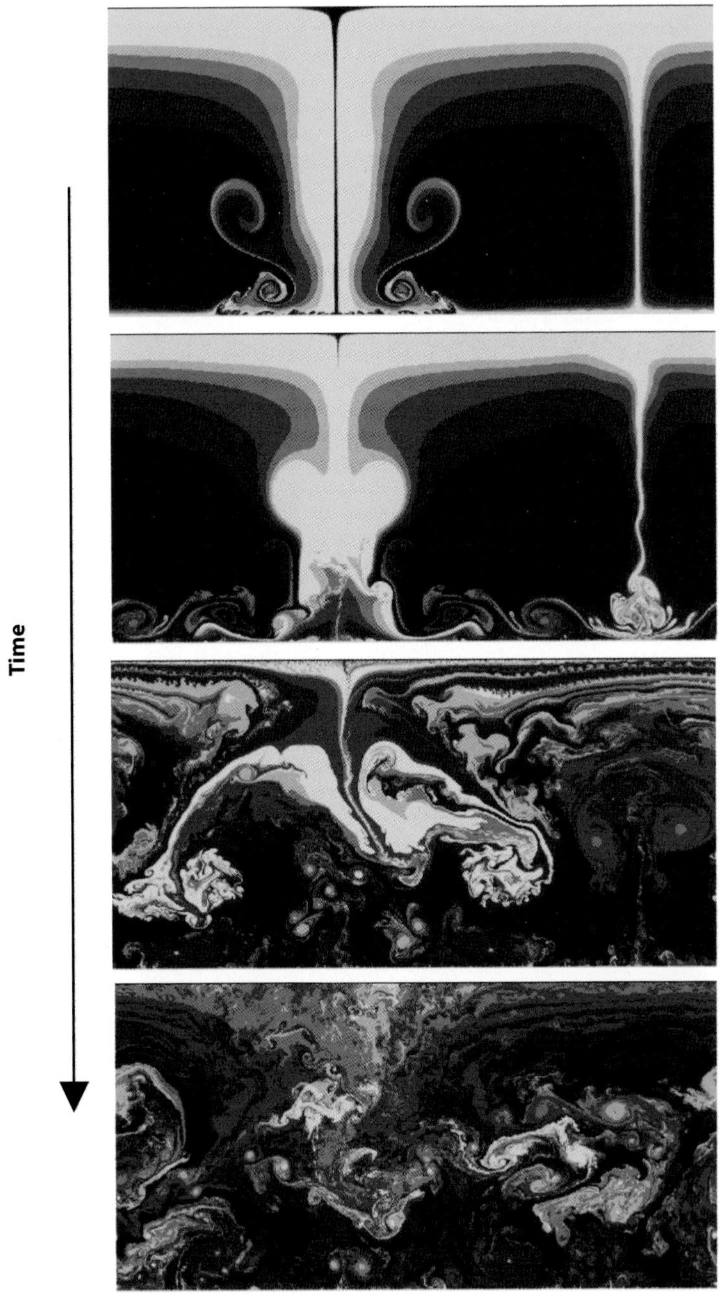

Plate 8 Four snapshots of entropy from an anelastic simulation of thermal convection in a 2D box with Ra $= 2 \times 10^{12}$, Pr $= 1/10$, Ek $= 10^{-9}$, and $N_\rho = 5$. Here dark colors represent cold fluid and light colors hot fluid. (This material is reproduced from Glatzmaier (2005a) with permission from *Taylor and Francis Group*, LLC, a division of *Informa plc*.)

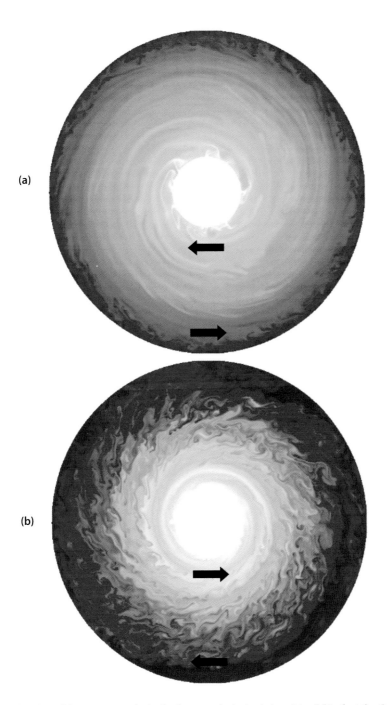

Plate 9 Snapshots of the entropy perturbation for the two anelastic simulations, (a) and (b), of rotating thermal convection in an annulus described in Section 13.3.2. Dark (light) colors represent cold (hot) fluid. The arrows show the angular velocity relative to the rotating frame, which rotates counterclockwise in the inertial frame. (This material is reproduced from Glatzmaier et al. (2009) with permission from *Taylor and Francis Group*, LLC, a division of *Informa plc*.)

Snapshot of entropy

Time averaged angular velocity

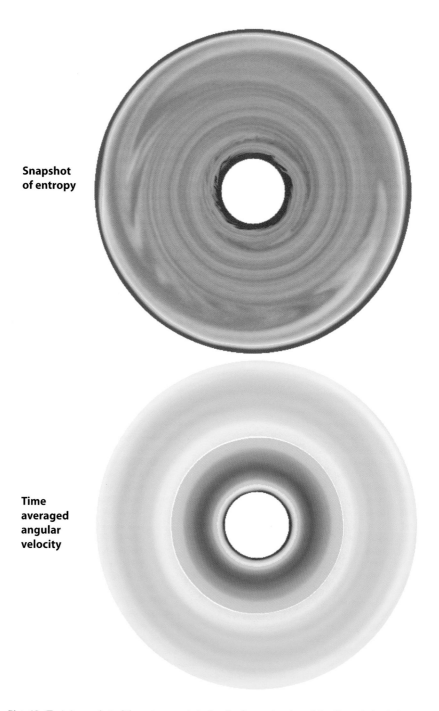

Plate 10 (Top) A snapshot of the entropy perturbation for the continuation of the $N_\rho = 3$ simulation illustrated in Fig. 12.3 but with Ra increased to 10^{11}, Pr decreased to 0.5, and now rotating with Ek = 2×10^{-8}. Reds represent hot fluid; blues represent cold fluid. (Bottom) Time-averaged differential rotation. Reds and yellows represent counterclockwise angular velocity (i.e., prograde flow) and blues represent clockwise (retrograde) flow, both relative to the counterclockwise rotating frame.

Radial component of the magnetic field

At surface   At core-mantle boundary

Geomagnetic field (1980) up to degree 12

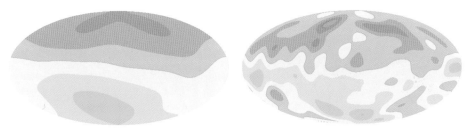

G–R simulation plotted up to degree 12

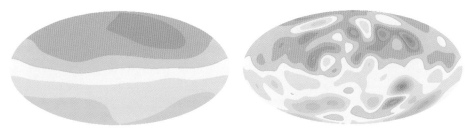

G–R simulation up to degree 95

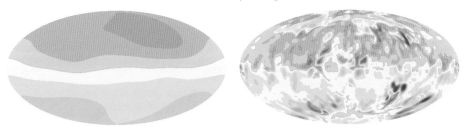

Plate 11 A snapshot of the radial component of the magnetic field simulated with the Glatzmaier-Roberts geodynamo model. The fields are plotted on equal-area (Hammer) projections of the entire core-mantle boundary and of what would be the Earth's surface at two different spatial resolutions: a coarse resolution (up to spherical harmonic degree 12) and a higher resolution (up to degree 95). These are compared to the Earth's field in the year 1980 on both surfaces up to degree 12. Blue represents inward-directed field and yellows and reds outward-directed field. The intensities (colors) are scaled the same, except that each of the three surface images have been multiplied by 10 to produce color intensities similar to those at the core-mantle boundary. (This material is reproduced from Roberts & Glatzmaier (2000) with permission from the *American Physical Society*.)

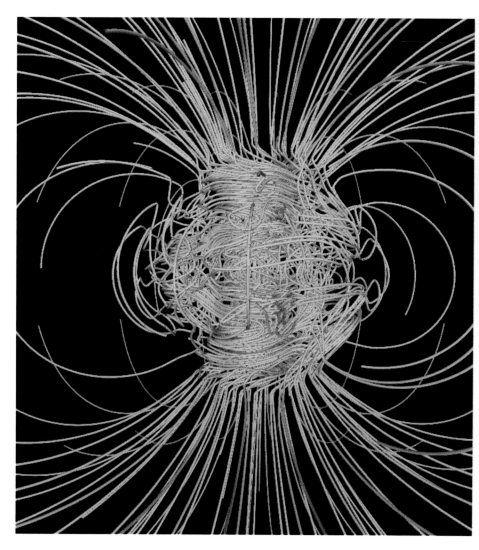

Plate 12 A snapshot of the magnetic field maintained in a geodynamo simulation illustrated as magnetic field lines. Gold field lines are directed outward and blue inward. (This material is reproduced from Glatzmaier & Roberts (1996a) with permission from *Elsevier Ltd.*)

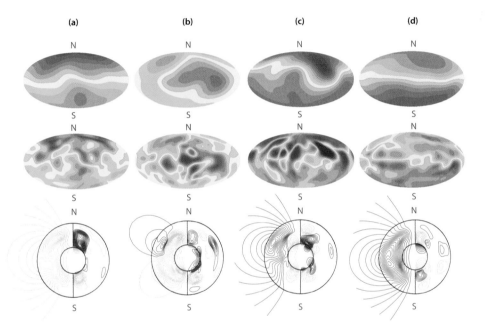

(a)	(b)	(c)	(d)

Plate 13 A sequence of snapshots, displayed at 3000-year intervals during a spontaneous simulated magnetic dipole reversal. (Bottom) The longitudinally averaged magnetic field plotted within the core. The small circle represents the inner-core boundary and the large circle is the core-mantle boundary. The poloidal field is shown as magnetic field lines on the left-hand sides of these plots (blue is clockwise and red is counterclockwise). The toroidal field direction and intensity are represented as contours (not magnetic field lines) on the right-hand sides (red is eastward and blue is westward). (Middle and Top) The radial component of the field on the core-mantle boundary and at what would be the surface, plotted as described in Fig. 13.13. (This material is reproduced from Glatzmaier et al. (1999) with permission from the *Nature Publishing Group*.)

Simulated banded zonal winds

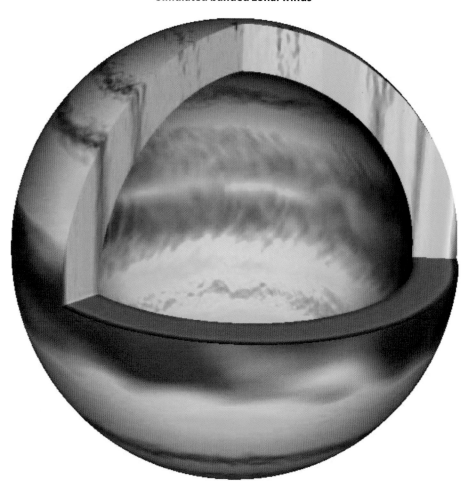

Plate 14 A snapshot of the longitudinal flow from a Saturn dynamo simulation. Reds and yellows are prograde flow relative to the rotating frame and blues are retrograde. (This material is reproduced from Stanley & Glatzmaier (2009) with permission from *Springer Science + Business Media BV.*)

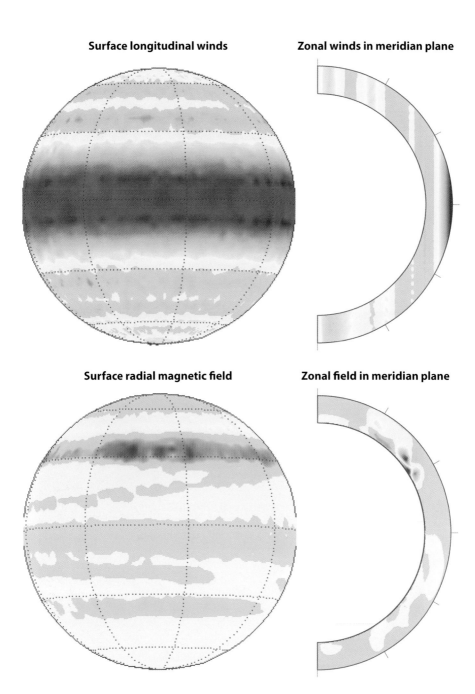

Surface longitudinal winds

Zonal winds in meridian plane

Surface radial magnetic field

Zonal field in meridian plane

Plate 15 A snapshot of a Saturn dynamo simulation, as in Fig. 13.17. (Top, left) The differential rotation (zonal winds) at the surface; red and yellow represent prograde flow, blues are retrograde. (Top, right) The longitudinal average of the zonal winds below the surface; red and yellow are prograde, blue is retrograde. (Bottom, left) The radial component of the magnetic field at the surface; yellow represents outward-directed field, blue is inward. (Bottom, right) The longitudinal average of the toroidal field below the surface; red and yellow are eastward-directed, blue is westward. (This material is reproduced from Stanley & Glatzmaier (2009) with permission from *Springer Science + Business Media BV*.)

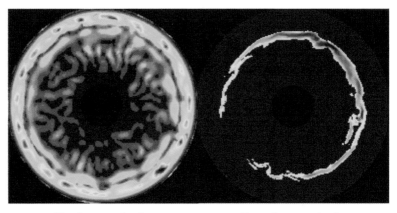

Kinetic energy density   **Magnetic energy density**

Plate 16a A snapshot of the kinetic and magnetic energy densities in the equatorial plane of a 3D dynamo simulation of a giant planet like Saturn. The kinetic energy is greatest near the surface and decays with depth; the magnetic energy peaks in a narrow layer at roughly 20% of the radius below the surface.

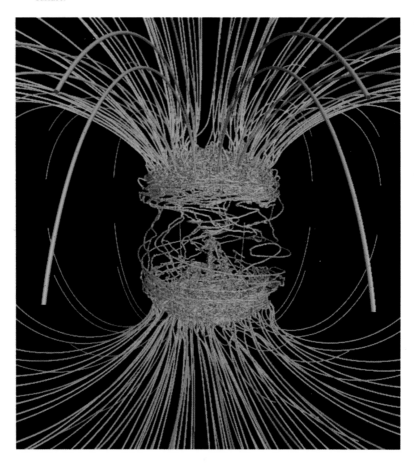

Plate 16b A snapshot of the magnetic field, illustrated with magnetic field lines, maintained in the Saturn dynamo simulation (Fig. 13.17). Gold field lines are directed outward and blue inward. (This material is reproduced from Stanley & Glatzmaier (2009) with permission from *Springer Science + Business Media BV.*)

PART 2
Additional Numerical Methods

.

Chapter Eight

Time Integration Schemes

The explicit Adams-Bashforth scheme for integrating the equations in time, which is described (Eq. 2.18) and employed in Part 1, is relatively simple and efficient since at each time step the time derivatives are needed for only the current and previous time steps. This works fairly well and is second-order accurate, although the accumulated error is proportional to only Δt^2. Many more accurate time integration schemes do exist, which are higher order but computationally more expensive. As an example, in this chapter we describe fourth-order accurate Runge-Kutta and predictor-corrector schemes. However, the stability constraints on the size of the Δt due to linear diffusion (Eq. 2.19) and due to the nonlinear advection (Eq. 4.7) for explicit schemes are still present. In many cases, the former is more severe than the latter. Therefore, we also describe time integration schemes that allow larger time steps (and therefore fewer steps for a given amount of simulated time) by treating the linear diffusion terms implicitly. The nonlinear terms, however, couple all the modes and so would be extremely expensive to treat implicitly; therefore they are usually treated explicitly. This "semi-implicit" scheme greatly improves the efficiency of the code.

8.1 FOURTH-ORDER RUNGE-KUTTA SCHEME

As described in Section 2.4, the explicit second-order Adams-Bashforth time integration scheme for, say, temperature, T, is illustrated in Eq. 2.18 where G_t represents the time derivative at time t. Now, instead, approximate the integral from time t to $t + \Delta t$ of the temperature time derivative with a weighted average of four evaluations of the time derivative: one at the current time t, two at the midpoint time $t + \Delta t/2$, and one at the new time $t + \Delta t$. That is, let the temperature equation be represented as

$$T_{t+\Delta t} = T_t + \frac{\Delta t}{6} \left(G_1 + 2G_2 + 2G_3 + G_4 \right),$$

the vorticity equation be

$$\omega_{t+\Delta t} = \omega_t + \frac{\Delta t}{6} \left(H_1 + 2H_2 + 2H_3 + H_4 \right),$$

and the solution of the Poisson equation for the streamfunction for a given vorticity be represented as

$$\psi_{t+\Delta t} = \psi(\omega_{t+\Delta t}).$$

For the standard fourth-order Runge-Kutta scheme (e.g., Press et al., 1992) applied to our problem the designated functions are

$$G_1 \equiv G(T_t, \ \psi(\omega_t)),$$

$$H_1 \equiv H(T_t, \ \omega_t, \ \psi(\omega_t)),$$

$$G_2 \equiv G(T_t + \frac{\Delta t}{2} G_1, \ \psi(\omega_t + \frac{\Delta t}{2} H_1)),$$

$$H_2 \equiv H(T_t + \frac{\Delta t}{2} G_1, \ \omega_t + \frac{\Delta t}{2} H_1, \ \psi(\omega_t + \frac{\Delta t}{2} H_1)),$$

$$G_3 \equiv G(T_t + \frac{\Delta t}{2} G_2, \ \psi(\omega_t + \frac{\Delta t}{2} H_2)),$$

$$H_3 \equiv H(T_t + \frac{\Delta t}{2} G_2, \ \omega_t + \frac{\Delta t}{2} H_2, \ \psi(\omega_t + \frac{\Delta t}{2} H_2)),$$

$$G_4 \equiv G(T_t + \Delta t G_3, \ \psi(\omega_t + \Delta t H_3)),$$

$$H_4 \equiv H(T_t + \Delta t G_3, \ \omega_t + \Delta t H_3, \ \psi(\omega_t + \Delta t H_3)).$$

One way to program this without involving a significant amount of code modification or additional disk space would be to add arrays tem1, omg1, and psi1 with the same dimensions as tem, omg, and psi, and to add an array here called rk defined as

$$\text{rk(irk, 1)} = 0, \ \frac{\Delta t}{2}, \ \frac{\Delta t}{2}, \ \Delta t,$$

$$\text{rk(irk, 2)} = 0, \ 1, \ 1, \ 1,$$

$$\text{rk(irk, 3)} = 1, \ 2, \ 2, \ 1,$$

for irk = 1,2,3,4. Then, just inside the loop over time steps, add a loop that cycles a counter, irk, from 1 to 4. Inside this loop first set tem1 = tem, omg1 = omg, and psi1 = psi if irk = 1 or, if irk > 1, set tem1 = tem + rk(irk,1)*dtemdt1, omg1 = omg + rk(irk,1)*domgdt1, and call the tridiagonal solver to get psi1 given omg1. Then continue with the original code that computes the linear and nonlinear contributions to dtemdt1 and domgdt1, but now using tem1, omg1, and psi1. After this, still within the loop over irk, set dtemdt2 = rk(irk,2)*dtemdt2 + rk(irk,3)*dtemdt1 and set domgdt2 = rk(irk,2)*domgdt2 + rk(irk,3)*domgdt1. This ends the loop over irk. The final step in updating the solution, per time step, is setting tem = tem + Δt/6*dtemdt2, omg = omg + Δt/6*domgdt2, and calling the tridiagonal solver to update psi using the updated omg. The remaining diagnostics and data storage are the same as in the original code, except that now there is no need to store the time derivatives, dtemdt2 and domgdt2, in the data files.

Of course, the price one pays for the additional accuracy is that each numerical time step takes about four times as much computer time as an Adams-Bashforth step. The type of problem and needed spatial and temporal resolutions will determine if a fourth-order method like this one is worth the price.

For a review of other Runge-Kutta schemes and a comparison of their accuracies and efficiencies see Carpenter & Kennedy (1994).

8.2 SEMI-IMPLICIT SCHEME

Consider again the time integration for temperature, T, as illustrated in Eq. 2.18 where G_t represents the time derivative at time t. An alternative integration scheme approximates the midpoint time derivative as

$$\left(\frac{\partial T}{\partial t}\right)_{t+\Delta t/2} = \alpha G_{t+\Delta t} + (1-\alpha)G_t$$

and therefore

$$T_{t+\Delta t} - \alpha \Delta t G_{t+\Delta t} = T_t + (1-\alpha)\Delta t G_t . \tag{8.1}$$

If α were set to 0, the scheme would be fully explicit and numerically unstable. If α were set to 1, the scheme would be fully implicit and numerically stable but not very accurate. In general, if $\alpha \geq 1/2$, the scheme is unconditionally stable, i.e., there is no limit (like Eq. 2.19) on the size of Δt to maintain numerical stability. However, if Δt were set to a relatively large value, Eq. 8.1 would reduce to

$$G_{t+\Delta t} \approx -\frac{(1-\alpha)}{\alpha}G_t ;$$

that is, the "solution" would erroneously change sign every time step. Therefore, the choice of Δt must also be based on accuracy.

The scheme depicted in Eq. 8.1 is accurate to second order only for $\alpha = 1/2$, which is called the *Crank-Nicolson* time integration scheme. Since a value of α only slightly less than one half would result in a numerical instability, sometimes when the solution is very time-dependent α is set to something greater than one half to ensure numerical stability at the cost of a little accuracy. For example, when starting a simulation from small random perturbations one might choose to set α to something greater than one half until the nonlinear terms have established a statistical equilibrium, after which it can be set to one half. If numerical instabilities occur, it may be better to achieve stability by reducing Δt (well below the CFL constraint), keeping α at one half. On the other hand, for very high Rayleigh number simulations, and therefore high Reynolds number, the linear diffusion terms play a relatively minor role at most length scales compared to the nonlinear terms; for such cases making the diffusion terms more implicit to avoid numerical instabilities may be a wise choice.

As mentioned, implementing an implicit scheme for the nonlinear terms is usually too expensive because all n modes are coupled and therefore would need to be updated simultaneously in one huge expression (per z-level) or iteratively with many cycles per time step. The linear diffusion terms, on the other hand, are decoupled in n and so we can easily implement a "semi-implicit" scheme, which treats the linear diffusion terms implicitly and the nonlinear terms explicitly (Durran, 1998). For example, the semi-implicit time integration scheme for mode n of the temperature equation 2.10 would be

$$\left[1 - \alpha\Delta t \left(\frac{\partial^2}{\partial z^2} - \left(\frac{n\pi}{a}\right)^2\right)\right] T_n(z, t + \Delta t) =$$

$$\left[1 + (1 - \alpha)\Delta t \left(\frac{\partial^2}{\partial z^2} - \left(\frac{n\pi}{a}\right)^2\right)\right] T_n(z, t)$$

$$+ \frac{\Delta t}{2} \left(3\left[-(\mathbf{v}\cdot\nabla)T\right]_{n,z,t} - \left[-(\mathbf{v}\cdot\nabla)T\right]_{n,z,t-\Delta t}\right) \tag{8.2}$$

and the constraint on the time step would be due to just the CFL condition, Eq. 4.7.

Although the horizontal Laplacian operator on the left side of Eq. 8.2 decouples in n, the vertical Laplacian couples all z-levels and therefore needs to be constructed as a matrix operator. Recalling the finite-difference approximation for $\partial^2/\partial z^2$, Eq. 2.16, the kth z-level row of the matrix operation representing the left side of Eq. 8.2, for a given mode n, is

$$\text{sub}(k)T_{k-1} + \text{dia}(k)T_k + \text{sup}(k)T_{k+1} ,$$

where

$$\text{sub}(k) = \text{sup}(k) = -\frac{\alpha\Delta t}{\Delta z^2}$$

and

$$\text{dia}(k) = 1 + \alpha\Delta t \left(\frac{2}{\Delta z^2} + \left(\frac{n\pi}{a}\right)^2\right) .$$

The right-hand side ($\text{rhs}(k)$) of Eq. 8.2 is easily calculated for each k because the T_n at the current time step and the nonlinear terms at the current and previous steps are already known. Note also that if a time-dependent "thermal forcing" term were employed, as discussed in Section 6.2, that term could also be treated semi-implicitly.

Schematically, this "collocation" equation (for $N_n = 5$) looks just like Eq. 2.21, where now the output $\text{sol}(k)$ is $T_{k,n}$ at the new time step. Rows $k = 1$ and N_z correspond to the bottom and top boundary conditions on T_n: $T_{1,n} = T_{N_z,n} = 0$ for $n > 0$. Therefore, the matrix elements in those two rows are the same as those listed in Eqs. 2.22. For $n = 0$, $\text{rhs}(1) = 1$ if convection is being simulated or $\text{rhs}(N_z) = 1$ if the gravity wave boundary condition is desired.

This Poisson equation is solved every numerical time step for each mode n using the same subroutine as listed in Appendix A. Note that if the code periodically checks the CFL condition and therefore occasionally changes the size of the time step Δt (according to Eq. 4.10), the matrix operator for each n needs to be reinverted since it includes Δt.

To be able to ignore the diffusive time step constraint equation 2.19 (with Pr in the denominator if Pr > 1), both the temperature and vorticity equations (and the magnetic vector potential equation if included) need to be treated semi-implicitly. The semi-implicit integration of the vorticity equation can be handled in a similar manner.

Now consider a few coding issues. Within the loop over time steps, start by setting the time derivatives at the current step for the temperature and vorticity to $(1 - \alpha)$ times all the linear terms, including, for example, the temperature-dependent buoyancy term in the vorticity equation. Then compute and add in the nonlinear terms using either the Galerkin method described in Chapter 4 or the spectral-transform method described in Chapter 10. Next update the temperature, one mode n at a time, by constructing the rhs(k) vector of the matrix operator, calling the tridiagonal matrix solver, and then overwriting the $T_n(z)$ with the sol(k). Then, using the updated $T_n(z)$, add α times the buoyancy term to the corresponding rhs(k) of the vorticity equation and update the vorticity, one n at a time. This treats the buoyancy term implicitly. The streamfunction is finally updated the original way using its matrix operator and the updated vorticity for the rhs(k).

8.3 PREDICTOR-CORRECTOR SCHEMES

A predictor-corrector time integration scheme approximates an implicit scheme. It has two (or more) steps per numerical time step. The predictor step evaluates the time derivative at the current time step and makes the first approximation to the updated function using a chosen explicit scheme. The corrector step then evaluates the time derivative at the new time step using this first approximation to the function and uses it in an implicit scheme in place of the actual time derivative at the new time step. This requires two evaluations of the (nonlinear) time derivative per numerical time step, but does not require a matrix solution method.

For example, consider again the second-order accurate Adams-Bashforth scheme (Eq. 2.18), which is explicit, and the second-order accurate Crank-Nicolson scheme (Eq. 8.1 for $\alpha = 1/2$), which is implicit. The predictor step first computes the time derivative of, say, temperature T_t at time t (which we again call G_t) and uses it to get the first approximation to the temperature at time $t + \Delta t$ using the Adams-Bashforth scheme:

$$T^*_{t+\Delta t} = T_t + \frac{\Delta t}{2} (3G_t - G_{t-\Delta t}) . \tag{8.3a}$$

Then the corrector step computes the time derivative of $T^*_{t+\Delta t}$ at time $t + \Delta t$ (i.e., $G^*_{t+\Delta t}$), as it would if this were our usual second-order Adams-Bashforth scheme updating T to time $t + 2\Delta t$. However, now this $G^*_{t+\Delta t}$ is instead used to get a better estimate of T at time $t + \Delta t$ by using it in the Crank-Nicolson scheme to get the updated $T_{t+\Delta t}$:

$$T_{t+\Delta t} = T_t + \frac{\Delta t}{2} \left(G^*_{t+\Delta t} + G_t \right) . \tag{8.3b}$$

If even greater accuracy were desired, one could do additional corrector steps each numerical time step. The number of additional steps could be determined by checking when either the average or maximum value of

$$\left| \frac{T^k_{t+\Delta t} - T^{k-1}_{t+\Delta t}}{T^k_{t+\Delta t}} \right|$$

drops below some prespecified tolerance, where here k is the iteration count for the additional iteration steps. However, this scheme is still only second-order accurate; so it would be better to reduce the Δt than to do many additional iterations.

Many other explicit and implicit time integration schemes exist, which could be used alone or in some combination within a predictor-corrector scheme. For example, in addition to the second-order Adams-Bashforth scheme we have been using, there is a family of higher order *explicit* Adams-Bashforth schemes obtained using a polynomial interpolation. The third- and fourth-order explicit Adams-Bashforth schemes are, respectively,

$$T_{t+\Delta t} = T_t + \frac{\Delta t}{12} \left(23G_t - 16G_{t-\Delta t} + 5G_{t-2\Delta t}\right) ,$$

$$T_{t+\Delta t} = T_t + \frac{\Delta t}{24} \left(55G_t - 59G_{t-\Delta t} + 37G_{t-2\Delta t} - 9G_{t-3\Delta t}\right) .$$

Likewise, there is a family of *implicit* Adams-Moulton schemes. The second-order Adams-Moulton scheme is what we have been calling the Crank-Nicolson scheme; the third- and fourth-order implicit Adams-Moulton schemes are, respectively,

$$T_{t+\Delta t} - \frac{5\Delta t}{12} G_{t+\Delta t} = T_t + \frac{\Delta t}{12} \left(8G_t - G_{t-\Delta t}\right) ,$$

$$T_{t+\Delta t} - \frac{9\Delta t}{24} G_{t+\Delta t} = T_t + \frac{\Delta t}{24} \left(19G_t - 5G_{t-\Delta t} + G_{t-2\Delta t}\right) .$$

A semi-implicit scheme (Section 8.2) could be designed that uses any of the explicit Adams-Bashforth schemes to update the nonlinear terms and any of the implicit Adams-Moulton schemes (of the same order) to update the linear terms.

Alternatively, a predictor-corrector scheme could be designed that uses, for example, the fourth-order accurate Adams-Bashforth scheme in the predictor step,

$$T^*_{t+\Delta t} = T_t + \frac{\Delta t}{24} \left(55G_t - 59G_{t-\Delta t} + 37G_{t-2\Delta t} - 9G_{t-3\Delta t}\right) , \tag{8.4a}$$

and the fourth-order accurate Adams-Moulton scheme in the corrector step,

$$T_{t+\Delta t} = T_t + \frac{\Delta t}{24} \left(9G^*_{t+\Delta t} + 19G_t - 5G_{t-\Delta t} + G_{t-2\Delta t}\right) . \tag{8.4b}$$

Again, one could use the check mentioned above to determine how many additional iteration steps would be needed to attain a prespecified accuracy using a constant Δt. However, it would be more efficient to use an *automatic time step adjustment* with the predictor-corrector scheme to maintain a given accuracy. The idea is use an estimate of the error after one predictor-corrector iteration to decide if the predictor-corrector step should be done over using a smaller Δt (if the estimated error exceeds some prespecified tolerance) or a larger Δt (if the error is too small). For this fourth-order predictor-corrector scheme (Eq. 8.4) the error estimate on the updated $T_{t+\Delta t}$ after on iteration step is

$$T^{true}_{t+\Delta t} - T_{t+\Delta t} \approx \frac{1}{14} \left(T_{t+\Delta t} - T^*_{t+\Delta t}\right) .$$

Again, this check would need to be an average or maximum value over all grid points and/or modes of T.

So, which time integration scheme should one use? It depends on the compromise that needs to be made between desired accuracy and computational expense and how that choice balances similar choices made about the spatial discretization, resolution, and geometry (Chapters 9 and 10) and how physically realistic the actual model equations are (Chapters 11–13). In any case, it is better to begin with a simple scheme (as we do in Part 1), get it working, and then improve the time integration later if and when needed. For a given problem, one needs to consider the additional computational time required for some schemes due to the additional times the nonlinear terms are computed per numerical time step and balance this with using a smaller Δt in a lower order accurate scheme. In practice, comparing the accuracy, efficiency, and memory requirements of several test runs with several different schemes at the desired spatial resolution and model parameters is usually the best way to make these choices. Having said all that, the Adams-Bashforth-Moulton predictor-corrector schemes, either the second-order accurate version, Eq. 8.3, or the fourth-order accurate version, Eq. 8.4, would be a good choice, especially with the automatic time step adjustment. If the stability constraint on the time step due to linear diffusion (e.g., Eq. 2.19) is more severe than the CFL constraint (e.g., Eq. 4.7), i.e., the diffusivities are not small, it may be worth implementing a semi-implicit scheme that treats the linear terms implicitly with a Crank-Nicolson scheme (or a higher-order Adams-Moulton scheme) and uses the predictor-corrector scheme only for updating the nonlinear terms.

8.4 INFINITE PRANDTL NUMBER: MANTLE CONVECTION

Thermal convection within the mantle of a terrestrial planet, like the Earth, is dominated by huge viscous forces to the extent that the inertial, Coriolis, and magnetic Lorentz forces in the momentum equation are negligible. The Prandtl number (ratio of viscous to thermal diffusivities) is on average about 10^{23} and so is approximated as being infinite in studies of geodynamics. Therefore, mantle convection is controlled by the balance among buoyancy, pressure gradient, and viscous forces. Convection in magma chambers is also typically studied using the infinite Prandtl number approximation and as a double-diffusive process (Hansen & Yuen (1995); Chapter 7). Note that the typical fluid velocity in the Earth's liquid core is estimated to be a few tens of kilometers per year, which is much smaller than typical velocities in the Earth's atmosphere or ocean but is about a million times larger than typical mantle convection (and tectonic plate) velocities. The slow creeping fluid flow of mantle convection is called "solid-state convection".

It is interesting that viscosity is usually the challenge in most studies of convection in planets and stars. However, unlike studies of convection in the liquid iron cores of terrestrial planets or in the atmospheres of planets and stars or in the deep fluid interiors of giant planets and stars, for which making the model's viscosity as small as possible is the challenge, the challenge for mantle convection is that the viscous diffusivity is highly temperature- and strain-rate-dependent. Viscosity in a terrestrial mantle can vary by several orders of magnitude from one location to another because of the large variations in temperature that are needed to produce sufficient buoyancy forces to drive the convection. The viscous diffusivity is also not just a scalar, but a tensor, dependent on the amplitude and direction of the local rate of strain. Therefore, the viscous force is a complicated nonlinear term. As shown in Chapter 12, a spectral model can accommodate a depth-dependent scalar viscosity, which would represent its horizontally averaged value; but if one wants a viscosity that also varies in the horizontal direction in a nonlinear time-dependent manner, a local numerical method (e.g., finite-difference) is recommended, especially if large gradients of viscosity can develop. In addition, nonlinear viscous heating (Eqs. 1.8 and 12.31), which we have so far neglected, is important in mantle convection; see Eqs. 12.31 and 12.32.

Here we show how the current model can easily be modified to study mantle convection (also called "geodynamics"). Consider the vorticity equation (2.11) in the limit of an infinite Pr:

$$\left(\frac{\partial^2 \omega_n}{\partial z^2} - \left(\frac{n\pi}{a} \right)^2 \omega_n \right) = -\text{Ra} \left(\frac{n\pi}{a} \right) T_n . \tag{8.5}$$

This equation is a Poisson equation and therefore can be solved like the streamfunction equation (2.12). At each time step, first update the temperature semi-implicitly as in Eq. 8.2. With the updated T_n, solve for ω_n via Eq. 8.5. Then use the updated ω_n to solve for ψ_n via Eq. 2.12. Each of these three steps, for every time step and each mode n, involves a tridiagonal matrix solution; only the temperature equation (8.2), however, has a time derivative and a nonlinear term.

We discuss the infinite Prandtl number approximation further at the end of Section 10.6.2 for a 3D spectral density-stratified spherical-shell model of mantle convection (Glatzmaier, 1988). However, readers interested in simulating mantle convection, especially with more realistic visco-elasto-plastic rheologies and phase transitions, are advised to check the numerical methods outlined in, for example, Ismail-Zadeh & Tackley (2010) and Gerya (2010). For a very comprehensive review of mantle convection studies for the Earth, for other terrestrial planets, and for satellites of giant planets check Schubert et al. (2001).

SUPPLEMENTAL READING

Durran (1998)
Ferziger & Perić (1997)
Gerya (2010)
Ismail-Zadeh & Tackley (2010)
Peyret (2002)
Press et al. (1992)
Schubert et al. (2001)

EXERCISES

1. *Runge-Kutta scheme*
   Modify a computer code developed in Part 1 by replacing the Adams-Bashforth time integration scheme with the fourth-order Runge-Kutta scheme outlined in Section 8.1.

2. *Semi-implicit scheme*
   Modify a computer code developed in Part 1 by replacing the Adams-Bashforth time integration scheme with the second-order semi-implicit scheme outlined in Section 8.2.

3. *Predictor-corrector scheme*
   Modify a computer code developed in Part 1 by replacing the Adams-Bashforth time integration scheme with the second-order predictor-corrector scheme outlined in Section 8.3.

4. *Infinite Prandtl number scheme*
   Modify a computer code developed in Part 1 by replacing the Adams-Bashforth time integration scheme with the infinite Prandtl number scheme outlined in Section 8.4.

COMPUTATIONAL PROJECTS

1. *Comparing the Runge-Kutta and Adams-Bashforth schemes*
   Compare the accuracy of the fourth-order Runge-Kutta scheme (Section 8.1) with the second-order Adams-Bashforth scheme (Part 1) by running the convection scenario illustrated on the left in Fig. 4.2 and tabulated in Table 4.1 with both schemes using the same spatial resolution and a series of different time step sizes Δt. Compare the computer time required per numerical time step by each scheme to obtain a desired degree of accuracy.

2. *Comparing the semi-implicit and Adams-Bashforth schemes*
   Compare a convection simulation produced with an explicit Adams-Bashforth code with one that employs the semi-implicit scheme (Section 8.2). Confirm that when using the same value for Δt (which will need to be less than both the diffusion limit and the CFL limit) and the same spatial resolution (i.e., the same N_n and N_z) that the two codes produce very nearly the same solution. This comparison is easier when a steady-state case is tested. Then test how the solutions from these two codes compare as the Δt for the semi-implicit code is increased above the diffusion limit but below the CFL limit. Also test how the value of α affects the results. Confirm that an α less than one half causes a numerical instability.

3. *Comparing the predictor-corrector and Adams-Bashforth schemes*
   Compare a convection simulation produced with an explicit Adams-Bashforth code with one that employs the second-order predictor-corrector scheme outlined in Section 8.3. Use the same value for Δt and the same spatial resolution.

4. *Comparing the semi-implicit and infinite Prandtl number schemes*
   Compare an infinite Prandtl number simulation (Section 8.4) with a series of
   semi-implicit finite Prandtl number simulations (Section 8.2) with the same
   Rayleigh number and aspect ratio but with increasing values of the Prandtl
   number. At what value of Pr does the finite Prandtl number solution look
   nearly identical to the infinite Prandtl number solution?

5. *Comparing internal heating to bottom heating*
   Using an infinite Prandtl number model compare the basic structure of the
   thermal plumes for a case with no internal heating (i.e., just bottom heating
   via the isothermal boundary conditions) to that for a case with no bottom
   heating (i.e., a zero vertical temperature gradient at the bottom boundary) and
   a prescribed amount of uniform internal heating such that the total heat flow
   through the top boundary is comparable to that for the case with no internal heating. The bottom thermal boundary condition for the internally heated
   case will require the temperature perturbation for all n, including $n = 0$, to
   be modified in a way that forces the diffusive heat flux through the bottom
   boundary to vanish.

Chapter Nine

Spatial Discretizations

In Part 1 we chose to treat the horizontal direction with a spectral method and the vertical direction with a finite-difference method on a uniform grid. For some problems it is desirable to be able to employ a spatial resolution that varies with position. In this chapter we introduce two ways of doing this within a finite-difference method: using a nonuniform grid and mapping to a new coordinate variable. We then outline how one can simulate the convection and gravity wave problems described in Part 1 either by using finite differences in both directions or by using a spectral method in both directions.

So many other methods have been developed and are described in many papers and books. They all have advantages and disadvantages, which depend on the particular details of the problem to be solved. After being introduced to the basic methods in this book, one should explore and compare methods like finite-volume, finite-element, spectral-element, compact finite-difference, arbitrary Lagrangian-Eulerian, particle-in-cell (marker-in-cell), and adaptive mesh refinement.

Related to the choice of spatial discretization is the design of spatial decomposition of the computation among processors on a massively parallel computer. We provide a very brief introduction to parallel processing at the end of this chapter.

9.1 NONUNIFORM GRID

In Part 1 we employ uniformly spaced grid levels in the vertical direction for the second-order accurate finite-difference method. However, as the Rayleigh number increases the depths of the thermal and viscous boundary layers at the bottom and top boundaries decrease, requiring smaller (constant) Δz (i.e., a larger N_z) to resolve them. In addition, v_z decreases as one approaches the bottom or top boundary because these boundaries are impermeable (and fixed in space); so using smaller Δz in these regions usually does not affect the CFL condition. Therefore, it can be advantageous to prescribe a nonuniform vertical grid, one that smoothly varies from a minimum Δz at the boundaries to a maximum Δz at mid-depth. There may also be interfaces within the fluid domain where better spatial resolution is needed; for example, at the interface between the stable and unstable regions in the case illustrated on the right in Fig. 6.2.

A general finite-difference method can be derived using the process, mentioned in Section 2.3, of adding and subtracting two Taylor series expansions, but now with a prescribed grid, z_k, for which the grid spacing, $z_k - z_{k-1}$, depends on the level index k. Consider three grid levels, z_{k-1}, z_k, z_{k+1}, such that $\Delta z_- \equiv z_k - z_{k-1}$

and $\Delta z_+ \equiv z_{k+1} - z_k$. Write two Taylor series about z_k, one for f_{k-1} and the other for f_{k+1}. That is,

$$f_{k-1} = f_k - \left(\frac{\partial f}{\partial z}\right)_k \Delta z_- + \frac{1}{2}\left(\frac{\partial^2 f}{\partial z^2}\right)_k \Delta z_-^2 + O(\Delta z_-^3) \qquad (9.1)$$

and

$$f_{k+1} = f_k + \left(\frac{\partial f}{\partial z}\right)_k \Delta z_+ + \frac{1}{2}\left(\frac{\partial^2 f}{\partial z^2}\right)_k \Delta z_+^2 + O(\Delta z_+^3). \qquad (9.2)$$

Subtracting Eq. 9.1 from 9.2 and dropping terms of order $(\Delta z_-^3 + \Delta z_+^3)$ and smaller produces an expression for $(\partial f/\partial z)_k$ in terms of $(\partial^2 f/\partial z^2)_k$. Likewise, adding Eqs. 9.1 and 9.2 produces an expression for $(\partial^2 f/\partial z^2)_k$ in terms of $(\partial f/\partial z)_k$, again dropping the higher order terms. Substituting the latter into the former gives a finite-difference approximation to $(\partial f/\partial z)_k$. Then substituting this into the latter gives a finite-difference approximation to $(\partial^2 f/\partial z^2)_k$. The result is that the first derivative of a function f on a general grid z_k is

$$\left(\frac{\partial f}{\partial z}\right)_k = a(k, -1)f_{k-1} + a(k, 0)f_k + a(k, 1)f_{k+1} \qquad (9.3)$$

and its second derivative is

$$\left(\frac{\partial^2 f}{\partial z^2}\right)_k = b(k, -1)f_{k-1} + b(k, 0)f_k + b(k, 1)f_{k+1}, \qquad (9.4)$$

where the arrays a and b (dimensioned as $(1 : N_z, -1 : 1)$) are

$$a(k, -1) = \frac{-\Delta z_+}{\Delta z_-(\Delta z_- + \Delta z_+)},$$

$$a(k, 0) = \frac{\Delta z_+ - \Delta z_-}{\Delta z_- \Delta z_+},$$

$$a(k, 1) = \frac{\Delta z_-}{\Delta z_+(\Delta z_- + \Delta z_+)},$$

$$b(k, -1) = \frac{2}{\Delta z_-(\Delta z_- + \Delta z_+)},$$

$$b(k, 0) = \frac{-2}{\Delta z_- \Delta z_+},$$

$$b(k, 1) = \frac{2}{\Delta z_+(\Delta z_- + \Delta z_+)}. \qquad (9.5)$$

Note that these reduce to the corresponding centered finite-difference expressions for a uniform grid (Eqs. 2.15 and 2.16) when $\Delta z_- = \Delta z_+$.

Assuming the nonuniform grid is time-independent, these coefficients only need to be calculated once and then stored. If derivatives are needed on a boundary, "ghost points" could be established at a grid point outside the boundary the same distance from the boundary as the adjacent grid level within the fluid. The value of a variable on a ghost point is typically a function of its values at the boundary and at one grid point inside the boundary and is determined by what it needs to be to satisfy the boundary condition with a centered finite-difference formula. See the discussion on the use of ghost points in Section 11.2. An example is also discussed in Section 9.3.

A convenient nonuniform set of z-levels that provides a smooth transition from high spatial resolution (small Δz) near the boundaries to coarse resolution (large Δz) at mid-depth is a Chebyshev grid defined as

$$z_k = \frac{1}{2}\left(1 - \cos\left(\frac{(k-1)\pi}{(N_z - 1)}\right)\right) \tag{9.6}$$

with, again, $k = 1 \rightarrow N_z$ (Fig. 9.1).

Note that when using a nonuniform grid the time step constraints require treating Δz as a function of z. That is, the (nondimensional) diffusive constraint is now

$$\Delta t < \frac{(\Delta z_{MIN})^2}{4},$$

assuming an explicit treatment of the diffusion terms and constant diffusion coefficients, and the CFL constraint is

$$\Delta t < \left(\frac{\Delta z}{v_z}\right)_{MIN}.$$

Also note that if graphics data were prepared and stored while the solution is being generated (as described in Section 5.1), first interpolating it onto a uniform grid before storing it in the output file might be beneficial. That way the same graphics postprocessor can be used for different computational grids.

9.2 COORDINATE MAPPING

An alternative way of achieving higher spatial resolution in different regions of the fluid domain is to map (or "project") a nonuniformly spaced set of grid levels in the original independent variable, z, onto a new independent variable, say, ζ, that is uniformly spaced. This requires all derivatives with respect to z in the original equations to be written as derivatives with respect to ζ.

For example, let $0 \leq \zeta \leq 1$, like the nondimensional z, and define a Chebyshev mapping of z as a function of ζ:

$$z(\zeta) = \frac{1}{2}\left(1 - \cos(\pi\zeta)\right).$$

The discrete version of this is Eq. 9.6. Then the first derivative of, say, T with respect to z at level k is

$$\left(\frac{\partial T}{\partial z}\right)_k = \frac{2}{\pi\,\sin(\pi\zeta_k)}\left(\frac{\partial T}{\partial \zeta}\right)_k$$

$$\approx \frac{2}{\pi\,\sin(\pi\zeta_k)}\left(\frac{T_{k+1} - T_{k-1}}{2\,\Delta\zeta}\right) \tag{9.7}$$

and the second derivative is

$$
\begin{aligned}
\left(\frac{\partial^2 T}{\partial z^2}\right)_k &= -\frac{4}{\pi} \frac{\cos(\pi \zeta_k)}{\sin^3(\pi \zeta_k)} \left(\frac{\partial T}{\partial \zeta}\right)_k + \frac{4}{\pi^2 \sin^2(\pi \zeta_k)} \left(\frac{\partial^2 T}{\partial \zeta^2}\right)_k \\
&\approx -\frac{4}{\pi} \frac{\cos(\pi \zeta_k)}{\sin^3(\pi \zeta_k)} \left(\frac{T_{k+1} - T_{k-1}}{2\,\Delta \zeta}\right) \\
&\quad + \frac{4}{\pi^2 \sin^2(\pi \zeta_k)} \left(\frac{T_{k+1} - 2T_k + T_{k-1}}{\Delta \zeta^2}\right),
\end{aligned}
\tag{9.8}
$$

where, again, $k = 1 \rightarrow N_z$ and now $\Delta \zeta$ is a constant $1/(N_z - 1)$. All z-derivatives in the temperature, vorticity, and streamfunction equations, as described in Part 1, would need to be replaced with corresponding ζ-derivatives like those of Eqs. 9.7 and 9.8. These two equations would have problems at $k = 1$ and N_z where $\sin(\pi \zeta_k) = 0$; however, a prescribed constant-temperature boundary condition would avoid this problem.

Similar methods use an independent variable other than height or radius for the "vertical" direction to achieve a more natural distribution of vertical levels or to accommodate topographic or bathymetric bottom boundaries or top boundaries defined by a specified pressure or total column mass. For example, 1D stellar evolution models typically use the integrated mass from the center of the star as the vertical coordinate, so radius becomes a dependent variable. Some ocean circulation models employ "isopycnal coordinates," which use density (relative to an adiabatic density profile) as the vertical coordinate. Atmospheric general circulation models often use the hydrostatic pressure (normalized by the surface pressure) as the vertical coordinate. In these cases, the equations need to be written in terms of the new independent variable instead of height or radius.

9.3 FULLY FINITE-DIFFERENCE

As mentioned in Section 2.3, a spectral solution converges much faster to the actual solution as the number of modes employed increases compared to a finite-difference method as the number of grid points increases. However, the programming required is typically much simpler for a finite-difference method. In addition, for a computer code written using parallel processing on a cluster of processors, a local numerical method, like finite-difference, requires much less communication between processors because usually only "nearest neighbor communication" is needed (Section 9.5.2). That is, the spatial domain is typically divided into cells (or grid points); finite-difference derivatives at a given cell depend only on the values in the adjacent cells. A spectral method, on the other hand, usually requires "global communication", for which every processor needs to send and receive data from every other processor every numerical time step (Glatzmaier & Clune, 2000) because all spectral modes are needed to compute spatial derivatives and these modes are distributed over the all the processors. So, a simple parallel finite-difference method with enough grid points to obtain comparable accuracy might be more computationally efficient at some high spatial resolution than a spectral method. Also, a spectral method, which is inherently nonlocal, can have more difficulty than a local method with a solution that is supposed to have large amplitude flows at one end of the domain and none at the other end. Therefore, for one or more of these reasons, a finite-difference method could be preferred over a spectral method. Many books have been written describing local methods like the finite-difference and finite-volume methods (e.g., Patankar, 1980; Ferziger & Perić, 1997; Griebel et al., 1967; Durran, 1998; Slingerland & Kump, 2011). High-order compact finite-difference methods (e.g., Rai & Moin, 1991; Lele, 1992; Yu et al., 1994; Durran, 1998; Gamet et al., 1999; Liao, 2008; Takahashi, 2012) provide a nice compromise between a spectral method and a strictly local method, providing high numerical accuracy without some of the disadvantages of a spectral method. Spatial derivatives are obtained with this method via an implicit coupling among all the grid levels in, for example, the vertical direction; they can easily be solved at each time step using a tridiagonal solver (Section 2.5).

Here we outline one way of treating both the vertical and horizontal directions with simple finite-difference methods. One could solve Eqs. 1.15–1.17 for T, p, v_x, and v_z without introducing the vorticity or streamfunction. However, here we choose, as we do in Part 1, to define a streamfunction (Eqs. 2.5–2.7), which automatically satisfies mass conservation, and to solve for vorticity via the curl of the momentum conservation equation, 2.4. We use mass conservation, Eq. 1.15, to write the nonlinear advection terms in Eqs. 1.17 and 2.4 in their "conservative" forms, i.e., as divergences of temperature and vorticity fluxes, respectively. Recall that since we are assuming that the $\nabla \cdot \mathbf{v} = 0$, mathematically $\mathbf{v} \cdot \nabla T = \nabla \cdot (T\mathbf{v})$. However, the finite-difference approximations of these two forms are not exactly equal. It is usually more accurate and stable to compute the finite-difference divergence of a nonlinear product than to compute the nonlinear product of $\mathbf{v}$ and a gradient of T or ω. Therefore, our working set of equations is

$$\omega = -\nabla^2 \psi, \tag{9.9}$$

$$\frac{\partial \omega}{\partial t} = -\nabla \cdot \omega \mathbf{v} - \text{RaPr} \frac{\partial T}{\partial x} + \text{Pr} \nabla^2 \omega, \tag{9.10}$$

$$\frac{\partial T}{\partial t} = -\nabla \cdot T \mathbf{v} + \nabla^2 T. \tag{9.11}$$

These equations need to be updated on a 2D set of grid points. Let the horizontal grid points be defined as $x_i = (i-1)\Delta x$ for $i = 1 \rightarrow N_x$ and the vertical grid points again be defined as $z_k = (k-1)\Delta z$ for $k = 1 \rightarrow N_z$. To obtain comparable accuracy in both directions, we choose the (constant) grid size in the x-direction, $\Delta x = a/(N_x - 1)$, to be as close as possible to the grid size in the z-direction, $\Delta z = 1/(N_z - 1)$. That is, we choose the number of grid cells in the x-direction, $(N_x - 1)$, to equal the nearest integer to $(N_z - 1)a$. Of course, one could use nonuniform grids (Section 9.1) or coordinate mappings (Section 9.2) in both the x- and z-directions.

Since the set of equations 9.9–9.11 has Laplacian operators on each of the three dependent variables, the system requires three boundary conditions on each of the four boundaries. As described in Part 1, the impermeable boundary condition forces ψ to vanish on all four boundaries; and this combined with the stress-free boundary condition forces ω to vanish on all four boundaries. That is, $\psi_{i,1} = \psi_{i,N_z} = \psi_{1,k} = \psi_{N_x,k} = 0$ and likewise for ω. Therefore, ψ and ω should be updated only on grid points interior to the boundaries. The isothermal boundary conditions on the bottom and top boundaries require $T_{i,1} = 1$ and $T_{i,N_z} = 0$ (or vice versa if gravity waves are desired instead of thermal convection). Therefore, temperature should not be updated on the bottom and top boundaries. It does, however, need to be updated on the insulating side boundaries. The vanishing horizontal gradient of temperature on these boundaries requires ghost points on which $T_{0,k} = T_{2,k}$ and $T_{N_x+1,k} = T_{N_x-1,k}$. An example of ghost points on the top and bottom boundaries is discussed in Section 11.2. As in that section, here the ghost points do not actually need to be allocated; instead, first- and second-order horizontal derivatives of T on, say, the left side boundary, $(1, k)$, should be calculated as

$$\left(\frac{\partial T}{\partial x}\right)_{1,k} = \frac{T_{2,k} - T_{0,k}}{2\Delta x} = 0, \tag{9.12}$$

$$\left(\frac{\partial^2 T}{\partial x^2}\right)_{1,k} = \frac{T_{2,k} - 2T_{1,k} + T_{0,k}}{\Delta x^2} = \frac{2(T_{2,k} - T_{1,k})}{\Delta x^2}. \tag{9.13}$$

Similar formulas should be used on the right side boundary, (N_x, k).

Equations 9.10 and 9.11 can be integrated in time using the explicit Adams-Bashforth scheme employed in Part 1. However, since all dependent variables are already in grid space, the nonlinear terms are much simpler to compute in this fully finite-difference method compared to the spectral Galerkin method. The advection of temperature at grid point (i, k) in Eq. 9.11 is, making use of Eq. 2.15,

$$-[\nabla \cdot T\mathbf{v}]_{i,k} = -\left[\frac{(Tv_x)_{i+1,k} - (Tv_x)_{i-1,k}}{2\,\Delta x} + \frac{(Tv_z)_{i,k+1} - (Tv_z)_{i,k-1}}{2\,\Delta z}\right] \tag{9.14}$$

and similarly for the advection of vorticity in Eq. 9.10.

Recall that temperature needs to be updated on the side boundaries. However, instead of writing the advection of temperature there as in Eq. 9.14, it can be written as $-\mathbf{v} \cdot \nabla T$, where only the $-v_z \partial T / \partial z$ part survives because of our side boundary conditions; and this part does not require ghost points. Using Eq. 2.16, the thermal diffusion term is

$$[\nabla^2 T]_{i,k} = \left[\frac{T_{i+1,k} - 2\,T_{i,k} + T_{i-1,k}}{\Delta x^2} + \frac{T_{i,k+1} - 2\,T_{i,k} + T_{i,k-1}}{\Delta z^2}\right]. \tag{9.15}$$

This term does require a ghost-point treatment on the side boundaries like that of Eq. 9.13.

The buoyancy term in Eq. 9.10 is simply

$$-\left[\text{Ra Pr}\frac{\partial T}{\partial x}\right]_{i,k} = -\text{Ra Pr}\left[\frac{T_{i+1,k} - T_{i-1,k}}{2\,\Delta x}\right].$$

The Laplacian terms in Eqs. 9.9 and 9.10 are formulated like that for temperature in Eq. 9.15.

Note that if $\Delta x = \Delta z$, the finite-difference thermal diffusion operator (Eq. 9.15) would make a positive contribution to the temperature (in Eq. 9.11) if the temperature were less than the average of the temperatures of its four nearest neighbors. Likewise, the operator would try to decrease the temperature if it were larger than the average of its nearest neighbors. That is, thermal diffusion tries to smooth the spatial distribution of temperature, which is needed to balance the tendency for nonlinear advection of temperature (Eq. 9.14) to shear the temperature profile into small scales.

Now consider how Eq. 9.9 can be solved each time step to update ψ with the updated ω. There are several methods for accomplishing this, usually involving an iterative relaxation process to converge to a numerical solution that satisfies some prescribed tolerance. A simple method for solving this Poisson equation is the Jacobi relaxation method, which iterates on the finite-difference representation of Eq. 9.9 until the error in the numerical solution is within some set tolerance. Using Eq. 9.15 for ψ instead of T, Eq. 9.9 at grid point (i, k) for iteration from step n to $n + 1$ is

$$\psi_{i,k,n+1} = c1 \, (\psi_{i+1,k,n} + \psi_{i-1,k,n}) + c2 \, (\psi_{i,k+1,n} + \psi_{i,k-1,n}) + c3 \, \omega_{i,k}, \quad (9.16)$$

where the coefficients are

$$c1 \equiv \frac{\Delta z^2}{2(\Delta x^2 + \Delta z^2)},$$

$$c2 \equiv \frac{\Delta x^2}{2(\Delta x^2 + \Delta z^2)},$$

$$c3 \equiv \frac{\Delta x^2 \, \Delta z^2}{2(\Delta x^2 + \Delta z^2)}.$$

Apply Eq. 9.16 to all internal grid points, keeping the boundary values equal to zero. Two arrays are needed for $\psi_{i,k}$, one for iteration n and the other for $n + 1$. Set the value of ψ on the first iteration to its value from the previous time step. Stop the iteration when the maximum value of

$$\frac{|\psi_{i,k,n+1} - \psi_{i,k,n}|}{(|\psi_{i,k,n+1}| + |\psi_{i,k,n}|)}$$

(over all i between 2 and $N_x - 1$ and k between 2 and $N_z - 1$) drops below some prescribed tolerance, say, 10^{-6}, or when the interaction count reaches some prescribed maximum number, say, 100. If and when the latter occurs, the maximum error should be printed and the simulation stopped. Note that a smaller Δt requires fewer iterations per time step; but the cost of a simulation is related to the product of the number of time steps and the average number of iterations per step.

One can try to accelerate the convergence each time step by incorporating a weighting factor. Let the right side of Eq. 9.16 be called $\psi_{i,k,*}$ and set

$$\psi_{i,k,n+1} = c \, \psi_{i,k,*} + (1 - c) \, \psi_{i,k,n},$$

where $0 < c < 2$. Try $c = 1.5$, for example, to see if the average number of iterations per time step decreases relative to $c = 1$.

Chebyshev polynomials

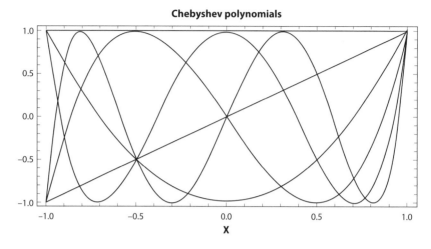

Figure 9.1 Plots of Chebyshev polynomials vs. x for degrees $m = 0 \rightarrow 5$.

9.4 FULLY SPECTRAL: CHEBYSHEV-FOURIER

Here we choose to simulate our convection or gravity wave problem using spectral methods in both the horizontal and vertical directions. We continue to use the Fourier method in the horizontal direction, which could be implemented by using the Galerkin method as outlined in Part 1 or by using the modified techniques described in Chapter 10. For the spectral method in the vertical direction we choose a Chebyshev collocation method. This is a much better choice than a Fourier method because of the impermeable bottom and top boundaries. Although an expansion in sines is employed in Section 3.4 for the onset of thermal convection, a highly non-linear solution has steep radial gradients near the bottom and top boundaries, which are difficult to resolve with just sines. Chebyshev expansions converge much faster for problems like this and can much more easily be used to impose fairly arbitrary boundary conditions.

9.4.1 Chebyshev Polynomials

Traditionally, the independent variable for Chebyshev polynomials has been called x, not to be confused with our horizontal coordinate. The Chebyshev polynomial (of the first kind) of degree m on $-1 \le x \le 1$ is defined as

$$\mathrm{T}_m(x) = \cos(m \arccos(x)) . \tag{9.17}$$

Plots of $\mathrm{T}_m(x)$ for $m = 0 \rightarrow 5$ are displayed in Fig. 9.1. Note that the degree of the Chebyshev polynomial equals to number of "zero crossings" the function has between $-1 \le x \le 1$ and that these nodes become more concentrated at the two boundaries of this domain as the degree increases. Also note that $\mathrm{T}_m(1) = 1$ and $\mathrm{T}_m(-1) = (-1)^m$.

Given that

$$T_0(x) = 1 \tag{9.18a}$$

and

$$T_1(x) = x, \tag{9.18b}$$

higher degree polynomials can be obtained from a recursion relation:

$$T_{m+1}(x) = 2x T_m(x) - T_{m-1}(x). \tag{9.18c}$$

Taking the first derivative of Eqs. 9.18, we have

$$\frac{dT_0}{dx} = 0, \tag{9.19a}$$

$$\frac{dT_1}{dx} = 1 \tag{9.19b}$$

and for higher degrees the recursion relation gives

$$\frac{dT_{m+1}}{dx} = 2T_m + 2x\frac{dT_m}{dx} - \frac{dT_{m-1}}{dx}. \tag{9.19c}$$

Taking the second derivative of Eqs. 9.18, we have

$$\frac{d^2T_0}{dx^2} = 0, \tag{9.20a}$$

$$\frac{d^2T_1}{dx^2} = 0 \tag{9.20b}$$

and for higher degrees the recursion relation gives

$$\frac{d^2T_{m+1}}{dx^2} = 4\frac{dT_m}{dx} + 2x\frac{d^2T_m}{dx^2} - \frac{d^2T_{m-1}}{dx^2}. \tag{9.20c}$$

Now we choose the Chebyshev-Gauss-Lobatto grid points:

$$x_k = \cos\left(\frac{(N_z - k)\pi}{N_z - 1}\right),$$

where again $1 \le k \le N_z$. Substituting this into Eq. 9.17 gives

$$T_m(x_k) = \cos\left(\frac{m(N_z - k)\pi}{N_z - 1}\right).$$

The Chebyshev-Gauss-Lobatto grid points can be written in a more general form:

$$z_k = z_1 + \frac{D}{2}\left(1 + \cos\left(\frac{(N_z - k)\pi}{N_z - 1}\right)\right),$$

where the depth of the fluid domain, D, is $z_{N_z} - z_1$ with z_1 being the bottom boundary and z_{N_z} the top boundary. Here we continue to use nondimensional variables; so $z_1 = 0$ and $z_{N_z} = 1$, which makes $D = 1$. Therefore,

$$z_k = \frac{1}{2}\left(1 + \cos\left(\frac{(N_z - k)\pi}{N_z - 1}\right)\right),$$

which is equivalent to Eq. 9.6; and so here we write the Chebyshev polynomials as functions of z_k,

$$T_m(z_k) = \cos\left(\frac{m(N_z - k)\pi}{N_z - 1}\right),$$

which should be stored as, say, cheb(m, k).

The point of all this is that an arbitrary function, say, $F(z_k)$, can be approximated as a series of (orthogonal) Chebyshev polynomials,

$$F(z_k) = \left(\frac{2}{N_z - 1}\right)^{1/2} \sum_{m=0}^{N_z-1} {}'' f_m T_m(z_k), \tag{9.21}$$

where the Chebyshev transform provides the coefficients

$$f_m = \left(\frac{2}{N_z - 1}\right)^{1/2} \sum_{k=1}^{N_z} {}'' F(z_k) T_m(z_k). \tag{9.22}$$

The double primes on the summations in Eqs. 9.21 and 9.22 mean that the first and last terms in those summations need to be multiplied by $1/2$.

The first derivative of a Chebyshev polynomial with respect to z is simply

$$\left(\frac{dT_m}{dz}\right)_k = \frac{2}{D}\left(\frac{dT_m}{dx}\right)_k, \tag{9.23}$$

where the right side is obtained from Eqs. 9.19. Likewise, the second derivative with respect to z is

$$\left(\frac{d^2T_m}{dz^2}\right)_k = \left(\frac{2}{D}\right)^2 \left(\frac{d^2T_m}{dx^2}\right)_k, \tag{9.24}$$

where here the right side is obtained from Eqs. 9.20. These should be stored in arrays, say, dcheb(m, k) and d2cheb(m, k), respectively. Using these one can compute the first and second derivatives of a function F with respect to z when the Chebyshev coefficients, f_m, of F are known. That is,

$$\left(\frac{dF}{dz}\right)_k = \left(\frac{2}{N_z - 1}\right)^{1/2} \sum_{m=0}^{N_z-1} {}'' f_m \left(\frac{dT_m}{dz}\right)_k \tag{9.25}$$

and

$$\left(\frac{d^2F}{dz^2}\right)_k = \left(\frac{2}{N_z - 1}\right)^{1/2} \sum_{m=0}^{N_z-1} {}'' f_m \left(\frac{d^2T_m}{dz^2}\right)_k. \tag{9.26}$$

An alternative way of taking derivatives of functions expanded in Chebyshev polynomials is convenient in some situations. It applies a backward recursion relation directly on the Chebyshev coefficients of the function. Let the coefficients of the first derivative of a function F with respect to z be called df_m. That is,

$$\left(\frac{dF}{dz}\right)_k = \left(\frac{2}{N_z - 1}\right)^{1/2} \sum_{m=0}^{N_z-1} {}'' df_m T_m(z_k), \tag{9.27}$$

where, for the two highest degree coefficients,

$$df_{N_z-1} = 0 \tag{9.28a}$$

and

$$df_{N_z-2} = \frac{2(N_z - 1)}{D} f_{N_z-1} \tag{9.28b}$$

and, for degrees $m = (N_z - 3) \rightarrow 0$,

$$df_m = df_{m+2} + \frac{4(m + 1)}{D} f_{m+1}. \tag{9.28c}$$

Again, the depth D is set to one in our nondimensional equations. The coefficients for the second derivative, ddf_m, can be computed in the same manner after computing the df_m, which would be used in the recursion relation instead of f_m. Note that for the second derivative $ddf_{N_z-2} = 0$.

We can also compute the integral of a function over the vertical domain. This is useful for calculating the volume integral of, for example, kinetic energy as discussed in Section 5.3 or for imposing an integral constraint on the system in place of a boundary condition. The integral of a Chebyshev polynomial is

$$\int T_m(z) \, dz = \frac{D}{2} \left[\frac{m \, T_{m+1}}{m^2 - 1} - \left(\frac{2 \, z}{D} - 1 \right) \frac{T_m}{m - 1} \right].$$

Now define Q_m to be the definite integral of a Chebyshev polynomial over the entire vertical domain, i.e.,

$$Q_m \equiv \int_0^D T_m(z) \, dz \,,$$

which equals

$$Q_m = \left\{ \begin{array}{ll} \frac{-D}{(m^2 - 1)} & \text{even } m \\ 0 & \text{odd } m \end{array} \right\}. \tag{9.29}$$

Then, using Eq. 9.21,

$$\int_0^D F(z_k) \, dz = \left(\frac{2}{N_z - 1} \right)^{1/2} \sum_{m=0}^{N_z-1}{}'' f_m Q_m \,. \tag{9.30}$$

More information about Chebyshev polynomials and their use can be found in several books. For example, check Fox & Parker (1968) and Boyd (2001).

9.4.2 Chebyshev collocation

With this brief introduction to Chebyshev expansions on a discrete Chebyshev grid, we now consider the Chebyshev collocation method (e.g., Peyret, 2002). An important thing to note is that we are *not* using a Galerkin method in the vertical direction as we have been doing for the Fourier method in the horizontal direction. The reason for this is that when we convert our model from a box to an annulus or to a spherical shell (Chapter 10) or use a density-stratified reference state (Chapter 12), the equations will have terms involving the product of two dependent variables (i.e., a quadratic nonlinear term) times a coefficient that is also a function of the vertical coordinate (height z or radius r), which makes these terms more than just quadratic nonlinear in the vertical coordinate. As a result, a Galerkin method would be much more complicated and computationally inefficient. The Chebyshev collocation method projects the solution onto Chebyshev space much more easily and efficiently.

Consider again the semi-implicit time integration scheme described in Section 8.2. However, instead of applying the finite-difference operators to an equation like 8.2, now expand the temperature coefficient, T_n, in Chebyshev polynomials as in Eq. 9.21. That is,

$$T_n(z_k, t) = \left(\frac{2}{N_z - 1}\right)^{1/2} \sum_{m=0}^{N_z-1}{}'' T_{nm}(t) \mathrm{T}_m(z_k). \tag{9.31}$$

Be careful here not to confuse the temperature coefficient, $T_{nm}(t)$, with the Chebyshev polynomial, $\mathrm{T}_m(z_k)$. Then, using the precalculated second-order derivatives of Chebyshev polynomials (Eq. 9.24),.

$$\frac{\partial^2 T_n}{\partial z^2} = \left(\frac{2}{N_z - 1}\right)^{1/2} \sum_{m=0}^{N_z-1}{}'' T_{nm}(t) \left(\frac{d^2 \mathrm{T}_m}{dz^2}\right)_k. \tag{9.32}$$

Substituting Eqs. 9.31 and 9.32 into the linear terms of Eq. 8.2 gives us

$$\left(\frac{2}{N_z - 1}\right)^{1/2} \sum_{m=0}^{N_z-1}{}'' \left[\mathrm{T}_m(z_k) - \alpha \Delta t \left(\left(\frac{d^2 \mathrm{T}_m}{dz^2}\right)_k - \left(\frac{n\pi}{a}\right)^2 \mathrm{T}_m(z_k)\right)\right] T_{nm}$$

$$= \mathrm{rhs}_n(z_k), \tag{9.33}$$

where $\mathrm{rhs}_n(z_k)$ is the (known) right-hand side of Eq. 8.2 and T_{nm} is the updated temperature solution in Fourier(n)-Chebyshev(m) space at $(t + \Delta t)$. This solution is obtained by a collocation method, which forces Eq. 9.33 (for a given n) to be satisfied on the $N_z - 2$ internal z-grid levels corresponding to rows $k = 2 \to N_z - 1$ of the $N_z \times N_z$ matrix operator. The rows corresponding to $k = 1$ and N_z force the bottom and top boundary conditions to be satisfied, respectively.

Boundary conditions are easily formulated in Chebyshev series. For our problem the temperature is a constant at the bottom and top boundaries; so the $k = 1$ row of the matrix operator is simply composed of the Chebyshev polynomial values $\mathrm{T}_m(z = 0)$ and the $k = N_z$ row is composed of the $\mathrm{T}_m(z = 1)$ values for matrix columns $m = 0 \to N_z - 1$. In addition, for $n > 0$, $\mathrm{rhs}_n(z = 0)$ and $\mathrm{rhs}_n(z = 1)$ need to be set to zero. For $n = 0$, $\mathrm{rhs}_0(z = 0) = 1$ and $\mathrm{rhs}_0(z = 1) = 0$ if thermal convection is being simulated or vice versa if gravity waves are being simulated. Remember to multiply every element of the matrix operator by $(2/(N_z - 1))^{1/2}$, including those in the bottom and top rows. In addition, remember to multiply every element in the first ($m = 0$) and last ($m = N_z - 1$) columns of the matrix operator by $1/2$.

For example, a schematic of matrix Eq. 9.33 for $N_z = 5$ and some horizontal mode number $n > 0$ is

$$C \begin{bmatrix} T_{10}/2 & T_{11} & T_{12} & T_{13} & T_{14}/2 \\ A_{20n}/2 & A_{21n} & A_{22n} & A_{23n} & A_{24n}/2 \\ A_{30n}/2 & A_{31n} & A_{32n} & A_{33n} & A_{34n}/2 \\ A_{40n}/2 & A_{41n} & A_{42n} & A_{43n} & A_{44n}/2 \\ T_{50}/2 & T_{51} & T_{52} & T_{53} & T_{54}/2 \end{bmatrix} \begin{bmatrix} T_{n0} \\ T_{n1} \\ T_{n2} \\ T_{n3} \\ T_{n4} \end{bmatrix} = \begin{bmatrix} 0 \\ rhs_{n2} \\ rhs_{n3} \\ rhs_{n4} \\ 0 \end{bmatrix},$$

where the internal matrix elements are denoted as

$$A_{kmn} \equiv T_m(z_k) - \alpha \Delta t \left(\left(\frac{d^2 T_m}{dz^2} \right)_k - \left(\frac{n\pi}{a} \right)^2 T_m(z_k) \right).$$

The Chebyshev functions for degree m at z-level k are $T_{km} \equiv T_m(z_k)$; the coefficients for the updated temperature of Chebyshev degree m and Fourier mode n are T_{nm}; the right-hand side elements for z-level k and Fourier mode n are $rhs_{nk} \equiv rhs_n(z_k)$; and the constant is $C \equiv (2/(N_z - 1))^{1/2}$.

This $N_z \times N_z$ matrix equation is then solved using a standard matrix solver, for example, LAPACK routines *dgetrf* and *dgetrs*. The *dgetrf* routine factors the matrix operator using LU decomposition with partial pivoting and the result is stored in the original array for each Fourier mode n. This needs to be done at the beginning of the run and at any time the size of the time step, Δt, is changed during the simulation. During each numerical time step, the *dgetrs* routine uses the factored matrix operator to solve the system by doing simple lower and upper triangular matrix multiplications on the current $rhs_n(z_k)$ vector; this produces the solution vector, T_{nm}.

With the updated temperature solution now in Chebyshev-Fourier space, use Eqs. 9.31 and 9.32 to transform it and its second derivative back to z-space. These are needed to construct the rhs of the temperature equation for the next time step. Depending on how one chooses to compute the nonlinear advection of temperature, its first derivative may also need to be computed and transformed to z-space. Recall, the Chebyshev coefficients for the first and second derivatives can be calculated via Eqs. 9.27 and 9.28.

After updating the temperature, a similar matrix solution procedure, for each $n > 0$, is required for the semi-implicit vorticity equation after including the buoyancy term,

$$\text{Ra Pr} \left(\alpha T_n(z_k, t + \Delta t) + (1 - \alpha) T_n(z_k, t) \right),$$

in its rhs vector. Then, with the updated vorticity $\omega_n(z_k, t + \Delta t)$, use a similar matrix solution procedure to solve the Poisson equation for the streamfunction, $\psi_n(z_k, t + \Delta t)$.

The nonlinear terms for the next (Adams-Bashforth) time step in the temperature and vorticity equations can be calculated using the Galerkin method described in Chapter 4 or the spectral-transform method described in Chapter 10. Any z-derivatives can be calculated by first transforming the function to Chebyshev space using Eq. 9.22 and then transforming its derivative back to z-space using Eq. 9.25.

The Chebyshev transforms that are performed on both the linear and nonlinear terms are much more efficient when using Fast Fourier Transforms (FFTs). This is done by first converting the Chebyshev polynomials to quarter-wave cosine functions (the "preprocess"), then doing a fast Fourier transform (Cooley & Tukey, 1965), and then converting back from the Fourier to the Chebyshev representation (the "postprocess"). This may seem like much more work but the efficiency of the FFT makes this absolutely worthwhile (see, e.g., Press et al., 1992; Boyd, 2001). However, when using a fast Chebyshev transform, Eqs. 9.27 and 9.28 should be used to compute the first and second derivatives. Running the FFTs in parallel, i.e., performing the FFTs in z simultaneously over all (or a subset) of the horizontal modes n, makes this spectral method even faster. A recommended FFT package is "FFTW," the "Fastest Fourier Transform in the West" (*www.fftw.org*). Fourier transforms are discussed further in Chapter 10.

The rest of the fully spectral code, the diagnostics, the CFL condition check, and the data processing for output files can remain as is.

9.5 PARALLEL PROCESSING

Parallel processing is necessary, even for 2D problems, to take full advantage of today's powerful computers in order to produce simulations that approach the turbulent conditions of planetary and stellar interiors. However, even a cursory summary of parallel processing techniques would be beyond the scope of this book. There are several comprehensive books on this topic (e.g., Chapman et al., 2007; Snir et al., 1998) and tutorials on the Internet (for example, https://computing.llnl.gov/tuto rials/openMP/ and https://computing.llnl.gov/tutorials/mpi/). Here we provide just a brief introduction to the basic ideas, some issues that need to be addressed, and choices that need to be made when designing a parallel code.

The basic goal is to either reduce the "wallclock time" required to run a given job or to be able to increase the spatial and temporal resolution of the job without significantly increasing the wallclock time. The former goal exists when the job is CPU-limited; the latter exists when the job is memory-limited. Usually the latter is the case for problems described in this book.

Because of the limitations in reducing communication times within a computer processor, combining tens of thousands of processors all working concurrently on a problem is much more efficient than improving the speed and memory of a single processor. This requires considerable communication among the processors every time step while the job is running. The computation and memory can be divided in many different ways: the challenge is to design a code that is efficient, i.e., all processors are working nearly all the time. There are two basic styles of parallel programming. One is *Open Multiprocessing* (OpenMP), which is a specification for doing "shared memory multiprocessing"; that is, all processes have direct (high-speed) access to the memory. The other is *Message Passing Interface* (MPI), which is a specification for doing "distributed memory multiprocessing"; that is, each processor (or node) has direct access to only the memory attached to it.

A code's performance is typically measured by how well it scales as the number of processors increases. For a job that is CPU-limited the performance is measured by its *strong scaling efficiency*, which is the number of processors, N_p, times the wallclock time required for N_p processors to complete the job divided by the wallclock time required for one processor to complete the same job, i.e., with the same spatial and temporal resolutions and for the same number of time steps. For a job that is memory-limited the performance is measured by its *weak scaling efficiency*, which is the wallclock time required for N_p processors to complete a job divided by the wallclock time required for one processor to complete a job that requires N_p times fewer computations. That is, for weak scaling we would like the required wallclock time to be nearly same among jobs for which the amount of computation is proportional to the number of processors used. This is typically the case for parallelization schemes that employ nearest-neighbor communication, and not typically for schemes that require global communication. In addition, a code is efficiently parallelized when the time the average process spends communicating with other processes is a small fraction of the time it spends computing.

OpenMP and MPI both require the programmer to add to a code statements that determine how computations are distributed among the available processors and how results are communicated between processors. Each style needs to have its libraries loaded with the compiled parallel code, which can be written in, for example, Fortran, C, or C++. OpenMP and MPI can run on most computer architectures and operating systems.

9.5.1 OpenMP

OpenMP is simpler than MPI to incorporate into a code but may be less effective in speeding up a code. There are many OpenMP directives that control where in the code the work is distributed among processors (i.e., "multithreaded") and where it is done by a single processor (i.e., the "master thread"). Usually the main objective is to make large "do loops" that are executed many times and run much faster by distributing their iterations over the available "threads" (i.e., processes). The programmer needs to identify which loops to parallelize by adding statements (preprocessor directives) just before each loop that list those variables in the loop that are "shared" among the threads and those that are "private," i.e., are temporary and used by only one of the threads. The OpenMP library uses this information to design how the iterations are divided among the processors, which would then be run concurrently. The programmer must be sure all statements within the loop are independent of the others for this to work. This works best for loops with many iterations, require a large amount of computational work per iteration, and are executed many times during the run.

See an example of an OpenMP loop in Appendix D.

In addition, a hybrid parallel code can be designed that uses both OpenMP and MPI. For example, a parallel computer can have many "nodes," each of which has several processors, each of which has several "cores." All the cores on one node share the memory on that node; and so OpenMP can be used among the cores of one node, one thread per core. However, the cores on different nodes need to communicate via MPI.

9.5.2 MPI

MPI can run on a small cluster of processors, even on a laptop computer, all the way up to the most powerful supercomputers. It provides considerable flexibility and potential for speedup. The price for this is that the programmer needs to design and implement the details of how the processes share data storage and computations and how they communicate, instead of having a compiler configure most of these details as is done with OpenMP. Typically, every process sees exactly the same code but, by knowing its process ID number, it knows what computations it needs to perform, where all the data is distributed (stored in memory) on all the processors, and what data it needs to send to and receive from all the other processes each numerical time step.

An efficient MPI code is one that has a speedup that scales nearly linearly with the number of processes it uses. Of course the additional time needed to communicate data among the processes prevents perfect linear scaling; however, the more communication can be done while computations are being done the more efficient the parallel code. Therefore, it is important to design the interprocessor communication to avoid conflicts; for example, a processor should not be receiving data from several other processors at the same time.

Designing an efficient "domain decomposition" is also critical; but the design depends on the numerical method. If a local method (like finite differences) is used, the physical domain can be divided into many subdomains with the gridded data residing within each subdomain being stored in the memory associated with a separate processor (or core if the processors are multicore). In such a case, only the grid points on the perimeters of the subdomains need to communicate with their adjacent subdomains to compute derivatives; that is, only "nearest-neighbor communication" is required. A spectral method, on the other hand, usually requires "global communication"; that is, because the value of a variable or its derivative at a given location on the physical grid requires a summation over all the spectral modes, which are usually distributed over all the processors, every process needs to send data to and received data from every other process (Glatzmaier & Clune, 2000). However, even a fully finite-difference method would require some sort of global communication if a poison solver is used (for updating the streamfunction, for example) or if a semi-implicit time integration scheme is used.

Consider, for example, convection in a 2D box using a finite-difference method in the z-direction and a Fourier method in the x-direction with a spectral-transform method to calculate the nonlinear terms each time step as we do in Section 10.4. One could choose to distribute the data and computations over the Fourier modes; that is, every process is in charge of a relatively small number of Fourier modes (for each variable) but has all the z-level data for those modes. In this way the finite-difference derivatives in z can easily be calculated. However, the spectral-transform method requires FFTs over the Fourier modes for each z-level. This could be done using parallel FFTs, which would do the necessary interprocessor data transfers. Alternatively, one could transpose all the data from being distributed over Fourier modes to being distributed over z-levels (using, for example, an MPI_ALLTOALL command), then do the needed FFTs within processor to go from Fourier space to x-space, compute the nonlinear products in x-space, do the inverse FFTs back to Fourier space, and then transpose the data back to being distributed over Fourier modes. Although these two ways tend to require comparable amounts of computational time, the latter way is usually the preferred way of doing a spectral transform in parallel.

How many processes to use is also an important choice. The more used, the less wall-clock time per run; however, at some point the code will no longer scale well with the number of processes. "Coarse-grained" domain decomposition means the domain (physical or spectral) is divided into a relatively small number of subdomains; and so a large amount of computation is performed between interprocessor communications, which keeps the ratio of communication time to computation time small. "Fine-grained" domain decomposition, on the other hand, reduces the wall-clock execution time but may be less efficient at some number of parallel processes.

Another issue is how data is input in and output from the disk during a run. Since IO is usually done much less frequently than updating the time step, one could designate one process as the one that is in charge of IO. However, typically there is not enough memory per processor to store all the data of the problem. Therefore, when reading in the data the IO process needs to read in small chunks of it at a time into a temporary array and send it to the correct process before reading in the next chunk. Likewise, when writing out the data to the disk the IO process needs to receive small chunks of data at a time from each of the other processes. Alternatively, parallel IO could be employed, for which each processor simultaneously reads (writes) its data from (to) a separate data file on the disk. Of course a large number of data files then would be associated for a given job for each restart; additional sorting would be required if one chooses to restart with a different number of processes. The single designated IO process method does not have these disadvantages.

After this very brief introduction to parallel processing, one needs to read some of the suggested books and tutorials and write simple programs that test the various parallel processing directives and routines to gain a good understanding of basic tools. Then, before writing a parallel convection code, one should first write a serial code to test the chosen numerical method and to have a numerical solution to which the parallel solutions can be compared. However, converting a serial code to a parallel code is usually not straightforward. For example, data arrays usually need to be restructured since each process works with only a portion of the data. Debugging is also more challenging since many processes are running simultaneously and data is passed between processes; "Totalview" is a common parallel debugger that can keep tract of what each process is doing. For obvious reasons, it is usually good to debug using a very small number of processes.

See examples of some common MPI routines in Appendix E.

SUPPLEMENTAL READING

Boyd (2001)
Chapman et al. (2007)
Durran (1998)
Ferziger & Perić (1997)
Fox & Parker (1968)
Griebel et al. (1967)
Patankar (1980)
Peyret (2002)
Press et al. (1992)
Slingerland & Kump (2011)
Snir et al. (1998)

EXERCISES

1. *Finite-difference formulas on a nonuniform grid*
   Derive the coefficients defined in Eqs. 9.5 for the finite-difference formulas
   (9.3 and 9.4) on an arbitrary nonuniform grid.

2. *Test Chebyshev derivatives*
   Write a test program that computes the Chebyshev coefficients, f_m, df_m, and
   ddf_m, of an arbitrary function of your choice, $F(z_k)$, that has known analytic
   first and second derivatives with respect to z. Then Chebyshev transform these
   coefficients to z_k-space and compare the results to their known values. Test
   how the error depends on the spatial resolution N_z.

3. *Test Chebyshev integrals*
   Write a test program that computes the Chebyshev coefficients, f_m, and integral coefficients, Q_m, of some function, $F(z_k)$, that has a known analytic
   integral from $z = 0$ to 1. Test how the error in the integral via Eq. 9.30 depends on N_z.

4. *Test the Chebyshev transform method*
   Construct a function of your choice, $F(z_k)$, on a Chebyshev grid, z_k, for a
   given N_z. Use Eq. 9.22 to calculate its Chebyshev coefficients, f_m. Then do
   a reverse transform, Eq. 9.21, to calculate the approximation, $G(z_k)$, of the
   original z-dependent function. Show how the standard deviation of $G(z_k)$ relative to $F(z_k)$ depends on N_z for several choices of $F(z_k)$.

COMPUTATIONAL PROJECTS

1. *Finite-difference solutions on uniform and nonuniform grids*
   Use a nonuniform finite-difference grid in the vertical direction, as described
   in Section 9.1, for the case illustrated on the right side of Fig. 6.2. Choose
   the nonuniform grid to be a Chebyshev grid in the upper unstable region and
   another in the lower stable region, with the two overlapping at the interface.
   Make the number of z-levels in the upper region be one-quarter of the

total and the number in the lower region three-quarters of the total, i.e., proportional to the depths of the two regions. This "stacked" Chebyshev grid provides better resolution near the interface between the stable and unstable regions and in the boundary layers at the top and bottom boundaries. Compare the flow and temperature patterns obtained with the finite-difference method on this nonuniform z-grid with those obtained with the finite-difference method on a uniform z-grid having the same total number of z-levels. Use a semi-implicit time integration scheme for both methods and a Δt small enough to satisfy the CFL constraints for both methods.

2. *Finite-difference solutions on nonuniform and coordinate-mapped grids*
Simulate the problem described in the *Finite-difference solutions on uniform and nonuniform grids* project but now using Chebyshev coordinate mappings to uniform grids, ζ_k, in each of the two regions, as described in Section 9.2. Use the same number of ζ-levels in each region as z-levels for the nonuniform method. Compare the flow and temperature patterns obtained with the finite-difference method on this mapped coordinate method with those obtained with the finite-difference method on the nonuniform z-grid method, using semi-implicit schemes with the same Δt.

3. *Fully finite-difference method on nonuniform grids*
Construct a fully finite-difference model of convection using nonuniform grids in the x- and z-directions.

4. *Fully finite-difference method with a semi-implicit scheme*
Starting with the *Fully finite-difference method on nonuniform grids* project, treat the thermal and viscous diffusion with an implicit Crank-Nicolson time integration scheme.

5. *Fully finite-difference vs. Fourier/finite-difference solutions*
Compare a solution obtained using a fully finite-difference model of convection as described in Section 9.3 with that obtained using the original model of Part 1. Set the number of horizontal grid points, N_x, equal to the number of horizontal Fourier modes, N_n. Use either explicit or semi-implicit time integration schemes for both methods.

6. *Finite-difference and Chebyshev spectral solutions*
Simulate the problem described in the *Finite-difference solutions on uniform and nonuniform grids* project but now using a Chebyshev collocation method in each of the two regions, as described in Section 9.4.2. Use the same number of z-levels in each region as used in the finite-difference non-uniform grid method. Compare the flow and temperature patterns obtained with the spectral Chebyshev collocation method with those obtained with the finite-difference method on the nonuniform grid method, using semi-implicit schemes with the same Δt.

7. *Fully finite-difference with constant heat flux through top boundary*
Using a fully finite-difference method, implement a constant diffusive heat flux boundary condition at the top boundary, keeping the bottom boundary at a constant temperature. Note that ghost points will need to be employed above the top boundary and the temperature equation will need to be solved on that boundary level.

8. *Fully finite-difference with background flow through bottom and top boundaries*

Using a fully finite-difference method, implement a z-dependent background flow that enters through a small vent centered on the bottom boundary between $x = x_v$ and $x = a - x_v$, where x_v is a chosen (nondimensional) distance and a is the aspect ratio. The total mass flow rate through this vent always equals the total mass flow rate through the entire top boundary. Do this by adding the following prescribed term to the perturbation streamfunction, ψ:

$$
\begin{cases}
0 & \text{for } x \leq (1 - z)x_v \,, \\[2mm]
A\left(\dfrac{x - (1 - z)x_v}{a - 2(1 - z)x_v}\right) & \text{for } (1 - z)x_v \leq x \leq a - (1 - z)x_v \,, \\[2mm]
A/a & \text{for } x \geq a - (1 - z)x_v \,.
\end{cases}
$$

Show that mass is conserved. Prescribe a time-dependent amplitude, A, to simulate a volcanic eruption into a convecting atmosphere.

9. *Fully finite-difference simulations of internal gravity waves*

Repeat the *Internal gravity waves excited by a continuous central source* project described at the end the Chapter 6 using a fully finite-difference method.

10. *Fully finite-difference simulations with nested grid*

To obtain higher spatial resolution within a portion of the computational domain, construct a "nested grid" in a chosen region of the interior of a 2D box model of convection. Double the resolution by adding grid points between each pair of the original grid points and in the center of each original cell. That is, decrease Δx and Δz each by a factor of 2 within this region. During each numerical time step, first update the solution on the coarse grid and then, using the updated values on the boundaries of the fine grid as the current boundary conditions for the fine grid, solve the equations on the fine grid. The boundary conditions on the new grid points need to be interpolated from those on the original grid points. When the fine grid solution is obtained for a given time step, update the coarse solution on those original grid points that coincide with the grid points on the fine grid. Compare this nested simulation with one that runs only on the original coarse grid, using the same time step, which will need to be small enough to satisfy the constraints on the fine grid.

Chapter Ten

Boundaries and Geometries

We begin this chapter by outlining how one can implement "absorbing" top and bottom boundaries, which reduce the large-amplitude convectively driven flows within shallow boundary layers or the reflection of internal gravity waves off these boundaries in a stable stratification. Then we focus on the side boundaries, outlining how to replace the impermeable side boundary conditions with permeable periodic side boundary conditions. This permits fluid flow through these boundaries and nonzero mean flow, i.e., time-dependent horizontal flow that varies in the vertical direction but not in the horizontal direction. The model can then easily be converted from cartesian box geometry to polar annulus geometry with gravitational acceleration directed toward the center of the annulus. This changes the model from a small-scale regional model, which is not significantly influenced by global curvature, to a global model, albeit one that is still 2D. Then we describe how, with these modifications, it is straightforward to also replace the Galerkin method for calculating nonlinear terms (in a spectral model) with a spectral-transform method, which is much more efficient. Next we introduce "two and a half dimensional" (2.5D) geometry within a cartesian box geometry for which there are now three components of the fluid flow but all variables still depend on just two spatial coordinates; we also outline how a fully 3D cartesian box model could be constructed. We finish this chapter with a description of a model of convection in a fully 3D global spherical shell and how to easily reduce this to a 2.5D (axisymmetric) spherical-shell model. For this we represent the horizontal structures in terms of spherical harmonic expansions.

10.1 ABSORBING TOP AND BOTTOM BOUNDARIES

Convection in the Earth's outer fluid core is confined by impermeable top and bottom boundaries; however, planetary and stellar atmospheres do not have impermeable top boundaries. Therefore, an impermeable boundary condition is not always very realistic for the problem being simulated. Within an atmosphere, rising convective plumes should not be artificially forced to "splash" against a fixed impermeable boundary and be converted into a shallow, high-speed horizontal flow along such a boundary. Likewise, in a stable stratification, upward-propagating internal gravity waves should not be artificially reflected downward by such a boundary.

Several methods have been developed to reduce these problems. The methods attempt to simulate "open boundaries," through which mass and energy can flow without a periodicity constraint (see, for example, Durran, 1998). A simple example is the use of "absorbing boundaries," sometimes called "sponge boundaries." The idea is to extend the computational domain somewhat beyond the region of interest and to severely damp the flows within this "sponge layer," reducing the intensity of the disturbances flowing back into the region of interest. Here we introduce two ways to implement a sponge layer.

One way is to gradually increase the viscous and thermal diffusivities with distance beyond the (internal) boundary of the region of interest, i.e., with distance into the sponge layer. For example, to produce a sponge layer of thickness δ at both the top and bottom boundaries define the viscous diffusivity as

$$\nu(z) = \nu_o \left(1 + \frac{A}{2} \left[1 + \cos\left(\frac{\pi z}{\delta}\right) \right] \right) \qquad \text{for } 0 \leq z \leq \delta,$$

$$= \nu_o \qquad \text{for } \delta \leq z \leq (D - \delta),$$

$$= \nu_o \left(1 + \frac{A}{2} \left[1 + \cos\left(\frac{\pi (D - z)}{\delta}\right) \right] \right) \qquad \text{for } (D - \delta) \leq z \leq D$$

and likewise for the thermal diffusivity. The usual impermeable and stress-free boundary conditions are applied at $z = 0$ and D, which are the exterior edges of the sponge layers.

The amplitude of the relative increase, A, in diffusivities across the sponge layers, and the thickness of the layer, δ, need to be adjusted for a given scenario and desired effect. A reasonable first attempt would be to set $A = 10$ and $\delta = D/4$. To avoid a severe constraint on the size of the numerical time step (Eq. 2.19), the viscous and thermal diffusion should be treated implicitly (Section 8.2).

Another way to damp the disturbances within a sponge layer is to employ "Rayleigh damping" by adding a term to the right sides of the temperature and vorticity equations (2.10 and 2.11) that reduces the amplitude of the variable. For example, add

$$-\frac{\omega_n(z, t)}{\tau}$$

to the right side of the vorticity equation, where the time scale, τ, is defined as

$$\frac{1}{\tau} = \frac{A}{2} \left[1 + \cos\left(\frac{\pi z}{\delta}\right) \right] \qquad \text{for } 0 \leq z \leq \delta,$$

$$= 0 \qquad \text{for } \delta \leq z \leq (D - \delta),$$

$$= \frac{A}{2} \left[1 + \cos\left(\frac{\pi (D - z)}{\delta}\right) \right] \qquad \text{for } (D - \delta) \leq z \leq D.$$

A similar expression would be added to the right side of the temperature equation. Again, the values of the parameters A and δ need to be chosen for a desired effect and the usual impermeable and stress-free boundary conditions are applied at the bottom and top boundaries of the full domain. Of course, the solution should be analyzed only within $\delta \leq z \leq (D - \delta)$.

10.2 PERMEABLE PERIODIC SIDE BOUNDARIES

As discussed in Chapter 2, the Boussinesq approximation ($\nabla \cdot \mathbf{v} = 0$) in 2D allows us to define a streamfunction, ψ, the spatial derivatives of which give the two components of the fluid velocity (Eqs. 2.6). Now we wish to allow fluid to flow in and out of the side boundaries with the only constraint being that the time-dependent flow and temperature profiles in z on the two side boundaries always be identical or, instead of duplicating that location in x, the left and right sides should be treated as being adjacent to each other. That is, the side boundaries are *periodic*.

Periodic side boundaries for a finite-difference method in the x-direction are relatively easy to implement. Instead of imposing fixed boundary conditions on the sides, calculate the first- and second-order derivatives in x on the left side boundary, x_1, in terms of the function values at x_{N_x}, x_1, and x_2. Likewise, derivatives on the right side boundary, x_{N_x}, are written in terms of x_{N_x-1}, x_{N_x}, and x_1.

Periodic side boundaries for our spectral method in the x-direction requires full Fourier expansions in x over the length of the box, L (or the aspect ratio a if nondimensional), that corresponds to multiples of 2π:

$$\psi(x, z, t) = \sum_{m=-N_m}^{N_m} \psi_m(z, t)\, e^{2\pi i m x/L}.$$

Note, here m is the horizontal mode number instead of n, which is the symbol used up to this point. We make this switch in anticipation of the conversion (later in this chapter) to cylindrical and then spherical geometry for which m is the more frequently used symbol for the longitudinal mode number. In addition, the symbol n is the commonly used symbol for the polytropic index, which we use to describe density stratification in the vertical direction in Chapter 12. The symbol i in the above equation is the imaginary number $\sqrt{-1}$. The wavenumber is now $k_m = 2\pi m/L$ (or $2\pi m/a$ if nondimensional, instead of $\pi m/a$ as it has been for the box with impermeable side boundaries) so each mode will exactly span an integral number of wavelengths within the box, making the side boundaries periodic. Likewise, the temperature and vorticity are now represented as full Fourier functions in x.

For $\psi(x, z, t)$ to be real (opposed to a complex number) the coefficients $\psi_m(z, t)$ need to be complex with the constraint that the imaginary contributions in the summation cancel out. That is,

$$\psi_{-m} = \psi_m^*$$

(where the ψ_m^* is the complex conjugate of ψ_m); so for each pair of modes, m and $-m$ (for $m > 0$),

$$\psi_{-m}\, e^{-2\pi i m x/L} + \psi_m\, e^{2\pi i m x/L} = \left(\psi_m\, e^{2\pi i m x/L}\right)^* + \left(\psi_m\, e^{2\pi i m x/L}\right)$$

$$= 2\,\mathrm{Re}\left(\psi_m\, e^{2\pi i m x/L}\right)$$

(where here $\text{Re}(f)$ is the real part of a complex function f). For $m = 0$, ψ_0 is purely real. Therefore,

$$\psi(x, z, t) = \psi_0(z, t) + 2 \sum_{m=1}^{N_m} \text{Re}\left(\psi_m(z, t)\, e^{2\pi i m x/L}\right)$$

$$\equiv 2 \sum_{m=0}^{N_m}{}' \text{Re}\left(\psi_m(z, t)\, e^{2\pi i m x/L}\right), \tag{10.1}$$

where the single prime on the summation symbol means that the first term in the series (i.e., $m = 0$) is multiplied by $1/2$. Likewise,

$$T(x, z, t) = 2 \sum_{m=0}^{N_m}{}' \text{Re}\left(T_m(z, t)\, e^{2\pi i m x/L}\right) \tag{10.2}$$

and

$$\omega(x, z, t) = 2 \sum_{m=0}^{N_m}{}' \text{Re}\left(\omega_m(z, t)\, e^{2\pi i m x/L}\right). \tag{10.3}$$

Note that an alternative to complex variables is to use double summations; that is, each variable would need to be expanded in both sines and cosines. The relationship between a single summation with complex coefficients and a double summation of real coefficients can be seen by noting that the right side of Eq. 10.1 equals

$$\psi(x, z, t) = \sum_{m=0}^{N_m}{}' \left[\psi_{m,cos}(z, t) \cos(2\pi m x/L) + \psi_{m,sin}(z, t) \sin(2\pi m x/L)\right], \tag{10.4a}$$

where the two real coefficients for each value of m are

$$\psi_{m,cos}(z, t) \equiv 2\text{Re}\left(\psi_m(z, t)\right), \tag{10.4b}$$

$$\psi_{m,sin}(z, t) \equiv -2\text{Im}\left(\psi_m(z, t)\right). \tag{10.4c}$$

However, the complex variable formulation, which we use here, is recommended over the double summation because the "bookkeeping" is much simpler and programming languages easily provide for complex arrays and operations. Of course, the T_m, ω_m, and ψ_m arrays would need to be declared complex in the code. Note, the $m = 0$ modes would be included in these complex arrays but would have a zero imaginary part.

The two components of the fluid velocity are therefore

$$v_x = -\frac{\partial \psi}{\partial z} = -2 \sum_{m=0}^{N_m}{}' \text{Re}\left(\frac{\partial \psi_m}{\partial z}\, e^{2\pi i m x/L}\right) \tag{10.5}$$

and

$$v_z = \frac{\partial \psi}{\partial x} = 2 \sum_{m=1}^{N_m}{}' \text{Re}\left(\frac{2\pi i m}{L}\, \psi_m\, e^{2\pi i m x/L}\right). \tag{10.6}$$

Note that, unlike for impermeable side boundaries, now v_x has an x-independent ($m = 0$) part because fluid can flow through the sides. That is, the x-averaged horizontal velocity is $-\partial \psi_0 / \partial z$, which can be a function of z and t. The x-averaged vertical velocity, however, vanishes for all z and t because otherwise mass would not be conserved for this incompressible fluid in a box with impermeable bottom and top boundaries.

Likewise, the y-component of vorticity is now

$$\omega = -\nabla^2 \psi = -2 \sum_{m=0}^{N_m} {}' \, \text{Re} \left(\omega_m \, e^{2\pi i m x / L} \right), \tag{10.7}$$

where

$$\omega_m = -\left(\frac{2\pi m}{L} \right)^2 \psi_m + \frac{\partial^2 \psi_m}{\partial z^2}. \tag{10.8}$$

Therefore, since ψ_0 no longer necessarily vanishes, neither does ω_0.

Since the bottom and top boundaries are still impermeable, v_z vanishes at $z = 0$ and D; therefore, by Eq. 10.6, $\psi_m = 0$ at $z = 0$ and D for all $m > 0$. For the $m = 0$ mode we consider the total momentum in the x-direction (per length in the y-direction), which is

$$\int_0^D \int_0^L \rho_o v_x \, dx \, dz = -\rho_o L \int_0^D \frac{\partial \psi_0}{\partial z} \, dz = -\rho_o L \left(\psi_0(D) - \psi_0(0) \right).$$

Since the bottom and top boundaries are tangentially stress-free and there are no external forces, this total x-component of momentum needs to remain constant in time. For convenience, we choose a frame of reference in which it vanishes, which means that $\psi_0(D) = \psi_0(0)$ for all times; and we choose their arbitrary value to be 0. Therefore, for all $m \geq 0$,

$$\psi_m = 0 \quad \text{at} \quad z = 0 \text{ and } D.$$

In addition, the stress-free bottom and top boundary conditions ($\partial v_x / \partial z = 0$) mean that, by Eq. 10.5,

$$\frac{\partial^2 \psi_m}{\partial z^2} = 0 \quad \text{at} \quad z = 0 \text{ and } D.$$

Given these boundary conditions on ψ_m, Eq. 10.8 forces

$$\omega = 0 \quad \text{at} \quad z = 0 \text{ and } D.$$

These bottom and top boundary conditions on ψ_m and ω_m are the same as those for the impermeable-side-boundary case. Likewise the boundary conditions on T_m have not changed.

The z-derivatives in these equations can be computed by using a finite-difference method on a uniform grid (Section 2.3) or on a nonuniform grid (Sections 9.1, 9.2) or by using a Chebyshev spectral method (Section 9.4). The nonlinear terms can be computed via a Galerkin method (Section 4.2) or a spectral-transform method (Section 10.4), which is presented later in this chapter. The time integration can

Temperature   Streamfunction

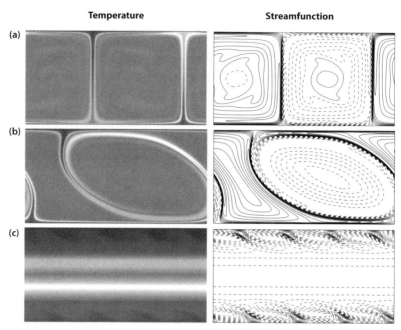

Figure 10.1 Three snapshots of a Boussinesq simulation with permeable periodic side
boundaries (see *Color Plate* 2 for a color version of this figure). The parameters are Ra $= 10^7$, Pr $= 1$, and $a = 2$. The snapshots are at (a) 0.01, (b)
0.017, and (c) 0.1 thermal diffusion times. The temperature perturbation is represented in the left column with maximum value (1) being dark red to minimum
value (0) in dark blue. The corresponding streamlines are plotted on the right
with broken lines representing counterclockwise flow. A second-order finite-
difference method on a uniform vertical grid ($N_z = 256$) is employed with a
full Fourier spectral method in the horizontal direction ($N_x = 512$, $N_n = 170$).
A spectral-transform method is used to compute the nonlinear terms. The solution is integrated in time with a semi-implicit time integration scheme using a
(nondimensional) numerical time step of $\Delta t = 10^{-7}$.

be done by using an explicit scheme (Sections 2.4, 8.1, 8.3), or by using a semi-
implicit scheme (Sections 8.2, 8.3). Of course, modifications need to be made to
all of these methods to account for the variables now being complex (or real but
expanded in both sines and cosines) and for the vorticity and streamfunction now
having nonzero x-independent ($m = 0$) modes.

An example of convection in a box with permeable periodic side boundaries
is presented in Fig. 10.1. This is a Boussinesq simulation with Rayleigh number
Ra $= 10^7$, Prandtl number Pr $= 1$, and an aspect ratio of $a = 2$. The figure shows
three snapshots during the simulation, which spans 0.1 of a thermal diffusion time
(D^2/κ). The temperature perturbation and the streamfunction are shown at each
snapshot. The first snapshot (Fig. 10.1a) is at nondimensional time 0.01 and shows
a pair of well-developed upwelling and downwelling plumes that initially appear
to be in steady state. However, it is an unstable equilibrium. Small noise (due to

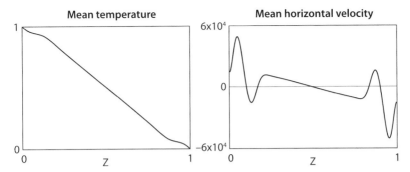

Figure 10.2 The horizontally averaged temperature and horizontal velocity profiles at the final snapshot ($t = 0.1$) illustrated in Fig. 10.1.

truncation errors) causes these plumes to slowly deform and then drift, the lower parts to the right and the upper parts to the left (in this case). The tilt in the flow pattern produces negative Reynolds stresses ($\rho \mathbf{v}_x \mathbf{v}_z$); that is, positive x-momentum is transported in the negative z-direction and negative x-momentum is transported in the positive z-direction (Fig. 10.1b, at time 0.017). The convergence of this non-linear momentum flux drives an x-independent (i.e., mean, $m = 0$) x-component of momentum to the left (i.e., negative momentum) in the upper part of the box and a mean x-component of momentum to the right in the lower part. The final con-figuration (Fig. 10.1c, at time 0.1) is a superposition of tilted plumes in extended convective boundary layers that are advected by the mean horizontal flow, which is sheared in the z-direction (Fig. 10.2). That is, fluid flows to the left (negative mean v_x) near the top boundary ($z = 1$) and to the right near the bottom bound-ary ($z = 0$). However, its direction reverses at five different heights. The greatest shears in the z-direction of the mean horizontal velocity ($\partial < v_x > /\partial z$) occur in the extended boundary layers. The pattern of the streamfunction in these boundary lay-ers (at the final configuration) is that of a characteristic Kelvin-Helmholtz (shear) instability (e.g., Kundu & Cohen, 2008). This boundary layer convection slightly flattens the mean temperature profile (Fig. 10.2) in these regions. Outside the boundary layers the z-derivatives of T and v_x are nearly constant, which mini-mizes both thermal and viscous diffusion in this middle region. Note, the directions in which the plumes initially drift, which ultimately determines the directions of the final mean flows in the upper and lower parts, are randomly determined by the noise. This example demonstrates how important it is to run a simulation for enough time, possibly a diffusion time or more depending on the Rayleigh number, before concluding that the solution is really in a stable steady state.

Alternatively, one may wish to suppress any mean horizontal flow within the computational box to avoid its dominance over the eddies or, for example, to simu-late only a small region within a larger aspect-ratio box that has impermeable side boundaries far from the computational domain. Such suppression of the mean flow can easily be done by maintaining $\omega_{m=0} = 0$ and $\psi_{m=0} = 0$ at all z and t. Likewise, the mean horizontal magnetic field could be suppressed by maintaining $A_{m=0} = 0$ and $J_{m=0} = 0$ (Chapter 11).

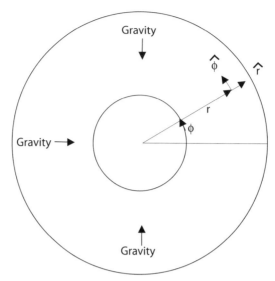

Figure 10.3 A sketch of an equatorial annulus with cylindrical coordinates r and ϕ and
         radial, $\hat{r}$, and longitudinal, $\hat{\phi}$, unit vectors indicated at the current position. The
         z-axis is directed outward, normal to this plane.

10.3 2D ANNULUS GEOMETRY

A model of a 2D box of fluid with periodic side boundaries can be converted into
a 2D annulus of fluid in which gravitational acceleration is everywhere directed toward the center as if representing the equatorial plane of a planet or star that has no
flow or variation normal to this plane. However, not until we add rotation (Section
13.3) is the name "equatorial plane" physically meaningful; until then we simply
mean an arbitrary plane containing the center of the body, the origin of our coordinate system. As can be seen in Fig. 10.3, we now switch from cartesian coordinates to cylindrical coordinates: r for cylindrical radius with N_r radial levels, ϕ for
longitude with N_ϕ longitudinal levels, and z now for the direction normal to the
plane of the flow. If we wished to work with nondimensional variables, the length
scale could be the radius of the top boundary, $D = r_{top}$; so the nondimensional
top radius would be unity and the bottom radius would be r_{bot}/r_{top}. An alternative
would be to choose the depth of the annulus, $D = r_{top} - r_{bot}$, as the length scale.
Here we work with dimensional variables.

10.3.1 2D Annulus: Equations

With the convective domain now representing much of the body's equatorial plane,
we can no longer assume that the reference state gravitational acceleration,
$\mathbf{g} = -\bar{g}\hat{r}$, has a constant amplitude. For a reference state that depends

only on r,

$$\nabla \cdot \mathbf{g} = -4\pi G \bar{\rho},$$

where G is the gravitational constant and $\bar{\rho}$ is the reference state density, which also should be a function of r, especially for stars and gas giant planets. Density stratification is considered in Chapter 12; here we continue to make the Boussinesq approximation and so the reference state density is assumed independent of r. Although we will work in cylindrical coordinates, we are interested in the equatorial plane within a *spherical* planet or star. Therefore we compute $\bar{g}(r)$ using the divergence operator in spherical coordinates assuming spherical symmetry:

$$\nabla \cdot \mathbf{g} = -\frac{1}{r^2} \frac{\partial (r^2 \bar{g})}{\partial r} = -4\pi G \bar{\rho}.$$

Consequently, the *amplitude* of the gravitational acceleration is

$$\bar{g}(r) = \frac{4\pi G}{r^2} \int_0^r \bar{\rho} r^2 dr, \qquad (10.9)$$

which, for constant $\bar{\rho}$, is

$$\bar{g}(r) = \frac{4\pi G \bar{\rho}}{3} r. \qquad (10.10)$$

Here we have assumed that the core below the bottom boundary has the same constant density as that of the convecting fluid within the annulus; this can easily be modified if desired by prescribing the total mass of the central core. Note that $\bar{g}$ now being a function of r means that one needs to choose a value of $\bar{g}$ to use for the definition of the Rayleigh number. For example, it could be the value at the bottom boundary or top boundary or it could be a radial or volumetric average; no one choice has become a standard. We will choose the value of $\bar{g}$ at mid-depth.

The reference state momentum equation still satisfies hydrostatic equilibrium, which appears the same in spherical and cylindrical coordinates:

$$\frac{d\bar{p}}{dr} = -\bar{g}\bar{\rho}. \qquad (10.11)$$

In Chapter 12 we consider a radially dependent $\bar{\rho}$; but here we continue to assume a constant reference state density, $\bar{\rho}$.

We choose to solve the fluid dynamics equations in cylindrical coordinates within the equatorial plane. Therefore the streamfunction, ψ, is defined by

$$\mathbf{v} = \nabla \times \psi \hat{z} = \frac{1}{r} \frac{\partial \psi}{\partial \phi} \hat{r} - \frac{\partial \psi}{\partial r} \hat{\phi};$$

and so the velocity components are

$$v_r = \frac{1}{r} \frac{\partial \psi}{\partial \phi}, \quad v_\phi = -\frac{\partial \psi}{\partial r}, \quad \text{and} \quad v_z = 0. \qquad (10.12\text{a,b,c})$$

The 2D assumption also constrains all derivatives with respect to z to vanish. Notice that $\nabla \psi \cdot \mathbf{v}$ vanishes in cylindrical coordinates; therefore contours of this ψ are also instantaneous "streamlines" of the flow for this annulus geometry. The vorticity in the z-direction is

$$\omega = -\nabla^2 \psi = -\left(\frac{\partial^2 \psi}{\partial r^2} + \frac{1}{r}\frac{\partial \psi}{\partial r} + \frac{1}{r^2}\frac{\partial^2 \psi}{\partial \phi^2} \right). \tag{10.13}$$

Now consider the boundary conditions. Our boundaries are impermeable $(v_r = 0)$; so

$$\frac{\partial \psi}{\partial \phi} = 0 \quad \text{at} \quad r_{bot} \quad \text{and} \quad r_{top}. \tag{10.14}$$

Therefore, as is done in Section 10.2, we set

$$\psi = 0 \quad \text{at} \quad r_{bot} \quad \text{and} \quad r_{top}. \tag{10.15}$$

Our boundaries are also stress-free, which now corresponds to a vanishing radial gradient of the angular velocity:

$$0 = \frac{\partial (v_\phi / r)}{\partial r} = -\frac{1}{r}\left(\frac{\partial^2 \psi}{\partial r^2} - \frac{1}{r}\frac{\partial \psi}{\partial r} \right) \quad \text{at} \quad r_{bot} \quad \text{and} \quad r_{top}. \tag{10.16}$$

These two conditions on each boundary (Eqs. 10.15 and 10.16) with Eq. 10.13 force the vorticity on the boundaries to be

$$\omega = -\frac{2}{r}\frac{\partial \psi}{\partial r} = \frac{2v_\phi}{r} \quad \text{at} \quad r_{bot} \quad \text{and} \quad r_{top}. \tag{10.17}$$

Therefore, unlike the box geometry for which the vorticity vanishes on the straight bottom and top boundaries, vorticity on the curved boundaries of the annulus equals twice the angular velocity there. Consequently, unlike the solution method for the box geometry, the vorticity equation in this cylindrical geometry cannot be solved before the streamfunction is updated via Eq. 10.13; instead, both need to be solved simultaneously with all four boundary conditions (Eqs. 10.15 and 10.16) placed on ψ.

As discussed in Section 10.2, these equations could be solved by using a finite-difference method in longitude, carefully constructing the continuous horizontal derivatives at $\phi = 0$ and 2π, or by using a full Fourier spectral method in longitude. We present the latter here, replacing $2\pi x/L$ in Section 10.2 with longitude ϕ (in radians):

$$T(r, \phi, t) = 2 \sum_{m=0}^{N_m}{}' \operatorname{Re}\left(T_m(r, t)\, e^{im\phi} \right),$$

$$\omega(r, \phi, t) = 2 \sum_{m=0}^{N_m}{}' \operatorname{Re}\left(\omega_m(r, t)\, e^{im\phi} \right),$$

$$\psi(r, \phi, t) = 2 \sum_{m=0}^{N_m}{}' \operatorname{Re}\left(\psi_m(r, t)\, e^{im\phi} \right).$$

Then, by Eqs. 10.12, we have

$$v_r = \frac{2}{r} \sum_{m=1}^{N_m}{}' \text{Re}\left(im\psi_m\, e^{im\phi}\right) \tag{10.18}$$

and

$$v_\phi = -2 \sum_{m=0}^{N_m}{}' \text{Re}\left(\frac{\partial\psi_m}{\partial r}\, e^{im\phi}\right). \tag{10.19}$$

Substituting these expressions into the temperature, vorticity, and streamfunction equations (noting the orthogonality of the Fourier functions) gives the following set of *dimensional* spectral equations for the annulus:

$$\frac{\partial T_m}{\partial t} = -[\mathbf{v}\cdot\nabla T]_m + \kappa\left(\frac{\partial^2 T_m}{\partial r^2} + \frac{1}{r}\frac{\partial T_m}{\partial r} - \frac{m^2}{r^2}T_m\right), \tag{10.20}$$

$$\frac{\partial \omega_m}{\partial t} = -[\mathbf{v}\cdot\nabla\omega]_m - \frac{\bar{g}\alpha}{r}im\,T_m + \nu\left(\frac{\partial^2 \omega_m}{\partial r^2} + \frac{1}{r}\frac{\partial \omega_m}{\partial r} - \frac{m^2}{r^2}\omega_m\right), \tag{10.21}$$

$$\omega_m = -\left(\frac{\partial^2 \psi_m}{\partial r^2} + \frac{1}{r}\frac{\partial \psi_m}{\partial r} - \frac{m^2}{r^2}\psi_m\right). \tag{10.22}$$

Since there is no term linear in ψ or ω in Eq. 10.20, we can update, at each numerical time step, all the T_m before the ω_m and ψ_m are updated. Any of the time integration schemes described in Chapter 8 could be used. Here we choose a second-order semi-implicit time integration scheme, which treats the linear terms implicitly and the nonlinear terms explicitly:

$$\left[T_m - \frac{\Delta t\,\kappa}{2}\left(\frac{\partial^2 T_m}{\partial r^2} + \frac{1}{r}\frac{\partial T_m}{\partial r} - \frac{m^2}{r^2}T_m\right)\right]_{t+\Delta t} =$$
$$T_{m,t} + \frac{\Delta t\,\kappa}{2}\left(\frac{\partial^2 T_m}{\partial r^2} + \frac{1}{r}\frac{\partial T_m}{\partial r} - \frac{m^2}{r^2}T_m\right)_t -$$
$$\frac{\Delta t}{2}\left(3\,[\mathbf{v}\cdot\nabla T]_{m,t} - [\mathbf{v}\cdot\nabla T]_{m,t-\Delta t}\right). \tag{10.23}$$

This can be solved via the collocation method outlined in Section 8.2 using the usual temperature boundary conditions: for $m > 0$ $T_m = 0$ and for $m = 0$ $T_0 = 1$ at r_{bot} and 0 at r_{top} for convection or vice versa for internal gravity waves. Note that the solution, $T_{m,t+\Delta t}$, and the right-hand side are complex vectors. However, each $N_r \times N_r$ matrix operating on $T_{m,t+\Delta t}$ (left side of Eq. 10.23), if using either a finite-difference method or a Chebyshev spectral method in radius, is real (i.e., not complex). Therefore, the real and imaginary parts of T_m (for each mode m) could be updated separately via Eq. 10.23 and then combined again (at each time step) into the complex T_m.

Now consider ω_m and ψ_m. The impermeable boundary condition at r_{bot} and r_{top} (Eq. 10.14) does not require the axisymmetric part of the streamfunction, ψ_0, to vanish on the boundaries because the axisymmetric part ($m = 0$) of v_r vanishes everywhere (Eq. 10.18). However, we have arbitrarily set ψ_0 to zero on both boundaries (Eq. 10.15), which makes the total angular momentum in the z-direction (per length in the z-direction) be

$$\int_0^{2\pi} \int_{bot}^{top} \rho_o v_\phi \, r^2 dr \, d\phi = -2\pi\rho_o \int_{bot}^{top} \frac{\partial \psi_0}{\partial r} r^2 dr = 4\pi\rho_o \int_{bot}^{top} \psi_0 \, r \, dr.$$

(10.24)

Since the boundaries are stress-free and there are no external torques, total angular momentum should be conserved. Therefore, since we choose ψ_0 to be zero everywhere initially, this integral should vanish at all times, which is something that can be checked as a measure of accuracy. Note, if solving the system of equations in a rotating frame of reference (Chapter 13), the magnitude of Eq. 10.24, computed in the rotating frame, could be compared to the total solid-body angular momentum measured in the inertial frame to check how well angular momentum is being conserved. In summary, the boundary conditions at r_{bot} and r_{top} on all spectral modes $m \geq 0$ are

$$\psi_m = 0,$$

(10.25a)

$$\frac{\partial^2 \psi_m}{\partial r^2} - \frac{1}{r} \frac{\partial \psi_m}{\partial r} = 0.$$

(10.25b)

Sometimes, when employing stress-free boundary conditions with no external torques, one wishes to force the total angular momentum, Eq. 10.24, to exactly vanish at all times instead of being a nonzero, albeit small, time-dependent value due to numerical truncation errors. After demonstrating that indeed Eq. 10.24 does remain very small in magnitude for a reasonable number of time steps, one could apply the integral condition, Eq. 10.24, every time step instead of Eq. 10.25b for $m = 0$ at either r_{bot} or r_{top}. This would be especially straightforward and accurate if employing Chebyshev collocation in radius (Section 9.4.2) with the help of Eqs. 9.29 and 9.30.

10.3.2 2D Annulus: Direct Matrix Solution Method

After the T_m have been updated (Eq. 10.20) at a given numerical time step, the ω_m and ψ_m can be updated by combining them into a single (complex) vector for each mode m and solving Eqs. 10.21 and 10.22 simultaneously using a $2N_r \times 2N_r$ matrix operator. The top N_r rows of our operator represent the semi-implicit vorticity equation,

$$\left[\omega_m - \frac{\Delta t \, \nu}{2} \left(\frac{\partial^2 \omega_m}{\partial r^2} + \frac{1}{r} \frac{\partial \omega_m}{\partial r} - \frac{m^2}{r^2} \omega_m \right) \right]_{t+\Delta t} =$$

$$\omega_{m,t} + \frac{\Delta t \, \nu}{2} \left(\frac{\partial^2 \omega_m}{\partial r^2} + \frac{1}{r} \frac{\partial \omega_m}{\partial r} - \frac{m^2}{r^2} \omega_m \right)_t +$$

$$\frac{\Delta t \, \bar{g} \alpha \, im \, (T_{m,t+\Delta t} + T_{m,t})}{r} \frac{}{2} -$$

$$\frac{\Delta t}{2} \left(3 \, [\mathbf{v} \cdot \nabla \omega]_{m,t} - [\mathbf{v} \cdot \nabla \omega]_{m,t-\Delta t} \right),$$

(10.26)

and the bottom N_r rows represent the streamfunction equation,

$$\omega_m + \left(\frac{\partial^2 \psi_m}{\partial r^2} + \frac{1}{r} \frac{\partial \psi_m}{\partial r} - \frac{m^2}{r^2} \psi_m \right) = 0. \qquad (10.27)$$

Defining the top half of the solution vector as the radial values (or Chebyshev modes) of ω_m and the bottom half as the values of ψ_m make the top-right quadrant of the matrix operator completely zeros since the vorticity equation has no linear terms involving ψ_m; the bottom two quadrants would both be nonzero since the streamfunction equation involves linear terms in both ω_m and ψ_m.

Notice how updating the temperature first allows us to effectively treat the buoyancy term in Eq. 10.26 implicitly. That is, the average value of the temperatures at time t and $t + \Delta t$ is used and included in the right-hand-side vector. If, on the other hand, a modified problem introduced linear terms in the temperature equation involving ω_m or ψ_m, all three equations (10.23, 10.26, and 10.27) could be solved simultaneously using a $3N_r \times 3N_r$ matrix operator for each mode m.

Now consider the boundary conditions (Eqs. 10.25), which involve only ψ. The impermeable conditions ($\psi_m = 0$) can be imposed on the top and bottom rows of the top-right quadrant, instead of forcing the vorticity equation to be satisfied on the boundaries. The stress-free conditions involve radial derivatives, which require ghost points in the streamfunction equation on the top and bottom boundaries if using a finite-difference method. If, on the other hand, a Chebyshev spectral method were employed, this boundary condition could easily be represented as a series of Chebyshev coefficients (Section 9.4). In addition, element 1 and elements N_r through $2N_r$ in the right-hand-side vector need to be set to zero.

The nonlinear terms can be computed each time step via a Galerkin method; however, employing a spectral-transform method (Section 10.4) is recommended instead.

10.3.3 2D Annulus: Influence Matrix Solution Method

An alternative method of solving the coupled vorticity-streamfunction system, Eqs. 10.26 and 10.27 with boundary conditions 10.25, without solving one $2N_r \times 2N_r$ matrix equation per mode per time step is to use the *influence matrix method* (e.g., Peyret, 2002), which instead requires two $N_r \times N_r$ matrix solutions. The point here is to take advantage of the linear properties of the system, the nonlinear terms being part of the known forcing term.

To simplify the notation, we write Eq. 10.26 as

$$L_\omega \omega = rhs \qquad (10.28)$$

and Eq. 10.27 as

$$L_\psi \psi = -\omega, \qquad (10.29)$$

where L_ω and L_ψ are the linear operators with up to second-order derivatives in radius and here $\omega \equiv \omega_m(r, t + \Delta t)$ and $\psi \equiv \psi_m(r, t + \Delta t)$. The four boundary conditions are

$$\psi = 0 \quad \text{at} \quad r_{bot} \quad \text{and} \quad r_{top} \qquad (10.30)$$

and

$$\frac{\partial^2 \psi}{\partial r^2} - \frac{1}{r}\frac{\partial \psi}{\partial r} = 0 \quad \text{at} \quad r_{bot} \quad \text{and} \quad r_{top}. \tag{10.31}$$

The first step is to solve Eqs. 10.28 and 10.29 for a temporary solution, (ω_1, ψ_1), that satisfies the boundary conditions:

$$\omega_1 = 0 \quad \text{at} \quad r_{bot} \quad \text{and} \quad r_{top}$$

and

$$\psi_1 = 0 \quad \text{at} \quad r_{bot} \quad \text{and} \quad r_{top},$$

i.e., the same boundary condition as on ψ, (Eqs. 10.30). Since both equations have their own separate boundary conditions, this is easily done by first solving Eq. 10.28 to get ω_1 as a function of radius and then using this ω_1 in Eq. 10.29 to get ψ_1 as a function of radius.

The second step is to solve the homogeneous version of Eq. 10.28,

$$L_\omega \omega_2 = 0,$$

with boundary conditions

$$\omega_2 = 0 \quad \text{at} \quad r_{top}$$

and

$$\omega_2 = 1 \quad \text{at} \quad r_{bot}.$$

Then use this ω_2 to solve Eq. 10.29 to get ψ_2, again using boundary conditions

$$\psi_2 = 0 \quad \text{at} \quad r_{bot} \quad \text{and} \quad r_{top}.$$

The third step is to again solve the homogeneous version of Eq. 10.28,

$$L_\omega \omega_3 = 0,$$

but now with boundary conditions

$$\omega_3 = 1 \quad \text{at} \quad r_{top}$$

and

$$\omega_3 = 0 \quad \text{at} \quad r_{bot}.$$

Then use this ω_3 to solve Eq. 10.29 to get ψ_3, again using boundary conditions

$$\psi_3 = 0 \quad \text{at} \quad r_{bot} \quad \text{and} \quad r_{top}.$$

The actual solutions are then linear combinations of the three temporary solutions,

$$\omega = \omega_1 + c_1\omega_2 + c_2\omega_3$$

and

$$\psi = \psi_1 + c_1\psi_2 + c_2\psi_3,$$

which satisfies the first set of desired boundary conditions on ψ, Eqs. 10.30. The second set of desired boundary conditions on ψ, Eqs. 10.31, requires

$$\left(\frac{\partial^2 \psi_1}{\partial r^2} - \frac{1}{r}\frac{\partial \psi_1}{\partial r}\right) + c_1\left(\frac{\partial^2 \psi_2}{\partial r^2} - \frac{1}{r}\frac{\partial \psi_2}{\partial r}\right) + c_2\left(\frac{\partial^2 \psi_3}{\partial r^2} - \frac{1}{r}\frac{\partial \psi_3}{\partial r}\right) = 0 \quad (10.32)$$

at r_{bot} and r_{top}. The coefficients, c_1 and c_2, are then easily determined from this set of two algebraic equations, Eqs. 10.32, which is called the *influence matrix*, since everything else in these two equations is now known (after taking the radial derivatives of the temporary solutions ψ_1, ψ_2, and ψ_3 at the two boundaries).

Note, the homogeneous solutions, (ω_2, ψ_2) and (ω_3, ψ_3), do not depend on time; that is, they do not depend on the time-dependent forcing term, rhs. Therefore, they and their contributions to Eqs. 10.32 need to be solved only once and stored, before the time integration begins. Thus, as mentioned, this influence matrix method replaces the one $2N_r \times 2N_r$ matrix solution per mode per time step (required for the direct matrix method) with two $N_r \times N_r$ matrix solutions.

A modification to this method would be to apply

$$\left(\frac{\partial^2 \psi_1}{\partial r^2} - \frac{1}{r}\frac{\partial \psi_1}{\partial r}\right) = \left(\frac{\partial^2 \psi_2}{\partial r^2} - \frac{1}{r}\frac{\partial \psi_2}{\partial r}\right) = \left(\frac{\partial^2 \psi_3}{\partial r^2} - \frac{1}{r}\frac{\partial \psi_3}{\partial r}\right) = 0 \quad (10.33)$$

at r_{bot} and r_{top} in steps 1–3, respectively, instead of

$$\psi_1 = \psi_2 = \psi_3 = 0 \quad (10.34)$$

at r_{bot} and r_{top}. Then the influence matrix, from which c_1 and c_2 can be obtained, would be

$$\psi_1 + c_1\psi_2 + c_2\psi_3 = 0.$$

However, if applying the boundary conditions 10.33 within the matrix solvers were inconvenient (for example, if using a finite-difference method in radius), the former way of using boundary conditions 10.34 in the matrix solvers and using Eqs. 10.32 as the influence matrix may be the better choice. This could also be a reason for choosing this influence matrix method over the direct matrix method (Section 10.3.2).

10.3.4 2D Annulus: Simulation

As an example, consider a Boussinesq simulation of thermal convection in a 2D annulus with central gravity having a Rayleigh number $\text{Ra} = 10^{11}$, Prandtl number $\text{Pr} = 0.2$, and $r_{bot}/r_{top} = 0.2$. A Chebyshev-Fourier spectral method (Section 9.4) is used with a semi-implicit time integration scheme (Section 8.2) and a spectral-transform method to compute the nonlinear terms (Section 10.4). The spatial resolution for this case is $N_\phi = 4096$ and $N_r = 1537$, i.e., Fourier modes up to 1365 (Section 10.4) and Chebyshev modes up to 1536. The code was run in parallel using MPI (Section 9.5.2) on 512 processors.

Figure 10.4 is a snapshot of the temperature perturbation near the beginning of simulation. As seen in this image, one plume is dominant at this early stage; it brings cold (blue) fluid down on one side of the hot bottom boundary and takes

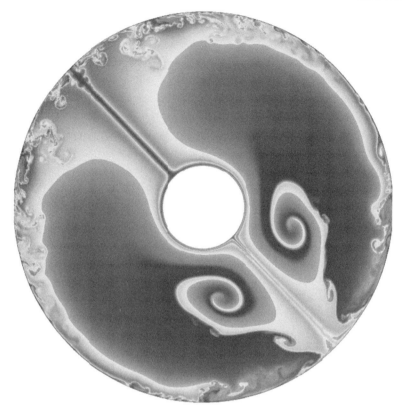

Figure 10.4 A snapshot of temperature for a Boussinesq simulation of thermal convection
in a 2D annulus (see *Color Plate* 3 for a color version of this figure). The
parameters for this case are Ra $= 10^{11}$, Pr $= 0.2$, and $r_{bot}/r_{top} = 0.2$. Yellows
represent hot fluid and blues represent cold fluid relative to the background
temperature.

heated (yellow) fluid up from the other side. This dipolar flow pattern is the most
efficient method of transporting heat from the bottom to the top boundary before
shear instabilities break up this pattern into small-scale turbulence. However, there
continues to be a tendency for a large-scale dipolar flow composed of many small-
scale turbulent eddies. The Reynolds number Re (Section 5.3) is about 10^5. The
initial single-plume dipolar flow pattern, seen here only at the beginning of the
simulation, is maintained as a permanent feature in simulations with much lower
Rayleigh and Reynolds numbers.

10.4 SPECTRAL-TRANSFORM METHOD

The Galerkin method (described in Part 1) for calculating the nonlinear terms every
time step certainly works and nicely illustrates how energy is transferred between

modes. However, the amount of computation increases rapidly with the number of modes. Since convection is turbulent in planetary and stellar interiors, more realistic simulations require significant spatial resolution, i.e., a large number of spectral modes. It was shown years ago (Eliasen et al., 1970; Orszag, 1970) that a spectral-transform method is much more efficient than a Galerkin method when the number of modes exceeds roughly 10. The objective here is to obtain the spectral coefficients for a nonlinear product of two dependent variables, i.e., a quadratic nonlinear term. The process is to transform the two variables from spectral (mode) space to physical (grid) space, simply multiply them together in physical space, and then transform that product back to spectral space. Although more programming is involved in this method, it is much faster than the Galerkin method when fast Fourier transforms (FFTs; Cooley & Tukey, 1965) are employed. Note, this method is often also called a "pseudo-spectral" method; this name, however, is usually reserved for the method that does the time integration in physical space and goes to spectral space and back each time step to accurately compute the spatial derivatives (instead of using something like a finite-difference method in physical space). In contrast, the spectral-transform method, which is described here for the horizontal direction, does the time integration in spectral space. As mentioned, Canuto et al. (1988), Boyd (2001), and Peyret (2002) review spectral methods.

Consider a function $F(x)$ on $0 \le x \le L$ that is periodic, i.e., $F(0) = F(L)$ and likewise each of its derivatives at $x = 0$ equals that at $x = L$. If modeling an annulus instead of a box, let $F(\phi)$ be on $0 \le \phi \le 2\pi$. As usual, define a discrete uniform grid, $x_j = L(j-1)/N_x$, where $j = 1 \to N_x$ for a box or, for an annulus, define $\phi_j = 2\pi(j-1)/N_\phi$ with $j = 1 \to N_\phi$. Note that in the computations (opposed to the graphics) the function at $x = L$ (or at $\phi = 2\pi$) is not allocated. Note also that one could choose to start counting grid points at $j = 0$ and end at $N_x - 1$, in which case the $(j-1)$ above would be replaced by j. Here we choose to start counting at 1.

As in Sections 10.2 and 10.3, we want all functions in physical space to be real; so the complex Fourier coefficients f_m of the real function F are constrained by $f_{-m} = f_m^*$. That is, for a periodic box the complex-to-real Fourier transform, which would be used to transform variables from spectral to physical space, is

$$F(x_j, z_k, t) = \sum_{m=-N_m}^{N_m} f_m(z_k, t)\, e^{2\pi i m x_j / L}$$

$$= \sum_{m=-N_m}^{N_m} f_m(z_k, t)\, e^{2\pi i m (j-1)/N_x}$$

$$= 2 \sum_{m=0}^{N_m}{}' \mathrm{Re}\left(f_m(z_k, t)\, e^{2\pi i m (j-1)/N_x} \right) \qquad (10.35)$$

and for an annulus is

$$F(\phi_j, r_k, t) = 2 \sum_{m=0}^{N_m}{}' \mathrm{Re}\left(f_m(r_k, t)\, e^{2\pi i m (j-1)/N_\phi} \right). \qquad (10.36)$$

The inverse transform (real-to-complex) for a periodic box, which would be used to transform a nonlinear product from physical to spectral space, is

$$f_m(z_k, t) = \frac{1}{N_x} \sum_{j=1}^{N_x} F(x_j, z_k, t)\, e^{-2\pi i m(j-1)/N_x} \tag{10.37}$$

and for an annulus is

$$f_m(r_k, t) = \frac{1}{N_\phi} \sum_{j=1}^{N_\phi} F(\phi_j, r_k, t)\, e^{-2\pi i m(j-1)/N_\phi}. \tag{10.38}$$

Setting up this spectral-transform method for calculating the nonlinear terms each time step is relatively straightforward. The first question is how many grid points (N_x or N_ϕ) should there be for a given number of spectral coefficients (N_m). If one were transforming linear terms back to spectral space (not the nonlinear products of variables), N_x would need to be no larger than $2N_m + 1$, i.e., the number of different m modes from $-N_m$ to $+N_m$. If, however, two modes both with $m = N_m$ were transformed to physical space, their product would involve a contribution from $m = 2N_m$, as can be seen from Eqs. 4.1 or simply by considering

$$e^{im_1\phi} e^{im_2\phi} = e^{i(m_1+m_2)\phi},$$

where $m_1 = m_2 = N_m$. Adding this nonlinear term to the solution each time step would contaminate the solution, which is only represented by modes with $|m| \leq N_m$. Such a solution is said to be "aliased." The problem is that additional energy is added to the solution each time step, which can cause the numerical solution to "blow up." Phillips (1959) solved this by using $4N_m + 1$ grid points in physical space (so the quadratic nonlinear term would be fully resolved there) and then kept only the part of this nonlinear term that is represented by $|m| \leq N_m$; this maintains a "de-aliased" solution. Orszag (1971a) later noticed that if just $3N_m + 1$ grid points were used in physical space, the aliasing would affect only those modes with $|m| > N_m$, which are being filtered out anyway (e.g., Boyd, 2001). Therefore, set

$$N_x \geq 3N_m + 1$$

and likewise for N_ϕ if an annulus.

The average amplitude of F is simply

$$\frac{1}{L} \int_0^L F(x)\, dx = f_0;$$

or, for an annulus, replace L with 2π and x with ϕ. Another convenient formula for calculating horizontally averaged variances and energies is

$$\frac{1}{L} \int_0^L F^2(x)\, dx = f_0^2 + 2 \sum_{m=1}^{N_m} |f_m|^2$$

and likewise for an annulus.

10.5 3D AND 2.5D CARTESIAN BOX GEOMETRY

A model that solves for all three components of the fluid flow (and magnetic field) but with dependence on only two of the three coordinate directions is said to be a *two and a half dimensional* (2.5D) model. Note, a "2.5D model" has also been defined by some authors as one that has one Fourier mode in the third coordinate direction; but here we define it as having no dependence in the third direction.

Here we briefly outline what modifications would need to be made to our 2D cartesian box model to convert it to a 2.5D model with cartesian coordinates x, y, and z. As in our 2D models, the gravitational acceleration is in the negative z-direction; but now fluid velocity is

$$\mathbf{v} = v_x\hat{\mathbf{x}} + v_y\hat{\mathbf{y}} + v_z\hat{\mathbf{z}} \tag{10.39}$$

with all velocity components and thermodynamic variables being functions of only x, z, and t. Such a configuration is mainly employed when the equations are solved in a rotating frame of reference (Chapter 13) with the axis of rotation in the z-direction so Coriolis forces, resulting from flows in the x-direction, drive flows in the y-direction.

Consider, for example, a small region of a sphere with the z-direction being the local radial direction (i.e., upward), the x-direction being the local colatitude direction (i.e., southward), and the y-direction being the local longitudinal direction (i.e., eastward). We are again assuming the size of this box is small relative to the radius of the spherical body so we can neglect the spherical curvature. Note that unless the body is rotating the south and east directions are arbitrary.

We start with the same 2D Boussinesq formulation that we employ in Part 1 and Section 10.2, vorticity (ω_y) and streamfunction (ψ), both in the y-direction, and temperature perturbation (T). To this we add another variable, the velocity in the y-direction (v_y), and another equation, the momentum equation in the y-direction. Therefore, the fluid velocity is

$$\mathbf{v} = \nabla \times \psi\,\hat{\mathbf{y}} + v_y\hat{\mathbf{y}}. \tag{10.40}$$

As can easily be seen, Eq. 10.40 forces mass conservation everywhere, Eq. 1.15. The y-component of vorticity is related to the streamfunction in the same way it is in the 2D formulation, Eq. 2.7. However, now vorticity also has components in the x- and z-directions:

$$\omega_x = -\frac{\partial v_y}{\partial z} \quad \text{and} \quad \omega_z = \frac{\partial v_y}{\partial x}. \tag{10.41a,b}$$

Using Eqs. 10.41 and that $\partial/\partial y = 0$, one can easily show that the vorticity equation in the y-direction is also the same as it is for a strictly 2D box, Eq. 2.4. That is, unless there is a Coriolis term (Chapter 13), the equation for $\partial\omega_y/\partial t$ is not affected by v_y. Likewise, the energy equation is unchanged, Eq. 1.17.

The additional equation, the y-component of the momentum equation (1.16), for this 2.5D formulation, assuming no Coriolis force, is simply

$$\frac{\partial v_y}{\partial t} = -(\mathbf{v} \cdot \nabla)v_y + \text{Pr}\,\nabla^2 v_y. \tag{10.42}$$

This equation needs boundary conditions. For all boundaries being impermeable and stress-free set

$$\frac{\partial v_y}{\partial x} = 0 \quad \text{at } x = 0 \text{ and } a$$

and set

$$\frac{\partial v_y}{\partial z} = 0 \quad \text{at } z = 0 \text{ and } 1.$$

Therefore, to satisfy the side boundary conditions, expand v_y and ω_x in $\cos(n\pi x/a)$ and ω_z in $\sin(n\pi x/a)$. Permeable side boundaries would require full Fourier expansions (Section 10.2).

Note, if v_y is initially zero everywhere, Eq. 10.42 shows that v_y will remain zero, i.e., the solution is strictly 2D. If, on the other hand, the initial v_y is set to be nonzero (for example, proportional to $\cos \pi z$), the x and z components of the flow will advect v_y around as v_y viscously diffuses away. Therefore, this 2.5D problem is not very interesting unless there is a Coriolis force (Chapter 13). For example, if the planetary rotation vector is in the z-direction, there is a Coriolis force driving v_y due to v_x and a Coriolis force driving v_x due to v_y.

In preparation for Section 10.6, consider an alternative to the "vorticity-stream-function" representation for this 2.5D box formulation. Instead of Eq. 10.40, let the divergence-free velocity be defined as

$$\mathbf{v} = \nabla \times \nabla \times W\hat{z} + \nabla \times Z\hat{z}, \tag{10.43}$$

where W is the *poloidal velocity scalar* and Z is the *toroidal velocity scalar*, both functions of x, y, z, and t if fully 3D or of x, z, and t if 2.5D. Like Eq. 10.40, Eq. 10.43 also forces mass conservation, i.e., makes the velocity divergence-free everywhere. In 3D, the three components of Eq. 10.43 are

$$v_x = \frac{\partial^2 W}{\partial x \partial z} + \frac{\partial Z}{\partial y}, \tag{10.44a}$$

$$v_y = \frac{\partial^2 W}{\partial y \partial z} - \frac{\partial Z}{\partial x}, \tag{10.44b}$$

$$v_z = -\nabla_H^2 W, \tag{10.44c}$$

where $\nabla_H^2 \equiv \partial^2/\partial x^2 + \partial^2/\partial y^2$. The three components of the curl of Eq. 10.43 are

$$(\nabla \times \mathbf{v})_x = -\frac{\partial \nabla^2 W}{\partial y} + \frac{\partial^2 Z}{\partial x \partial z},$$

$$(\nabla \times \mathbf{v})_y = \frac{\partial \nabla^2 W}{\partial x} + \frac{\partial^2 Z}{\partial y \partial z},$$

$$(\nabla \times \mathbf{v})_z = -\nabla_H^2 Z;$$

and the z-component of the double curl of Eq. 10.43 is

$$(\nabla \times \nabla \times \mathbf{v})_z = \nabla_H^2 \nabla^2 W.$$

Dropping all terms in Eqs. 10.44 with a $\partial/\partial y$ gives a 2.5D representation, which when compared to Eq. 10.40 shows that

$$\frac{\partial W}{\partial x} = -\psi \quad \text{and} \quad \frac{\partial Z}{\partial x} = -v_y. \tag{10.45a,b}$$

If W and Z were expanded in Fourier functions in x, which are the eigenfunctions of the Laplacian operator in x, it would be straightforward to update $W_n(z, t)$ via the z-component of the momentum equation and $Z_n(z, t)$ via the z-component of the curl of the momentum equation. The z-component of the double curl of the momentum equation is sometimes used instead of the z-component of the momentum equation to avoid having to solve for the pressure when updating $W_n(z, t)$. If a strictly 2D model were desired, the toroidal scalar, Z, would be set to zero and only the poloidal scalar, W, would be needed.

Note that it would also be straightforward to make a fully 3D cartesian box model with periodic side boundaries by representing the structures in both x and y by Fourier expansions and using, for example, either Chebyshev polynomial expansions or finite differences in the z-direction. One could also make the box periodic in all three dimensions by using Fourier expansions in all directions; or one could make all six boundaries impermeable by using Chebyshev polynomial expansions or finite differences in all directions.

We conclude by mentioning that, if a magnetic field (Chapter 11) were included, a 2.5D model would need to represent it with equations similar to Eqs. 10.40 and 10.43, where ψ would be replaced with the vector potential, A, v_y by B_y, and the poloidal and toroidal scalars by corresponding poloidal and toroidal magnetic scalars (Section 10.6).

10.6 3D AND 2.5D SPHERICAL-SHELL GEOMETRY

Consider a spherical shell of fluid (Fig. 10.5) with spherical coordinates r, θ and ϕ being the spherical radius, colatitude, and longitude. Colatitude is the polar coordinate, which is zero at the north geographic pole and increases southward to π radians at the south pole. Longitude is the azimuthal coordinate, which increases eastward from zero back to 2π radians. Fluid velocity is now

$$\mathbf{v} = v_r \hat{\mathbf{r}} + v_\theta \hat{\boldsymbol{\theta}} + v_\phi \hat{\boldsymbol{\phi}}, \tag{10.46}$$

where all the velocity components and the thermodynamic variables are functions of r, θ, ϕ, and t if fully 3D or functions of only r, θ, and t if 2.5D, i.e., axisymmetric. Here we describe a poloidal-toroidal representation of the equations that is solved using a spectral method involving spherical harmonic expansions for the horizontal structure and Chebyshev polynomial expansions for the radial structure (Glatzmaier, 1984). We present it here for a fully 3D model of convection and magnetic field generation (Chapter 11) in a spherical shell of a density-stratified fluid (Chapter 12) that is rotating (Chapter 13); then we explain how one can choose to use just the axisymmetric modes to make a simpler and less computationally expensive 2.5D model.

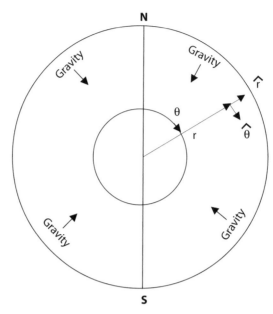

Figure 10.5 A sketch of a meridian plane in a 3D spherical shell with spherical coordinates r, θ, and ϕ and radial, $\hat{r}$, and colatitudinal, $\hat{\theta}$, unit vectors indicated at the current position. The $\hat{\phi}$ unit vector is directed into the plane on the right side of the axis and out of the plane on the left side. The $\hat{\theta}$ unit vector is everywhere directed southward.

The magnetic, density-stratification, and rotation terms are included in this section for convenience, before they are formally introduced and described in Chapters 11, 12, and 13. Therefore, reading the introductions to those chapters is encouraged before proceeding with this chapter. Alternatively, the magnetic, density-stratified, and rotation terms could be ignored the first time through this section. That is, the formulation reduces to a nonmagnetic, nonstratified, nonrotating problem when neglecting the magnetic (**B**) terms and equations, setting the background density ($\bar{\rho}$) to a constant, setting the entropy perturbation over the heat capacity (S/c_p) to the temperature perturbation over a constant background temperature ($T/\bar{T}$), and setting the planetary rotation rate (Ω) to zero.

10.6.1 Spherical Harmonic Expansions

Spherical harmonics, $Y_l^m(\theta, \phi)$, are a natural set of functions for this problem because they constitute a complete and orthogonal set of eigenfunctions of the horizontal Laplacian operator in θ and ϕ. That is,

$$\nabla_H^2 Y_l^m \equiv \frac{1}{r^2 \sin\theta} \frac{\partial}{\partial\theta}\left(\sin\theta \frac{\partial Y_l^m}{\partial\theta}\right) + \frac{1}{r^2 \sin^2\theta} \frac{\partial^2 Y_l^m}{\partial\phi^2}$$

$$= -\frac{l(l+1)}{r^2} Y_l^m, \tag{10.47}$$

where the eigenvalue is $-l(l + 1)/r^2$. This is convenient because all the diffusion operators involve the Laplacian operator. Although not something we will use, it is worth noting that the function

$$f(r, \theta, \phi) = \left(a_1 r^l + a_2 r^{-(l+1)}\right) Y_l^m(\theta, \phi)$$

is the solution to the Laplace equation

$$\nabla^2 f = 0.$$

in spherical coordinates.

Each spherical harmonic is a product of a complex Fourier function, $e^{im\phi}$, and an associated Legendre function (of the first kind), $P_l^m(\cos \theta)$. Two mode numbers define the Y_l^m, the spherical harmonic degree, l, and the spherical harmonic order, m, where $l \geq 0$ and $-l \leq m \leq l$. There are several choices for the normalization coefficient in front of this product. Here we adopt one common to physics, quantum mechanics in particular:

$$Y_l^m(\theta, \phi) \equiv \epsilon \left(\frac{2l + 1}{4\pi} \frac{(l - |m|)!}{(l + |m|)!}\right)^{1/2} P_l^{|m|}(\cos \theta) \, e^{im\phi}, \tag{10.48}$$

where ϵ, called the Condon-Shortley phase factor, is $(-1)^m$ for $m \geq 0$ and 1 for $m < 0$. Therefore, a spherical harmonic with a negative m is related to the complex conjugate of the corresponding positive m harmonic by

$$Y_l^{-m} = (-1)^m (Y_l^m)^*. \tag{10.49}$$

This normalization is chosen to make a convenient orthogonality integral,

$$\int_0^{2\pi} \int_0^\pi \left(Y_{l'}^{m'}\right)^* Y_l^m \sin \theta \, d\theta \, d\phi = \delta_{l'l}\delta_{m'm}.$$

A few examples of spherical harmonic functions are listed in Table 10.1. Recall that $\sin \theta = (1 - \cos^2 \theta)^{1/2}$; so these could be written as just functions of $\cos \theta$. The first routine listed in Appendix C is the way to compute the values of spherical harmonics for many degrees, orders, colatitudes, and longitudes, which is needed for a code that employs spherical harmonic expansions. The routine computes the value of a Legendre function, normalized over just the theta integral, $\overline{P}_l^m(\cos \theta)$, for a given degree, order, and colatitude. Then the value of a complex spherical harmonic for a given l, m, θ, and ϕ is

$$Y_l^m(\theta, \phi) = \frac{\epsilon}{\sqrt{2\pi}} \overline{P}_l^{|m|}(\cos \theta) \, e^{im\phi}.$$

Recursion relations exist for spherical harmonics that are very useful. For example,

$$\cos \theta \, Y_l^m = c_{l+1}^m Y_{l+1}^m + c_l^m Y_{l-1}^m \tag{10.50a}$$

and

$$\sin \theta \, \frac{\partial Y_l^m}{\partial \theta} = l \, c_{l+1}^m Y_{l+1}^m - (l + 1) \, c_l^m Y_{l-1}^m, \tag{10.50b}$$

Table 10.1 The First Few Spherical Harmonic Functions.

$$Y_0^0 = \sqrt{\frac{1}{4\pi}}$$

$$Y_1^0 = \sqrt{\frac{3}{4\pi}}\ \cos\theta$$

$$Y_1^1 = -\sqrt{\frac{3}{8\pi}}\ \sin\theta\ e^{i\phi}$$

$$Y_2^0 = \sqrt{\frac{5}{16\pi}}\ (3\cos^2\theta - 1)$$

$$Y_2^1 = -\sqrt{\frac{15}{8\pi}}\ \cos\theta\ \sin\theta\ e^{i\phi}$$

$$Y_2^2 = \sqrt{\frac{15}{32\pi}}\ \sin^2\theta\ e^{2i\phi}$$

$$Y_3^0 = \sqrt{\frac{7}{16\pi}}\ (5\cos^2\theta - 3)\cos\theta$$

$$Y_3^1 = -\sqrt{\frac{21}{64\pi}}\ (5\cos^2\theta - 1)\sin\theta\ e^{i\phi}$$

$$Y_3^2 = \sqrt{\frac{105}{32\pi}}\ \cos\theta\ \sin^2\theta\ e^{2i\phi}$$

$$Y_3^3 = -\sqrt{\frac{35}{64\pi}}\ \sin^3\theta\ e^{3i\phi}$$

where the coefficient, c_l^m, is defined as

$$c_l^m \equiv \left(\frac{(l+m)(l-m)}{(2l+1)(2l-1)}\right)^{1/2}. \tag{10.50c}$$

Using Eqs. 10.47 and 10.48, we also have

$$\sin\theta\ \frac{\partial}{\partial\theta}\left(\sin\theta\ \frac{\partial Y_l^m}{\partial\theta}\right) = m^2 Y_l^m - l(l+1)\ \sin^2\theta\ Y_l^m. \tag{10.50d}$$

Other useful relations can be constructed by combining these three relations.

Now consider how one expands scalar variables, like entropy (S), pressure (p), density (ρ), and temperature (T), in series of spherical harmonics to represent their horizontal structures. For example,

$$S(r, \theta, \phi, t) = \sum_{l=0}^{\infty} \sum_{m=-l}^{l} S_l^m(r, t)\ Y_l^m(\theta, \phi); \tag{10.51}$$

and we usually simplify the notation by writing the two summation symbols in Eq. 10.51 as just $\sum_{l,m}$. The complex coefficients, S_l^m, in Eq. 10.51 need to satisfy the constraint

$$S_l^{-m} = (-1)^m (S_l^m)^*; \tag{10.52}$$

so the double summation in Eq. 10.51 ends up being a purely real function. That is, using Eqs. 10.49 and 10.52, for every l and $m \neq 0$ combination in the double summation

$$S_l^{-m} Y_l^{-m} + S_l^m Y_l^m = (S_l^m Y_l^m)^* + (S_l^m Y_l^m) = 2\ \text{Re}\left(S_l^m Y_l^m\right).$$

Recall that S_l^0 and Y_l^0 are both purely real. Therefore, only the $m \geq 0$ modes are unique; and, like Eq. 10.1, the truncated version of Eq. 10.51 is

$$S(r, \theta, \phi, t) = \sum_{l=0}^{l_{max}} S_l^0(r, t) Y_l^0(\theta) + 2 \sum_{l=1}^{l_{max}} \sum_{m=1}^{l} \mathrm{Re}\left[S_l^m(r, t)\, Y_l^m(\theta, \phi)\right]$$

$$\equiv \sum_{l,m} S_l^m\, Y_l^m, \tag{10.53}$$

where the second line of Eq. 10.53 is the shorthand notation for the first line, which we will use from now on.

The inverse (real-to-complex) horizontal transform involves an inverse Fourier transform (if 3D) from r, θ, ϕ-space to r, θ, m-space using the discrete version defined by Eq. 10.38 or equivalently an FFT, which is much faster, and an inverse Legendre transform from r, θ, m-space to r, l, m-space, which can be accomplished (to machine accuracy) using an exact Gaussian quadrature method defined by

$$N_l^m(r_k, t) = \sum_{i=1}^{N_\theta} N^m(r_k, \theta_i, t)\, (-1)^m \sqrt{2\pi}\, G_i\, \overline{P}_l^m(\cos \theta_i). \tag{10.54}$$

Here, N_l^m represents the l, m coefficient of a nonlinear term. The N_θ Gaussian colatitudes (θ_i) and weights (G_i) can be calculated using the second routine listed in Appendix C. Usually, N_θ is chosen to be an even number. The θ_i are the colatitudes at which $\overline{P}_{N_\theta}^0(\cos \theta_i)$ vanish and the G_i are then

$$G_i = \frac{(2N_\theta + 3)}{(N_\theta + 1)^2}\left(\frac{\sin \theta_i}{\overline{P}_{N_\theta+1}^0(\cos \theta_i)}\right)^2.$$

A fast Legendre transform would make the spherical spectral transform much faster; however, although some approximate algorithms have been proposed, no stable exact fast Legendre transform has been devised yet. Fortunately, though, the multiply and add operations required for this discrete Legendre transform are typically very efficient on most modern computers.

However, a savings of roughly 50% can be made when doing the Legendre transform to θ-space from l-space by taking advantage of the fact that spherical harmonics, Y_l^m, are symmetric with respect to the equator $(\theta = \pi/2)$ if $(l - m)$ is even and they are antisymmetric (and vanish at $\theta = \pi/2$) if $(l - m)$ is odd. For example, the products in the summation in Eq. 10.53 need to be done only for the northern hemisphere $(\theta \leq \pi/2)$ because the contributions to the summation in the southern hemisphere are equal to those in the northern hemisphere (at the same latitude from the equator) for the even $(l - m)$ harmonics and the negative of those for odd $(l - m)$. Likewise, the reverse transform to l-space from θ-space can be done by forming the symmetric and antisymmetric parts of the nonlinear terms in θ-space. For example, the summation in Eq. 10.54 needs to be done only over $i = 1$ to $N_\theta/2$ if $N^m(r_k, \theta_i, t)$ is redefined as $N^m(r_k, \theta_i, t) + N^m(r_k, \theta_{N_\theta+1-i}, t)$ for even $(l - m)$ and as $N^m(r_k, \theta_i, t) - N^m(r_k, \theta_{N_\theta+1-i}, t)$ for odd $(l - m)$. Then a procedure similar to that for the transform to θ-space can be employed for this reverse transform from θ-space.

Note that Eq. 10.53 could also be represented with just real coefficients and real spherical harmonic functions written in terms of $\cos(m\phi)$ and $\sin(m\phi)$ as in Eqs. 10.4; however, as mentioned, the complex formulation is much easier to write and code.

Also, as illustrated in Eq. 10.53, these summations in practice need to be truncated; there are several different choices for exactly how to truncate them. A *triangular* truncation, which is illustrated in Eq. 10.53, sets the largest value of the spherical harmonic degree to l_{max} and includes all orders up to $m_{max} = l_{max}$. Then, to obtain an alias-free spherical harmonic transform, the number of colatitude grid points needs to be

$$N_\theta \geq \frac{3l_{max} + 1}{2} \qquad (10.55)$$

and the number of longitudinal grid points (for a spherical surface) needs to be $\geq 2N_\theta$. A triangular truncation scheme provides a relatively balanced amount of spatial resolution over a spherical surface. That is, as one chooses spherical harmonics (within the retained set) with a larger order m, the greatest latitudinal wavenumber, $(l_{max} - m)$, decreases. In other words, within a triangular truncation, increasing the longitudinal resolution is at the expense of decreasing latitudinal resolution, and vice versa.

Another common truncation scheme is a *rhomboidal* truncation, which strives to maintain the same amount of latitudinal resolution for all orders m. For this truncation scheme the maximum latitudinal wavenumber, i.e., the number of zeros between the north and south poles, is set to, say, N, which makes $l_{max} = N + m$. An alias-free spherical harmonic transform again requires the constraint Eq. 10.55; but now the number of longitudinal grid points (for a spherical surface) needs to be $\geq N_\theta$ and the largest l_{max}, i.e., the one for $m = m_{max}$, is $N + m_{max}$. However, unless noted, we will use a triangular truncation of spherical harmonics.

For vector variables that are divergence-free, like mass flux ($\bar{\rho}\mathbf{v}$) and magnetic field ($\mathbf{B}$), we use a *poloidal-toroidal* decomposition (Bullard & Gellman, 1954; Chandrasekhar, 1961), which, like the vorticity-streamfunction formulation, ensures mass conservation everywhere within either a Boussinesq or anelastic (Chapter 12) model. That is, we define the poloidal and toroidal mass flux scalars, W and Z, respectively, by writing the mass flux as

$$\bar{\rho}\mathbf{v} \equiv \nabla \times \nabla \times W\hat{\mathbf{r}} + \nabla \times Z\hat{\mathbf{r}} \qquad (10.56)$$

so that

$$\nabla \cdot \bar{\rho}\mathbf{v} = 0 \qquad (10.57)$$

everywhere. Note how this is similar to how we defined the mass flux in Eq. 10.43 for a Boussinesq fluid in a cartesian box with $\bar{\rho} = 1$. Note also that one could use the position vector, $\mathbf{r} = r\hat{\mathbf{r}}$, in Eq. 10.56 instead of the way we have chosen to define the poloidal and toroidal scalars with just the unit vector, $\hat{\mathbf{r}}$. Obviously either way would work; but the equations that follow in terms of W and Z would differ due to the extra factor of r.

We then expand these scalars, W and Z, in spherical harmonics:

$$W(r, \theta, \phi, t) = \sum_{l,m} W_l^m(r, t) \, Y_l^m(\theta, \phi), \tag{10.58}$$

$$Z(r, \theta, \phi, t) = \sum_{l,m} Z_l^m(r, t) \, Y_l^m(\theta, \phi). \tag{10.59}$$

Although the thermodynamic perturbations have time-dependent spherically symmetric modes, e.g., $S_0^0(r, t) \, Y_0^0$, because of Eq. 10.57 the spherically symmetric mass flux scalars, W_0^0 and Z_0^0, are set to zero. To include a spherically symmetric radial expansion or contraction or pulsation would require an additional equation (e.g., Gough, 1969).

Substituting Eqs. 10.58 and 10.59 into Eq. 10.56 gives the three components of mass flux in spherical coordinates,

$$\bar{\rho} v_r = \frac{1}{r^2} \sum_{l,m} l(l+1) \, W_l^m \, Y_l^m, \tag{10.60a}$$

$$\bar{\rho} v_\theta = \frac{1}{r \sin\theta} \sum_{l,m} \left[\frac{\partial W_l^m}{\partial r} \sin\theta \frac{\partial Y_l^m}{\partial \theta} + Z_l^m \frac{\partial Y_l^m}{\partial \phi} \right], \tag{10.60b}$$

$$\bar{\rho} v_\phi = \frac{1}{r \sin\theta} \sum_{l,m} \left[\frac{\partial W_l^m}{\partial r} \frac{\partial Y_l^m}{\partial \phi} - Z_l^m \sin\theta \frac{\partial Y_l^m}{\partial \theta} \right], \tag{10.60c}$$

and the divergence of the velocity,

$$\nabla \cdot \mathbf{v} = -h_\rho v_r = -\frac{h_\rho}{\bar{\rho} r^2} \sum_{l,m} l(l+1) \, W_l^m \, Y_l^m, \tag{10.61}$$

and the three components of the curl of mass flux,

$$(\nabla \times \bar{\rho} \mathbf{v})_r = \frac{1}{r^2} \sum_{l,m} l(l+1) \, Z_l^m \, Y_l^m, \tag{10.62a}$$

$$(\nabla \times \bar{\rho} \mathbf{v})_\theta = \frac{1}{r \sin\theta} \sum_{l,m} \left[\frac{\partial Z_l^m}{\partial r} \sin\theta \frac{\partial Y_l^m}{\partial \theta} + \left(\frac{l(l+1)}{r^2} W_l^m - \frac{\partial^2 W_l^m}{\partial r^2} \right) \frac{\partial Y_l^m}{\partial \phi} \right], \tag{10.62b}$$

$$(\nabla \times \bar{\rho} \mathbf{v})_\phi = \frac{1}{r \sin\theta} \sum_{l,m} \left[\frac{\partial Z_l^m}{\partial r} \frac{\partial Y_l^m}{\partial \phi} - \left(\frac{l(l+1)}{r^2} W_l^m - \frac{\partial^2 W_l^m}{\partial r^2} \right) \sin\theta \frac{\partial Y_l^m}{\partial \theta} \right], \tag{10.62c}$$

and the radial component of the double curl of the mass flux,

$$(\nabla \times \nabla \times \bar{\rho} \mathbf{v})_r = \frac{1}{r^2} \sum_{l,m} l(l+1) \left(\frac{l(l+1)}{r^2} W_l^m - \frac{\partial^2 W_l^m}{\partial r^2} \right) Y_l^m. \tag{10.63}$$

To obtain the horizontal components of vorticity, $\nabla \times \mathbf{v}$, from Eqs. 10.62 one would need to add terms involving h_ρ, which is the inverse density scale height defined as $h_\rho = d \ln \bar{\rho} / dr$.

Since the magnetic field vector, $\mathbf{B}$, also needs to be divergence-free (Chapter 11), we also describe it with poloidal (B) and toroidal (J) scalars:

$$\mathbf{B} \equiv \nabla \times \nabla \times \mathbf{B}\hat{r} + \nabla \times \mathbf{J}\hat{r} \tag{10.64}$$

so that

$$\nabla \cdot \mathbf{B} = 0 \tag{10.65}$$

everywhere. Then, like the mass flux scalars, the B and J scalars are expanded in spherical harmonics,

$$B(r, \theta, \phi, t) = \sum_{l,m} B_l^m(r, t) \, Y_l^m(\theta, \phi), \tag{10.66}$$

$$J(r, \theta, \phi, t) = \sum_{l,m} J_l^m(r, t) \, Y_l^m(\theta, \phi), \tag{10.67}$$

and the spherically symmetric scalars, B_0^0 and J_0^0, are set to zero. Replacing the mass flux vector, $\bar{\rho}\mathbf{v}$, with the magnetic field vector, $\mathbf{B}$, and the mass flux scalars, W and Z, with the magnetic scalars, B and J, respectively, in Eqs. 10.60 gives the three components of the magnetic field,

$$B_r = \frac{1}{r^2} \sum_{l,m} l(l+1) \, B_l^m \, Y_l^m, \tag{10.68a}$$

$$B_\theta = \frac{1}{r \sin \theta} \sum_{l,m} \left[\frac{\partial B_l^m}{\partial r} \sin \theta \frac{\partial Y_l^m}{\partial \theta} + J_l^m \frac{\partial Y_l^m}{\partial \phi} \right], \tag{10.68b}$$

$$B_\phi = \frac{1}{r \sin \theta} \sum_{l,m} \left[\frac{\partial B_l^m}{\partial r} \frac{\partial Y_l^m}{\partial \phi} - J_l^m \sin \theta \frac{\partial Y_l^m}{\partial \theta} \right]. \tag{10.68c}$$

Likewise, making these replacements in Eqs. 10.62 and 10.63 gives the curl of the magnetic field and the radial component of its double curl. Note that the electric current density is $\nabla \times \mathbf{B} / \mu$, where μ is the magnetic permeability.

Here is probably a good place to discuss the relationship between the poloidal magnetic field scalar, B_l^m, and *gauss coefficients*, which are used to describe the *potential* magnetic field at and above the top boundary of the convective dynamo (Section 13.6). By definition, within a potential field,

$$\mathbf{B} = -\nabla V, \tag{10.69}$$

where V is the scalar potential field. Therefore, $\nabla \times \mathbf{B} = 0$ and so electric current density and J_l^m are zero. The potential is expanded in spherical harmonics, but traditionally has been written as

$$V = r_{surf} \sum_{l=1}^{\infty} \sum_{m=0}^{l} \left(\frac{r_{surf}}{r}\right)^{l+1} \left(g_l^m \cos m\phi + h_l^m \sin m\phi\right) P_l^m(\cos\theta), \quad (10.70)$$

where here $P_l^m(\cos\theta)$ is the Schmidt quasi-normalized Legendre functions, which, for $m = 0$, are the same as the usual associated Legendre functions but for $m > 0$ are $(2(l-m)!/(l+m)!)^{1/2}$ times the usual associated Legendre functions. The g_l^m and h_l^m are the gauss coefficients and r_{surf} is the radius of the surface of the planet. By comparing the three components of Eq. 10.69 to those in Eqs. 10.68 one can see that for $m = 0$

$$g_l^0 = l \left(\frac{2l+1}{4\pi}\right)^{1/2} \left(\frac{r_{top}^l}{r_{surf}^{l+2}}\right) \text{Re}(B_l^0)_{top}, \quad (10.71a)$$

$$h_l^0 = 0; \quad (10.71b)$$

and for $m > 0$

$$g_l^m = (-1)^m l \left(\frac{2l+1}{2\pi}\right)^{1/2} \left(\frac{r_{top}^l}{r_{surf}^{l+2}}\right) \text{Re}(B_l^m)_{top}, \quad (10.71c)$$

$$h_l^m = -(-1)^m l \left(\frac{2l+1}{2\pi}\right)^{1/2} \left(\frac{r_{top}^l}{r_{surf}^{l+2}}\right) \text{Im}(B_l^m)_{top}. \quad (10.71d)$$

Here, r_{top} is the radius of the top of the convective dynamo and $(B_l^m)_{top}$ is the poloidal magnetic field scalar at r_{top}. For the Earth, r_{top} would be the core-mantle boundary radius; for a giant gas planet or star, r_{top} would be typically just r_{surf}.

The degree $l = 1$ gauss coefficients represent the magnetic dipolar part of the field; $l = 1, m = 0$ is the axial dipole and $l = 1, m = 1$ is the equatorial dipole. The magnitude of the "geocentric" magnetic dipole moment, m1, is

$$m1 = r_{surf}^3 \left[\left(g_1^0\right)^2 + \left(g_1^1\right)^2 + \left(h_1^1\right)^2\right]^{1/2}$$

$$= \left(\frac{3}{2\pi}\right)^{1/2} r_{top} \left[\frac{1}{2} \left(\text{Re}(B_1^0)_{top}\right)^2 + \left(\text{Re}(B_1^1)_{top}\right)^2 + \left(\text{Im}(B_1^1)_{top}\right)^2\right]^{1/2},$$

assuming the magnetic permeability, μ, is the free-space value. The SI and CGS units for this can be confusing. If the radii were in cm, the gauss coefficients in gauss, and B_l^m in gauss cm^2, the magnetic moment would be in gauss cm^3. If, on the other hand, the radii were in m and the gauss coefficients in nT, the magnetic moment would have units 10^{-2} A m^2, where here T is tesla and A is amperes. (Note, $1T = 10^4$ gauss and $1nT = 10^{-5}$ gauss.)

One more tool we need to describe is a set of algorithms (Glatzmaier, 1984) that is very useful for computing the full divergence and horizontal divergence of a vector, the radial component of the curl of a vector, and (in some codes) the radial component of the double curl of a vector:

$$\nabla \cdot \mathbf{A} = \sum_{l,m} \left[\frac{1}{r^2} \frac{\partial f_{1,l}^m}{\partial r} + (l+1)c_l^m f_{2,l-1}^m - l c_{l+1}^m f_{2,l+1}^m + im f_{3,l}^m \right] Y_l^m,$$

$$\nabla_H \cdot \mathbf{A} = \sum_{l,m} \left[(l+1)c_l^m f_{2,l-1}^m - l c_{l+1}^m f_{2,l+1}^m + im f_{3,l}^m \right] Y_l^m,$$

$$(\nabla \times \mathbf{A})_r = \sum_{l,m} \left[(l+1)c_l^m f_{3,l-1}^m - l c_{l+1}^m f_{3,l+1}^m - im f_{2,l}^m \right] Y_l^m,$$

$$(\nabla \times \nabla \times \mathbf{A})_r = \sum_{l,m} \left[\frac{l(l+1)}{r^4} f_{1,l}^m \right. \tag{10.72}$$

$$\left. + \frac{1}{r^2} \frac{\partial}{\partial r} \left(r^2 \left[(l+1)c_l^m f_{2,l-1}^m - l c_{l+1}^m f_{2,l+1}^m + im f_{3,l}^m \right] \right) \right] Y_l^m,$$

where c_l^m is defined in Eq. 10.50c. Here the arbitrary vector is

$$\mathbf{A} = A_r \hat{\mathbf{r}} + A_\theta \hat{\boldsymbol{\theta}} + A_\phi \hat{\boldsymbol{\phi}}$$

and the complex spherical harmonic coefficients, functions of r and t, are defined such that

$$r^2 A_r \equiv \sum_{l,m} f_{1,l}^m Y_l^m,$$

$$\frac{A_\theta}{r \sin \theta} \equiv \sum_{l,m} f_{2,l}^m Y_l^m,$$

$$\frac{A_\phi}{r \sin \theta} \equiv \sum_{l,m} f_{3,l}^m Y_l^m.$$

These algorithms are used when doing vector differential operations on nonlinear terms at the same time as they are being spectrally transformed from physical grid space back to spherical harmonic space.

Having discussed how the dependent variables can be expanded in finite series of spherical harmonics we mention here additional choices for selecting which spherical harmonics to use. Typically, all spherical harmonic degrees, l, and orders, m, within the limits of the selected truncation (e.g., triangular truncation) are used in the spherical harmonic expansions of the dependent variables. However, one could choose to impose a symmetry in longitude that effectively focuses all computational resources within a limited range in longitude for the purpose of reaching much higher spatial resolution without the much higher computational cost of simulating the entire spherical shell at that resolution. If only mode numbers m that are evenly divisible by N were used, there would be an N-fold symmetry in longitude. The Fourier expansions and transforms would then be done over just one longitudinal slice from $\phi = 0$ to $2\pi/N$. Of course, this imposed symmetry eliminates modes with wavelengths greater than $2\pi/N$, including flow through the axis of the coordinate system. This method could also be used for the 2D annulus if one wished to use extremely high spatial resolution while retaining the curvilinear geometry.

In addition, for a 3D spherical shell, one could impose symmetry across the equatorial plane by using only even values of $l - m$ for the thermodynamic scalars (e.g., S_l^m, p_l^m), the poloidal mass flux scalar (W_l^m), and the toroidal magnetic field scalar (J_l^m) and use only odd values of $l - m$ for the toroidal mass flux scalar (Z_l^m) and poloidal magnetic field scalar (B_l^m). This would allow radial and longitudinal components of the flow and a colatitudinal component of the field at the equator but preclude any flow in colatitude across the equator or any radial or longitudinal field at the equator. Usually these restrictions would be more unrealistic than those for an imposed longitudinal symmetry; therefore, imposing this equatorial symmetry is not advised.

10.6.2 The Equations for a Spherical Shell

If the model is anelastic, i.e., density-stratified (Chapter 12), the reference state is constructed as described in Section 12.2. Conservation of mass (Eq. 10.57) and magnetic flux (Eq. 10.65) are automatically satisfied by expressing mass flux and magnetic field in terms of poloidal and toroidal scalars, Eqs. 10.56 and 10.64.

Consider now the spherical harmonic decomposition of the internal energy equation, Eq. 12.34 (or its Boussinesq version, Eq. 1.14), which is needed to update the entropy perturbation $S_l^m(r, t)$ (or the temperature perturbation $T_l^m(r, t)$). The l, m-mode of the entropy form of the internal energy equation (12.34), for an adiabatic ($d\overline{S}/dr = 0$) reference state, is obtained by projecting the equation onto the set of spherical harmonics:

$$\frac{\partial S_l^m}{\partial t} = \bar{\kappa}\left[\frac{\partial^2 S_l^m}{\partial r^2} + \left(\frac{2}{r} + h_\kappa + h_\rho + h_T\right)\frac{\partial S_l^m}{\partial r} - \frac{l(l+1)}{r^2}S_l^m\right]$$
$$+ \left[-\mathbf{v}\cdot\nabla S + \frac{2\bar{\nu}}{\overline{T}}\left(e_{ij}e_{ij} - \frac{1}{3}(\nabla\cdot\mathbf{v})^2\right) + \frac{\bar{\eta}}{\mu\bar{\rho}\overline{T}}|\nabla\times\mathbf{B}|^2\right]_l^m. \quad (10.73)$$

Here $h_\kappa = d\ln\bar{\kappa}/dr$, $h_\rho = d\ln\bar{\rho}/dr$, and $h_T = d\ln\overline{T}/dr$; all of which vanish for a Boussinesq model. The last term in this equation represents the nonlinear advection of entropy and the nonlinear viscous and ohmic heating (Section 10.6.3). The prescribed reference state heating term, $\sqrt{4\pi}\ \overline{Q}/(\bar{\rho}\overline{T})$ (Eq. 12.33), should also be added to the right side of the spherically symmetric part ($l = 0, m = 0$) of Eq. 10.73.

The poloidal mass flux scalar, $W_l^m(r, t)$, could be updated via the radial component of the momentum equation, in which case one needs to simultaneously solve the divergence of the momentum equation, or just its horizontal part, for the pressure perturbation $P_l^m(r, t)$. Alternatively, one could use the radial component of the double curl of the momentum equation to avoid having to solve for the pressure perturbation; this, however, is a fourth-order differential equation in radius, which can be solved as two second-order differential equations (e.g., Jones & Kuzanyan, 2009); see Section 10.3.3.

Here we describe the former method. In theory, solving either the full or the horizontal divergence of the momentum equation with the radial component of the momentum equation should be equivalent because the full divergence equation can be constructed from the horizontal divergence and the radial derivative of the radial component equation. In practice, however, the full divergence of the anelastic (or Boussinesq) momentum equation is a diagnostic equation because of Eq. 10.57; that is, there is no time derivative. This forces a significant constraint on the system every numerical time step; whereas the time derivative in the horizontal divergence of the momentum equation produces small adjustments in the solution every time step and so tends to be more stable numerically and puts less energy into the smallest scales. Also, the full divergence equation requires a radial derivative of the nonlinear advection term; whereas the horizontal divergence equation does not (Eqs. 10.72). On the other hand, the horizontal divergence equation has a third-order radial derivative in the viscous term (Eq. 10.75), which the full divergence equation does not have. Here, however, we describe the method that uses the horizontal divergence of the momentum equation.

The l, m-mode of the radial component of the anelastic momentum equation (12.29) in spherical coordinates for an adiabatic reference state is

$$
\frac{\partial [\bar{\rho} v_r]_l^m}{\partial t} = \frac{l(l+1)}{r^2} \frac{\partial W_l^m}{\partial t}
$$

$$
= -\bar{\rho} \frac{\partial P_l^m}{\partial r} - \bar{g} \overline{\left(\frac{\partial \rho}{\partial S}\right)_p} S_l^m
$$

$$
+ \frac{2\Omega}{r} \left[(l+2) c_{l+1}^m Z_{l+1}^m - (l-1) c_l^m Z_{l-1}^m + im \frac{\partial W_l^m}{\partial r} \right]
$$

$$
+ \frac{\bar{\nu} l(l+1)}{r^2} \left[\frac{\partial^2 W_l^m}{\partial r^2} + \left(2h_\nu - \frac{h_\rho}{3} \right) \frac{\partial W_l^m}{\partial r} \right.
$$

$$
\left. - \left(\frac{4}{3} \left[\left(\frac{h_\rho}{r} + \frac{dh_\rho}{dr} \right) + h_\nu \left(\frac{3}{r} + h_\rho \right) \right] + \frac{l(l+1)}{r^2} \right) W_l^m \right]
$$

$$
+ \left[-(\nabla \cdot \bar{\rho} \mathbf{v} \mathbf{v})_r + ((\nabla \times \mathbf{B}) \times \mathbf{B})_r / \mu \right]_l^m , \qquad (10.74)
$$

where here P_l^m is the reduced pressure (Eq. 12.25), the thermodynamic derivative $\overline{(\partial \rho / \partial S)_p} = -\bar{\rho}/c_p$ for a perfect gas, and $h_\nu = d \ln \bar{\nu}/dr$. The terms on the left side of Eq. 10.74 are the pressure gradient, buoyancy, Coriolis, viscous, and the nonlinear advection and Lorentz terms. The projection onto spherical harmonics is straightforward for the pressure gradient, buoyancy, and viscous terms. The projection of the Coriolis term, however, requires the use of Eqs. 10.50 and therefore couples W_l^m with Z_{l+1}^m and Z_{l-1}^m. The Coriolis term also couples the real and imaginary parts of W_l^m for $m > 0$. The nonlinear terms couple all the l- and m-modes, which is why we compute these using a spectral-transform method (Section 10.6.3).

Equation 10.74 is solved simultaneously with the horizontal divergence of the anelastic momentum equation:

$$\frac{\partial}{\partial t}\left[\nabla_H\cdot\bar\rho\mathbf{v}\right]_l^m = -\frac{\partial}{\partial t}\left[\frac{1}{r^2}\frac{\partial}{\partial r}(r^2\bar\rho v_r)\right]_l^m = -\frac{l(l+1)}{r^2}\frac{\partial}{\partial t}\left(\frac{\partial W_l^m}{\partial r}\right)$$

$$= \frac{l(l+1)}{r^2}\bar\rho P_l^m + \frac{2\Omega}{r^2}\left[l(l+2)c_{l+1}^m Z_{l+1}^m + (l+1)(l-1)c_l^m Z_{l-1}^m\right.$$

$$\left. -im\left(\frac{\partial W_l^m}{\partial r} + \frac{l(l+1)}{r}W_l^m\right)\right]$$

$$+\frac{\bar\nu l(l+1)}{r^2}\left[-\frac{\partial^3 W_l^m}{\partial r^3} - (h_\nu - h_\rho)\frac{\partial^2 W_l^m}{\partial r^2}\right.$$

$$+\left(\frac{2h_\rho}{r} + \frac{\partial h_\rho}{\partial r} + h_\nu\left(\frac{2}{r} + h_\rho\right) + \frac{l(l+1)}{r^2}\right)\frac{\partial W_l^m}{\partial r}$$

$$\left. -\frac{l(l+1)}{r^2}\left(h_\nu + \frac{2}{3}h_\rho + \frac{2}{r}\right)W_l^m\right]$$

$$+\left[\nabla_H\cdot\left(-\nabla\cdot\bar\rho\mathbf{vv} + \frac{(\nabla\times\mathbf{B})\times\mathbf{B}}{\mu}\right)\right]_l^m. \tag{10.75}$$

We used the anelastic mass conservation (Eq. 10.57) in the first line in this equation in order to write the time derivative in terms of the poloidal mass flux scalar. Solving Eqs. 10.74 and 10.75 together each time step updates W_l^m and P_l^m. It is convenient to solve Eqs. 10.74 and 10.75 simultaneously because the pressure perturbation needs to be free to vary in time and space on the *fixed impermeable* boundaries. That is, no boundary condition should be placed on the pressure; the required number of boundary conditions can be all placed on the fluid velocity (Section 10.6.4).

The toroidal mass flux scalar, Z_l^m, is updated using the radial component of the curl of the momentum equation:

$$\frac{\partial}{\partial t}\left[(\nabla\times\bar\rho\mathbf{v})_r\right]_l^m = \frac{l(l+1)}{r^2}\frac{\partial Z_l^m}{\partial t}$$

$$= \frac{2\Omega}{r^2}\left[l(l+2)c_{l+1}^m\left(\frac{\partial W_{l+1}^m}{\partial r} + \frac{(l+1)}{r}W_{l+1}^m\right)\right.$$

$$\left. +(l+1)(l-1)c_l^m\left(\frac{\partial W_{l-1}^m}{\partial r} - \frac{l}{r}W_{l-1}^m\right) + imZ_l^m\right]$$

$$+\frac{\bar\nu l(l+1)}{r^2}\left[\frac{\partial^2 Z_l^m}{\partial r^2} + (h_\nu - h_\rho)\frac{\partial Z_l^m}{\partial r}\right.$$

$$\left. -\left(\left(\frac{2h_\rho}{r} + \frac{dh_\rho}{dr} + h_\nu\left(\frac{2}{r} + h_\rho\right)\right) + \frac{l(l+1)}{r^2}\right)Z_l^m\right]$$

$$+\left[\left(\nabla\times\left(-\nabla\cdot\bar\rho\mathbf{vv} + \frac{(\nabla\times\mathbf{B})\times\mathbf{B}}{\mu}\right)\right)_r\right]_l^m. \tag{10.76}$$

Here it is seen that the projection of the Coriolis term onto spherical harmonics couples Z_l^m with W_{l+1}^m and W_{l-1}^m.

The poloidal and toroidal magnetic field scalars (Eqs. 10.66 and 10.67) are updated in a similar manner using the magnetic induction equation (11.13). The radial component of the magnetic induction equation, projected onto spherical harmonics, provides the equation for updating B_l^m,

$$\frac{\partial\,[B_r]_l^m}{\partial t} = \frac{l(l+1)}{r^2}\frac{\partial B_l^m}{\partial t}$$

$$= \frac{\bar{\eta}l(l+1)}{r^2}\left(\frac{\partial^2 B_l^m}{\partial r^2} - \frac{l(l+1)}{r^2}B_l^m\right)$$

$$+ [(\nabla\times(\mathbf{v}\times\mathbf{B}))_r]_l^m\,, \tag{10.77}$$

and the radial component of the curl of the magnetic induction equation provides the equation for J_l^m,

$$\frac{\partial\,[(\nabla\times\mathbf{B})_r]_l^m}{\partial t} = \frac{l(l+1)}{r^2}\frac{\partial J_l^m}{\partial t}$$

$$= \frac{\bar{\eta}l(l+1)}{r^2}\left(\frac{\partial^2 J_l^m}{\partial r^2} + h_\eta\frac{\partial J_l^m}{\partial r} - \frac{l(l+1)}{r^2}J_l^m\right)$$

$$+ [(\nabla\times\nabla\times(\mathbf{v}\times\mathbf{B}))_r]_l^m\,, \tag{10.78}$$

where η is the magnetic diffusivity and $h_\eta = d\ln\bar{\eta}/dr$.

Notice that this method of updating the magnetic field differs from the 2D formulation described in Chapter 11 where we "uncurl" the magnetic induction equation to get an equation in terms of the magnetic vector potential. This difference affects the choice of magnetic boundary conditions, which constrain the total (integrated) amount of magnetic flux through a boundary in the 2D formulation instead of the local magnetic field on a boundary in the 3D formulation presented here (Section 10.6.4). Another important difference is that the magnetic field cannot decay away in the 2D *magnetoconvection* problems presented in Chapter 11 because the total magnetic flux through a boundary is prescribed and held constant. On the other hand, the magnetic boundary conditions we choose for the 3D problem allow the field to decay away if the convection is not able to generate enough new field to compensate for magnetic diffusion. When convection is able to maintain a dynamically self-consistent self-sustaining magnetic field, the mechanism is called a *magnetohydrodynamic dynamo* (Section 13.6). However, as Cowling (1934) proved, a magnetic field maintained by dynamo action needs to be fully 3D, i.e., dependent on all three dimensions. This usually involves a fully 3D fluid flow; but certain prescribed 2.5D flows that are spatially periodic can maintain 3D time-dependent magnetic fields (Roberts, 1972). However, a flow and field that are both 2.5D (Sections 10.6.7 and 13.4) with the same magnetic boundary conditions as the 3D case cannot be an MHD dynamo.

Another point worth discussing here is how this system of equations is modified for the *infinite Prandtl number approximation* (Section 8.4), which is employed in studies of mantle convection. In the limit of an infinite Prandtl number, which is very appropriate for mantle convection in terrestrial planets, the momentum equation reduces to an instantaneous balance among the pressure gradient, buoyancy, and viscous forces; the inertial, Coriolis, and Lorentz terms are negligible. Equations 10.74 and 10.75 are then independent of the toroidal mass flux scalar, Z_l^m (Glatzmaier, 1988). This toroidal scalar is the only dependent variable left in Eq. 10.76 and is therefore set to zero since it would diffuse away if ever excited. Equations 10.74 and 10.75 are then solved simultaneously as outlined in Section 10.6.3 but without the time derivatives, Coriolis terms, or nonlinear terms, as is done in Section 8.4 for a simple 2D cartesian box model of Boussinesq convection. If, however, tectonic plates are included, which have a rotational component due to plate boundary constraints, or if a more realistic nonlinear viscous term is employed, the toroidal mass flux scalar would couple with the poloidal scalar and pressure and therefore would need to be retained in the system of equations.

Bercovici et al. (1989) published the first Boussinesq benchmark comparison between two independently written, 3D spherical shell, infinite Prandtl number, mantle convection codes. Anelastic (Chapter 12) infinite Prandtl number simulations of thermal convection in a 3D spherical shell have been produced for studies of mantle convection in the Earth (e.g., Glatzmaier, 1988) and in Mars and Venus (e.g., Schubert et al., 1990). When the infinite Prandtl number approximation is valid, a simpler formulation, the "liquid anelastic" approximation, may also be valid (Jarvis & McKenzie, 1980; Anufriev et al., 2005).

10.6.3 Chebyshev Collocation and Spherical Spectral Transform

The Chebyshev collocation and spectral-transform methods that are described in Chapters 9 and 10, respectively, are now applied to a spherical shell. The complex spherical harmonic coefficients of the dependent variables, i.e., the thermodynamic perturbations and the poloidal and toroidal scalars for the mass flux and magnetic field (Section 10.6.1), are further expanded in a series of Chebyshev polynomials in radius $T_n(r_k)$. The Chebyshev radial grid levels are defined as

$$r_k = r_1 + \frac{D}{2}\left(1 + \cos\left(\frac{(N_r - k)\pi}{N_r - 1}\right)\right),$$

where the depth of the fluid shell, D, is $r_{N_r} - r_1$ with r_1 being the radius of the bottom boundary and r_{N_r} the top boundary. The Chebyshev grid levels are numbered from $k = 1$ to N_r and in spectral space the Chebyshev degrees (modes) are numbered from $n = 0$ to $N_r - 1$. The spherical harmonic coefficients of the entropy perturbation, for example, are

$$S_l^m(r_k, t) = \left(\frac{2}{N_r - 1}\right)^{1/2} \sum_{n=0}^{N_r - 1}{}'' S_{ln}^m(t)T_n(r_k), \tag{10.79a}$$

where the Chebyshev coefficients, S_{ln}^m, are complex and the Chebyshev polynomials, T_n, are real. The inverse Chebyshev transform from r, l, m-space to

n, l, m-space, like that described in Eqs. 9.21 and 9.22, is

$$S_{nl}^m(t) = \left(\frac{2}{N_r - 1}\right)^{1/2} \sum_{k=1}^{N_r}{}'' S_l^m(r_k, t) \mathrm{T}_m(r_k). \tag{10.79b}$$

However, instead of performing the transform 10.79b, we obtain the Chebyshev coefficients by solving semi-implicit matrix equations that update each spherical harmonic coefficient in ways similar to that described in Section 9.4.2. That is, the matrix equations force the equations described in Section 10.6.2 to be satisfied on all the interior Chebyshev levels, r_k for $k = 2$ to $N_r - 1$, and force the boundary conditions at levels r_1 and r_{N_r}. The matrix operators are composed of terms involving $\mathrm{T}_n(r_k)$ and their first and second radial derivatives. Here we will use the semi-implicit time integration scheme introduced in Section 8.2 and modified for Chebyshev collocation in Section 9.4.2, which treats the linear terms with a Crank-Nicolson scheme and the nonlinear term explicitly with an Adams-Bashforth scheme.

The nonlinear terms in Eqs. 10.73–10.78 are calculated each time step using a spectral-transform method (Section 10.4), but now in spherical coordinates. We use Eqs. 10.53 and 10.60–10.62 to construct the scalars,

$$S, \qquad \frac{\partial S}{\partial r}, \tag{10.80a,b}$$

$$r^2 \bar{\rho} v_r, \qquad r \sin\theta \, \bar{\rho} v_\theta, \qquad r \sin\theta \, \bar{\rho} v_\phi, \qquad r^2 (\nabla \times \bar{\rho} \mathbf{v})_r, \tag{10.80c,d,e,f}$$

$$\frac{\partial}{\partial r}(r^2 \bar{\rho} v_r), \qquad \sin\theta \frac{\partial}{\partial \theta}(r^2 \bar{\rho} v_r), \qquad \frac{\partial}{\partial \phi}(r^2 \bar{\rho} v_r), \tag{10.80g,h,i}$$

$$\frac{\partial}{\partial r}(r \sin\theta \, \bar{\rho} v_\theta), \qquad \frac{\partial}{\partial r}(r \sin\theta \, \bar{\rho} v_\phi), \tag{10.80j,k}$$

$$\frac{\partial}{\partial \phi}(r \sin\theta \, \bar{\rho} v_\theta), \qquad \frac{\partial}{\partial \phi}(r \sin\theta \, \bar{\rho} v_\phi), \tag{10.80l,m}$$

$$r^2 B_r, \qquad r \sin\theta \, B_\theta, \qquad r \sin\theta \, B_\phi, \tag{10.80n,o,p}$$

$$r^2 (\nabla \times \mathbf{B})_r, \qquad r \sin\theta \, (\nabla \times \mathbf{B})_\theta, \qquad r \sin\theta \, (\nabla \times \mathbf{B})_\phi, \tag{10.80q,r,s}$$

in grid space (one radius at a time) by first Legendre transforming their representations in r, l, m-space to r, θ, m-space and, if 3D (i.e., for $m > 0$) Fourier transforming them from r, θ, m-space to r, θ, ϕ-space (i.e., to fully physical grid space) using FFTs. These are used to produce the nonlinear terms,

$$S1_l^m \equiv \left[r^2 \bar{\rho} v_r S\right]_l^m, \tag{10.81a}$$

$$S2_l^m \equiv \left[\frac{\bar{\rho} v_\theta S}{r \sin\theta}\right]_l^m, \tag{10.81b}$$

$$S3_l^m \equiv \left[\frac{\bar{\rho} v_\phi S}{r \sin\theta}\right]_l^m, \tag{10.81c}$$

$$S4_l^m \equiv \left[\frac{2\bar{\nu}}{\overline{\overline{T}}}\left(e_{ij}e_{ij} - \frac{1}{3}(\nabla\cdot\mathbf{v})^2\right) + \frac{\bar{\eta}}{\mu\bar{\rho}\overline{T}}|\nabla\times\mathbf{B}|^2\right]_l^m, \tag{10.81d}$$

$$W1_l^m \equiv r^2\left[-(\nabla\cdot\bar{\rho}\mathbf{vv})_r + ((\nabla\times\mathbf{B})\times\mathbf{B})_r/\mu\right]_l^m, \tag{10.81e}$$

$$W2_l^m \equiv \left[-\frac{(\nabla\cdot\bar{\rho}\mathbf{vv})_\theta}{r\,\sin\theta} + \frac{((\nabla\times\mathbf{B})\times\mathbf{B})_\theta/\mu}{r\,\sin\theta}\right]_l^m, \tag{10.81f}$$

$$W3_l^m \equiv \left[-\frac{(\nabla\cdot\bar{\rho}\mathbf{vv})_\phi}{r\,\sin\theta} + \frac{((\nabla\times\mathbf{B})\times\mathbf{B})_\phi/\mu}{r\,\sin\theta}\right]_l^m, \tag{10.81g}$$

$$B1_l^m \equiv r^2\left[(\mathbf{v}\times\mathbf{B})_r\right]_l^m, \tag{10.81h}$$

$$B2_l^m \equiv \left[\frac{(\mathbf{v}\times\mathbf{B})_\theta}{r\,\sin\theta}\right]_l^m, \tag{10.81i}$$

$$B3_l^m \equiv \left[\frac{(\mathbf{v}\times\mathbf{B})_\phi}{r\,\sin\theta}\right]_l^m, \tag{10.81j}$$

which are first constructed in grid space using the (real) functions 10.80 and then (if 3D) Fourier transformed from r, θ, ϕ-space to r, θ, m-space using FFTs and then Legendre transformed from r, θ, m-space to r, l, m-space using Gaussian quadratures (Eq. 10.54). The complex spherical harmonic coefficients, 10.81, are then combined using Eqs. 10.72 to compute the additional divergence and curl operations needed for the nonlinear terms in Eqs. 10.73–10.78.

For example, the nonlinear advection term in Eq. 10.73 is computed as

$$[-\mathbf{v}\cdot\nabla S]_l^m = -\frac{1}{\bar{\rho}}[\nabla\cdot\bar{\rho} S\mathbf{v}]_l^m$$

$$= -\frac{1}{\bar{\rho}}\left[\frac{1}{r^2}\frac{\partial}{\partial r}S1_l^m + (l+1)c_l^m S2_{l-1}^m - l\,c_{l+1}^m S2_{l+1}^m + im\,S3_l^m\right].$$

Note that the radial derivative of $S1_l^m$ is computed using transforms to Chebyshev space and back (Eqs. 10.79) and recursion relations (Section 9.4.1) after the $S1_l^m$ functions are computed on all radial levels. The additional nonlinear viscous and ohmic heating terms in Eq. 10.73, $S4_l^m$, are computed using the spherical coordinate representation of the rate of strain tensor (e_{ij}, Eq. 1.4) and Eqs. 10.60–10.62 to calculate the velocity and magnetic field terms.

Likewise, the nonlinear terms in Eqs. 10.74–10.78 are

$$\left[-(\nabla\cdot\bar\rho\mathbf{vv})_r + \frac{((\nabla\times\mathbf{B})\times\mathbf{B})_r}{\mu}\right]_l^m = \frac{W1_l^m}{r^2},$$

$$\left[\nabla_H\cdot\left(-\nabla\cdot\bar\rho\mathbf{vv} + \frac{(\nabla\times\mathbf{B})\times\mathbf{B}}{\mu}\right)\right]_l^m$$
$$= (l+1)\,c_l^m\,W2_{l-1}^m - l\,c_{l+1}^m\,W2_{l+1}^m + im\,W3_l^m,$$

$$\left[\left(\nabla\times\left(-\nabla\cdot\bar\rho\mathbf{vv} + \frac{(\nabla\times\mathbf{B})\times\mathbf{B}}{\mu}\right)\right)_r\right]_l^m$$
$$= (l+1)\,c_l^m\,W3_{l-1}^m - l\,c_{l+1}^m\,W3_{l+1}^m - im\,W2_l^m,$$

$$[(\nabla\times(\mathbf{v}\times\mathbf{B}))_r]_l^m = (l+1)\,c_l^m\,B3_{l-1}^m - l\,c_{l+1}^m\,B3_{l+1}^m - im\,B2_l^m,$$

$$[(\nabla\times\nabla\times(\mathbf{v}\times\mathbf{B}))_r]_l^m = \frac{l(l+1)}{r^4}B1_l^m + \frac{1}{r^2}\frac{\partial}{\partial r}\left(r^2\left((l+1)\,c_l^m\,B2_{l-1}^m\right.\right.$$
$$\left.\left.-l\,c_{l+1}^m\,B2_{l+1}^m + im\,B3_l^m\right)\right).$$

Notice that $S2_l^m$, $W2_l^m$, $W3_l^m$, $B2_l^m$, and $B3_l^m$ need to be calculated up to degree $l = l_{max} + 1$ for each order m.

Consider again the matrix equations that use Chebyshev collocation to update the solution in time. Since the (linear) Coriolis terms in Eqs. 10.74–10.76 couple all degrees l of W_l^m and Z_l^m, these poloidal and toroidal mass flux scalars for all l for a given m would need to be updated simultaneously if the Coriolis terms were treated implicitly. This would require a complex block pentadiagonal matrix equation for each m, with block elements being $N_r \times N_r$ submatrices. Although this would be more accurate and stable, it would require a huge amount of computer memory, especially for large values of N_r and l_{max}. The odd degree W_l and even degree Z_l^m combine to form a set of equations that is decoupled from the even degree W_l and odd degree Z_l^m, which provides a little savings. Here, however, we treat the Coriolis terms explicitly with the nonlinear terms, which allows us to solve a larger number of much smaller matrix equations, one for each spherical harmonic mode l, m, each numerical time step. With this choice we could just compute the vector Coriolis forces in grid space with the nonlinear terms and project the combination back onto spherical harmonic space via the spectral-transform method instead of computing them the way they are expressed in Eqs. 10.74–10.76. However, we usually choose to compute them as expressed in these equations; but we set the Coriolis terms to zero for the ($l = l_{max}$) mode of these equations since for this mode there is no ($l_{max} + 1$) contribution to combine with the ($l_{max} - 1$) contribution.

Another issue is how Eqs. 10.74 and 10.75 are solved simultaneously for W_l^m and P_l^m each time step. Recall (Eq. 9.25) how the radial derivative of the poloidal mass flux scalar can be calculated via

$$\left(\frac{d\mathrm{W}_l^m}{dr}\right)_k = \left(\frac{2}{N_r - 1}\right)^{1/2}\sum_{n=0}^{N_r-1}{}'' \mathrm{W}_{ln}^m\left(\frac{d\mathrm{T}_n}{dr}\right)_k$$

using stored values of dT_n/dr on the N_r Chebyshev levels, r_k, and likewise the second- and third-order radial derivatives of W_l^m and the first-order derivative of P_l^m. The $2N_r \times 2N_r$ collocation matrix operator represents the terms at the new time step in the semi-implicit formulation of Eqs. 10.74 and 10.75; the $2N_r \times 1$ solution vector is composed of N_r Chebyshev modes W_{nl}^m and the N_r Chebyshev modes P_{nl}^m. The explicit parts of the time integration scheme compose the known $2N_r \times 1$ right-hand side vector. However, the actual method of solving the coupled system of Eqs. 10.74 and 10.75 simultaneously for W_l^m and P_l^m could be a direct matrix solution method similar to that described in Section 10.3.2 or an influence matrix solution method similar to that described in Section 10.3.3.

10.6.4 Spherical Shell Boundary Conditions

This combined system, for each spherical harmonic mode l, m, is fourth order in the radial coordinate because of the third-order derivative of W_l^m and first-order derivative of P_l^m and therefore requires four boundary conditions. However, for an impermeable boundary, which we assume here, it is not physical to place a boundary condition on the pressure. Therefore, besides being impermeable we constrain the flow to also be either stress-free or nonslip (or some linear combination of these). These four boundary conditions (e.g., impermeable and stress-free at both the bottom and top boundaries) are forced via their Chebyshev polynomial expansions on the four rows of the matrix operator that represent the bottom and top radii in each of the two equations. The equations are forced to be satisfied on all of the other (internal) rows of the matrix, as described in Section 9.4.2.

Boundary conditions on the velocity provide boundary conditions on the poloidal and toroidal mass flux scalars because of Eqs. 10.60 and the orthogonality of spherical harmonics. An impermeable boundary requires v_r to vanish there; therefore, W_l^m vanishes there. If an impermeable boundary is stress-free,

$$\frac{\partial}{\partial r}\left(\frac{v_\theta}{r}\right) = \frac{\partial}{\partial r}\left(\frac{v_\phi}{r}\right) = 0$$

there and therefore

$$\frac{\partial^2 W_l^m}{\partial r^2} - \left(\frac{2}{r} + h_\rho\right)\frac{\partial W_l^m}{\partial r} = 0$$

and

$$\frac{\partial Z_l^m}{\partial r} - \left(\frac{2}{r} + h_\rho\right)Z_l^m = 0.$$

If, on the other hand, a boundary is nonslip, v_θ and v_ϕ vanish there in addition to v_r and therefore

$$\frac{\partial W_l^m}{\partial r} = Z_l^m = 0.$$

Typical thermal boundary conditions are constant entropy (or constant temperature), over the boundary,

$$S_l^m \text{ (or } T_l^m) = \text{ constant,}$$

or constant diffusive heat flux through the boundary,

$$\frac{\partial S_l^m}{\partial r} \left(\text{or } \frac{\partial T_l^m}{\partial r}\right) = \text{constant.}$$

Typically, homogeneous (i.e., spherically symmetric) thermal boundary conditions are prescribed, which means that S_0^0 or its radial derivative is set to a constant and the S_l^m or $\frac{\partial S_l^m}{\partial r}$ for $l > 0$ are set to zero. If, on the other hand, a heterogeneous thermal boundary condition is desired, it can readily be produced by setting all or some subset of the S_l^m or $\frac{\partial S_l^m}{\partial r}$ to constants on the boundary.

There are also several choices for the magnetic boundary conditions. Typically the region above the top boundary is assumed to be a perfect insulator out to infinity; therefore, the magnetic field that extends into this region is a potential field. That is, there are no electric currents in this external region ($\mu \mathbf{J}_{ext} = \nabla \times \mathbf{B}_{ext} = 0$); and so $\mathbf{B}_{ext}$ can be written as the gradient of a potential (Section 11.1). In addition, the external field needs to decrease at least as fast as $1/r^3$ as $r \to \infty$ because there are no magnetic monopoles. We can therefore write the external magnetic field as

$$\mathbf{B}_{ext} \equiv \sum_{l,m} B_{l,ext}^m \nabla \left(\frac{Y_l^m}{r^{l+1}}\right), \tag{10.82}$$

where the coefficients, $B_{l,ext}^m$, are only functions of time and $B_{0,ext}^0 = 0$. The spherical components of Eq. 10.82 are

$$B_{r,ext} = -\sum_{l,m} B_{l,ext}^m \frac{(l+1)}{r^{l+2}} Y_l^m, \tag{10.83a}$$

$$B_{\theta,ext} = \sum_{l,m} B_{l,ext}^m \frac{1}{r^{l+2}} \frac{\partial Y_l^m}{\partial \theta}, \tag{10.83b}$$

$$B_{\phi,ext} = \sum_{l,m} B_{l,ext}^m \frac{1}{r^{l+2} \sin\theta} \frac{\partial Y_l^m}{\partial \phi}. \tag{10.83c}$$

This external potential field at the top boundary needs to exactly match the convectively generated field, Eqs. 10.68, at the top boundary at every time step, assuming μ is continuous across the top boundary. Therefore, at the top boundary

$$\left[J_l^m\right]_{top} = 0, \tag{10.84a}$$

$$\left[\frac{\partial B_l^m}{\partial r} + \frac{l}{r} B_l^m\right]_{top} = 0; \tag{10.84b}$$

and the external potential field can be calculated for $r \geq r_{top}$ at every time step using Eqs. 10.83 with

$$B_{l,ext}^m \equiv -l\, r_{top}^l \left[B_l^m\right]_{top}. \tag{10.85}$$

If the core below the convection zone were also assumed to be an insulator, the field there would be a potential field but one that is finite at $r = 0$. Consequently, this interior potential field can be written as

$$\mathbf{B}_{int} \equiv \sum_{l,m} \mathrm{B}_{l,int}^{m} \nabla \left(r^{l} Y_{l}^{m} \right), \tag{10.86}$$

where the coefficients, $\mathrm{B}_{l,int}^{m}$, are only functions of time and $\mathrm{B}_{0,int}^{0} = 0$. A derivation similar to the one above for the external field gives the following boundary conditions at the bottom boundary of the convecting shell:

$$\left[\mathrm{J}_{l}^{m} \right]_{bot} = 0, \tag{10.87a}$$

$$\left[\frac{\partial \mathrm{B}_{l}^{m}}{\partial r} - \frac{l+1}{r} \mathrm{B}_{l}^{m} \right]_{bot} = 0. \tag{10.87b}$$

Likewise, the inner potential field can be calculated for $r \leq r_{bot}$ at every time step using Eq. 10.86 with

$$\mathrm{B}_{l,int}^{m} \equiv \frac{l+1}{r_{bot}^{l+1}} \left[\mathrm{B}_{l}^{m} \right]_{bot}. \tag{10.88}$$

The other extreme case would be to approximate the core as a perfect conductor, which would not allow any magnetic field in (or out). Then

$$\left[\frac{\partial \mathrm{J}_{l}^{m}}{\partial r} \right]_{bot} = 0, \tag{10.89a}$$

$$\left[\frac{\partial^{2} \mathrm{B}_{l}^{m}}{\partial r^{2}} \right]_{bot} = 0. \tag{10.89b}$$

Instead of assuming the external or inner region is an insulator (zero electrical conductivity) or a perfect conductor (infinite electrical conductivity) one could specify a finite electrical conductivity in the region and solve the magnetic induction equation (10.77 and 10.78) there with no flow, i.e., a magnetic diffusion equation. This inner or external magnetic field would need to be solved simultaneously with the field in the convection zone, forcing continuity of the normal component of the magnetic field and of the tangential electric field at the interface. The tangential component of the magnetic field would also need to be continuous across the interface if the magnetic permeability, μ, were continuous and there were no surface current density on the interface. Likewise, if the permittivity, ϵ, were continuous and there were no electric charge density on the interface, the normal component of the electric field would also be continuous.

Here we should mention the different meanings of "core" used in various fields. Iron-rich cores in terrestrial planets, like the Earth, are usually described as a "solid inner core" surrounded by a "fluid outer core," or, if the planet is young enough, the core would be completely liquid. The "core" of a star is a dense gas and refers to the central region where nuclear burning occurs. The "silicate core" of a giant planet usually means the central region where elements heavier than hydrogen and helium have concentrated; it could be either solid or liquid.

Solving for the time-dependent magnetic field within a finitely conducting solid inner core, like the Earth's, is relatively straightforward. Such a core should be able to rotate as a solid body relative to the rotating frame of reference in response to the viscous, topographic, magnetic, and possibly gravitational torques acting on it (e.g., Buffett & Glatzmaier, 2000). The resulting time-dependent solid-body rotation is represented by spherical harmonic degree $l = 1$ of the colatitudinal and longitudinal components of the velocity. Of this, the "off-axis" rotations ($l = 1, m = 1$) are typically small compared to the "on-axis" rotations ($l = 1, m = 0$) if the body has a significant basic rotation rate; therefore in many cases only the on-axis solid-body rotations are computed. In either case, though, the only nonlinear term in the magnetic induction equation within a solid inner core is the induction due to this solid-body rotational velocity relative to the rotating frame of reference, which is relatively easy to compute knowing the moment of inertia of the inner core and the time-dependent torques on it.

The complication, for a spherical coordinate system, is the singularity at the origin, $r = 0$, where all but the $l = 1$ spherical harmonic contributions need to vanish. In addition, regularity conditions need to be satisfied that require the poloidal and toroidal magnetic scalars to be proportional to radius to the $(l + 1)$ power as the radius goes to zero. A convenient way to enforce this is to redefine the magnetic poloidal and toroidal scalars inside the inner core (Glatzmaier & Roberts, 1996a) as

$$b_l^m (r, t) \equiv x^{-(l+1)} B_l^m (r, t)$$

and

$$j_l^m (r, t) \equiv x^{-(l+1)} J_l^m (r, t),$$

where here $x \equiv r/r_{ICB}$ and r_{ICB} is the radius of the inner core boundary. The three components of the field are now

$$B_r = \frac{1}{r_{ICB}^2} \sum_{l,m} l(l + 1) x^{(l-1)} b_l^m Y_l^m, \tag{10.90a}$$

$$B_\theta = \frac{1}{r_{ICB} \sin \theta} \sum_{l,m} \left[x^{(l-1)} \left(x \frac{\partial b_l^m}{\partial r} + \frac{l + 1}{r_{ICB}} b_l^m \right) \sin \theta \frac{\partial Y_l^m}{\partial \theta} + x^l j_l^m \frac{\partial Y_l^m}{\partial \phi} \right],$$
$$\tag{10.90b}$$

$$B_\phi = \frac{1}{r_{ICB} \sin \theta} \sum_{l,m} \left[x^{(l-1)} \left(x \frac{\partial b_l^m}{\partial r} + \frac{l + 1}{r_{ICB}} b_l^m \right) \frac{\partial Y_l^m}{\partial \phi} - x^l j_l^m \sin \theta \frac{\partial Y_l^m}{\partial \theta} \right].$$
$$\tag{10.90c}$$

Corresponding changes to the magnetic induction equations for b_l^m and j_l^m also need to be made. These equations are solved simultaneously with those for B_l^m and J_l^m in the fluid core with the above mentioned continuity conditions at r_{ICB}. In addition, to reduce the radial resolution near the origin, we employ a half-Chebyshev grid in radius within the solid inner core by spanning the diameter of the inner core and using only even Chebyshev degrees.

Instead of modeling a solid core below a fluid *shell*, one may wish to model convection or internal gravity waves within a *full sphere*, i.e., within a body without a solid inner core. This is a challenge for codes employing a spherical coordinate system because of the singularity at the origin. It poses a similar challenge for a 2D cylinder without a solid inner core when using cylindrical coordinates. Not only would there be a severe CFL constraint on Δt due to a convergence of the grid at the origin, but regularity conditions similar to those described above for the magnetic field at the origin would exist for all variables, with the additional nonlinear complication of having nonvanishing fluid flows through the origin. A relatively simple way to avoid these issues would be to employ a different spatial discretization and a local numerical method. For example, a full sphere can be simulated within a 3D cartesian box or within a "cubed-sphere" grid, which better represents the outer spherical boundary. A more challenging and probably more accurate way would be to expand the radial dependencies of all variables in, say, *Worland polynomials* (Livermore et al., 2007; Boyd, 2011) instead of Chebyshev polynomials. Worland polynomials naturally satisfy the regularity conditions as a function of the spherical harmonic degree l; however, unlike the Chebyshev polynomials, they do not have a fast spectral transform. Another consideration is the relatively poor parallel processing efficiency of spectral methods in general compared to simpler local numerical methods (Section 9.5.2).

10.6.5 Spherical Shell Time Integration

Having discussed issues related to boundary conditions, now consider some issues related to the time integration. One is called the "pole problem," which is avoided when using spherical harmonic expansions on a sphere. That is, the CFL condition (Eq. 4.7) for horizontal flow on a spherical coordinate system would be a very severe constraint on the numerical time step if these equations were integrated in time in grid space because the longitudinal grid spacing, $r \sin\theta \, \Delta\phi$, is small near the poles, i.e., as $\theta \to 0$ or π. However, the spherical spectral-transform method integrates the equations in spherical harmonic space, which has a CFL condition based on a global average horizontal grid spacing:

$$\Delta t < \left(\frac{r^2}{l_{max}(l_{max} + 1)(v_\theta^2 + v_\phi^2)} \right)^{1/2}_{min}. \tag{10.91}$$

Notice how this spectral CFL constraint is essentially the minimum value of the smallest horizontal wavelength divided by the largest horizontal velocity at each radius.

Of course, there is also the CFL condition in the radial direction. Although the Chebyshev radial grid spacing decreases as either the bottom or top boundary is approached, the radial velocity also decreases as fluid approaches an impermeable boundary. However, the spectral CFL condition also depends on the Alfvén velocity (Eq. 11.35), which needs to be included in Eq. 10.91; and the radial component of this velocity, which is proportional to the radial component of the magnetic field, does not vanish at an impermeable boundary (unless it is a perfectly conducting boundary). Therefore, a strong magnetic field extending through a boundary can require the numerical time step to be significantly smaller than it would otherwise need to be for a weak field scenario.

Another time integration consideration is to treat the buoyancy term in the momentum equation implicitly without necessarily computing the energy and momentum equations simultaneously. This is accomplished at each time step by first updating the entropy while saving the old value so the average of the old and the updated values can be used in the buoyancy term when the momentum equation is updated.

Here we also just mention that this basic 3D spectral method for convection in a spherical fluid shell can also be adopted for a simple model of mantle convection (Glatzmaier, 1988). If viscosity is at most a function of radius and vorticity due to the interaction of tectonic plates is neglected, the toroidal mass flux scalar function (Z in Eq. 10.56) vanishes and the radial component of the curl of the momentum equation does not need to be solved. As mentioned in Section 8.4, the inertial, Coriolis, and magnetic forces are also neglected, which removes the time derivative and the nonlinear terms in the momentum equation. Only the energy equation has a time derivative (of entropy or temperature) and nonlinear terms.

10.6.6 Spherical Shell Energy Analysis

The thermal, kinetic, and magnetic energies and their spectra need to be calculated when analyzing simulations. For example, a measure of the total thermal energy relative to the adiabatic reference state is the volume integral over the convection zone of the perturbation entropy density:

$$\int_{shell} \bar{\rho}\bar{T}\,S\,dV = 4\pi \int_{bot}^{top} \bar{\rho}\bar{T}\,S_0^0 Y_0^0 r^2 dr$$

$$= \int_{bot}^{top} F(r)\,dr$$

$$= \left(\frac{2}{N_r - 1}\right)^{1/2} \sum_{n=0}^{N_r-1}{}'' f_n Q_n,$$

where

$$F(r) \equiv \sqrt{4\pi}\,\bar{\rho}(r)\bar{T}(r)S_0^0(r)r^2$$

and

$$f_n = \left(\frac{2}{N_r - 1}\right)^{1/2} \sum_{k=1}^{N_r}{}'' F(r_k)\mathrm{T}_n(r_k).$$

Q_n is defined by Eq. 9.29.

Unlike the anelastic thermal energy, which is linear in the perturbation, the kinetic and magnetic energies are nonlinear. The total kinetic energy within the convection zone relative to the rotating frame of reference is

$$\int_{shell} \frac{1}{2} \bar{\rho} v^2 dV = \int_{bot}^{top} F(r) dr$$

$$= \left(\frac{2}{N_r - 1} \right)^{1/2} \sum_{n=0}^{N_r - 1}{}'' f_n Q_n,$$

where now

$$F(r) \equiv \sum_{l=1}^{l_{max}} \sum_{m=0}^{l}{}' \frac{l(l+1)}{\bar{\rho}} \left[\frac{l(l+1)}{r^2} |W_l^m|^2 + \left| \frac{\partial W_l^m}{\partial r} \right|^2 + |Z_l^m|^2 \right] \qquad (10.92)$$

and f_n is defined as the Chebyshev coefficients of $F(r)$ as above. Recall that the square of the absolute value of a complex function is the product of that function times its complex conjugate. Also recall that the single prime on the summation symbol means that the first term in the series (i.e., $m = 0$, the axisymmetric part) is multiplied by $1/2$. The axisymmetric meridional circulation kinetic energy is obtained by summing over only the W_l^0 and $\partial W_l^0/\partial r$ terms; whereas the axisymmetric differential rotation kinetic energy is obtained by summing over only the Z_l^0 term.

The total magnetic energy within the convection zone is

$$\int_{shell} \frac{B^2}{2\mu} dV = \int_{bot}^{top} F(r) dr$$

$$= \left(\frac{2}{N_r - 1} \right)^{1/2} \sum_{n=0}^{N_r - 1}{}'' f_n Q_n,$$

where now

$$F(r) \equiv \sum_{l=1}^{l_{max}} \sum_{m=0}^{l}{}' \frac{l(l+1)}{\mu} \left[\frac{l(l+1)}{r^2} |B_l^m|^2 + \left| \frac{\partial B_l^m}{\partial r} \right|^2 + |J_l^m|^2 \right]. \qquad (10.93)$$

The axisymmetric poloidal magnetic energy is obtained by summing over only the B_l^0 and $\partial B_l^0/\partial r$ terms; whereas the axisymmetric toroidal magnetic energy is obtained by summing over only the J_l^0 term.

Instead of calculating the total energy, one could calculate and plot a two-dimensional energy spectrum on an l vs. m grid, i.e., the energy for each spherical harmonic mode; or one could sum over all the m values and plot the energy spectrum as a function of degree l; or sum over l and plot the energy spectrum as a function of order m. In addition, instead of plotting the volume-integrated energies, one could plot the energy densities as functions of radius.

10.6.7 2.5D Spherical Shell

Before attempting to construct a 3D spherical shell model, it is recommended that one get a 2.5D spherical shell model working. Such a model is easily obtained by using only spherical harmonic order $m = 0$ in the equations presented in this chapter, i.e., only the axisymmetric modes. This allows the fluid velocity and magnetic field to have ϕ components but makes all variables independent of longitude ϕ. It also makes all the spherical harmonic coefficients real instead of complex.

It is instructive to compare the streamfunction formulation, which we employ throughout most of this book, with the poloidal-toroidal formulation, which we recommend for a 2.5D spherical shell model since it can be extended to 3D. Mass flux in the streamfunction formulation of an anelastic 2.5D model in spherical coordinates is

$$\bar{\rho}\mathbf{v} = \nabla \times \left(\frac{\psi}{r \sin \theta} \hat{\boldsymbol{\phi}} \right) + \bar{\rho} v_\phi \hat{\boldsymbol{\phi}}.$$

Notice how now in spherical coordinates we need to include an $r \sin \theta$ in the curl operation in order to satisfy $\nabla \psi \cdot \mathbf{v} = 0$ and have contours of the streamfunction, ψ, represent the instantaneous longitudinally averaged streamlines of the meridional *mass flux*. On the other hand, a poloidal-toroidal formulation of mass flux is

$$\bar{\rho}\mathbf{v} = \nabla \times \nabla \times W\hat{\boldsymbol{r}} + \nabla \times Z\hat{\boldsymbol{r}}.$$

Expanding these vector operations and equating their spherical components reveals that the streamfunction is

$$\psi = -\sin\theta \frac{\partial W}{\partial \theta} = -\sum_l W_l^0 \sin\theta \frac{\partial Y_l^0}{\partial \theta} \tag{10.94}$$

and the ϕ component of the mass flux is

$$\bar{\rho} v_\phi = -\frac{1}{r}\frac{\partial Z}{\partial \theta} = -\frac{1}{r}\sum_l Z_l^0 \frac{\partial Y_l^0}{\partial \theta}. \tag{10.95}$$

This is similar to the relationships found for a 2.5D Boussinesq box model (Eqs. 10.45). Likewise, the 2.5D magnetic field is

$$\mathbf{B} = \nabla \times A_\phi \hat{\boldsymbol{\phi}} + B_\phi \hat{\boldsymbol{\phi}}$$

and

$$\mathbf{B} = \nabla \times \nabla \times B\hat{\boldsymbol{r}} + \nabla \times J\hat{\boldsymbol{r}},$$

which makes the ϕ component of the magnetic vector potential

$$A_\phi = -\frac{1}{r}\frac{\partial B}{\partial \theta} = -\frac{1}{r}\sum_l B_l^0 \frac{\partial Y_l^0}{\partial \theta} \tag{10.96}$$

and the ϕ component of the magnetic field

$$B_\phi = -\frac{1}{r}\frac{\partial J}{\partial \theta} = -\frac{1}{r}\sum_l J_l^0 \frac{\partial Y_l^0}{\partial \theta}. \tag{10.97}$$

Since here we have not included a factor of $r \sin \theta$ in the curl operation, contours of $(A_\phi r \sin \theta)$ represent the longitudinally averaged magnetic field lines in a meridian plane. Note that positive contours of ψ and of $(A_\phi r \sin \theta)$ represent clockwise circulation and clockwise magnetic field lines, respectively, in the $r - \theta$ meridian plane when the ϕ-direction (i.e., eastward) is viewed into the image; negative values are counterclockwise.

Another issue is the choice of an initial magnetic field for the 2.5D (or 3D) spherical shell problem. If, for example, both the inner core (below the bottom boundary) and the external region (above the top boundary) are perfect insulators, one might choose an initial magnetic field that is axisymmetric and purely poloidal inside the shell ($r_{bot} \leq r \leq r_{top}$). A potential field outside the shell would satisfy the insulating boundary conditions, Eqs. 10.84 and 10.87. A simple choice would be an axial dipole constructed by setting (at $t = 0$) $J_l^m = 0$ for all l and m and setting $B_l^m = 0$ for all l and m except $l = 1, m = 0$:

$$B_1^0(r) = a_1\, r^2 f_1(r) + \frac{a_2}{r} f_2(r). \tag{10.98}$$

The constants a_1 and a_2 are defined by the chosen values of the radial component of the field at the north geographic pole at the bottom and top boundaries, respectively:

$$a_1 \equiv \sqrt{\frac{\pi}{2}}\, B_r(\theta = 0, r = r_{bot}) \quad \text{and} \quad a_2 \equiv \sqrt{\frac{\pi}{2}}\, r_{top}^3 B_r(\theta = 0, r = r_{top}).$$

The radial functions are defined for $r < r_{bot}$ as

$$f_1(r) \equiv 1 \quad \text{and} \quad f_2(r) \equiv 0,$$

for $r > r_{top}$ as

$$f_1(r) \equiv 0 \quad \text{and} \quad f_2(r) \equiv 1,$$

and for $r_{bot} \leq r \leq r_{top}$ as

$$f_1(r) = 1 - 3x_1^2 + 3x_1^4 - x_1^6$$

and

$$f_2(r) = 1 - 3x_2^2 + 3x_2^4 - x_2^6,$$

where

$$x_1 \equiv \frac{r - r_{bot}}{r_{top} - r_{bot}} \quad \text{and} \quad x_2 \equiv \frac{r_{top} - r}{r_{top} - r_{bot}}.$$

Note, the current density associated with this initial field vanishes within the insulating inner and external regions but not necessarily within the shell.

It is important to understand, however, that the axisymmetric variables in a 2.5D model lack the maintenance of the nonaxisymmetric variables via nonlinear terms (e.g., Reynolds and Maxwell stresses and magnetic induction), which makes all the difference in 3D models. Many studies have been done using "mean-field" models and "intermediate dynamo" models; see Roberts (2007) for a review of the basic theory upon which these models are based. Simply put, they compensate for the lack of a three-dimensional influence on the axisymmetric flow and field by adding prescribed axisymmetric terms to the momentum and/or magnetic induction equations. Although the sensitivity to various types of prescribed forcing terms can be learned by this approach, the results cannot serve as predictions since the added terms are not self-consistently computed as they would be with a 3D model. These missing nonlinear contributions could be computed in a 3D simulation and saved as functions of r, θ, and t and then added to a 2.5D model as source terms; but of course this would provide no new information beyond that provided by the original 3D model. The effects on the axisymmetric (i.e., the longitudinally averaged) part of a simulation due to the 3D nonlinear terms is clearly seen by comparing 2.5D and 3D spherical shell simulations (with magnetic field, density stratification, and rotation) that are illustrated in Sections 13.4 and 13.6, respectively.

A final note for this section is that this 3D spectral method involving spherical harmonic and Chebyshev polynomial expansions and spectral transforms (Glatzmaier, 1984) has proven very efficient and accurate for many studies of stellar dynamos, giant planet dynamos, the geodynamo, mantle convection, flows within planetary satellites, and laboratory fluid dynamics experiments. However, if we wish to take advantage of the latest parallel computers, which have hundreds of thousands of processors, a change may be needed because of the rapid increase in computational resources required as one increases spatial resolution due to the global communication of spectral transforms (Section 9.5.2; Glatzmaier & Clune, 2000). Therefore, we may be on the verge of going back to simpler, less accurate, methods like finite element, which require only nearest neighbor communication and so will likely outperform the spectral method at some high spatial resolution even though requiring more nodes than would be needed to obtain comparable accuracy with a spectral method. A compromise might be a spectral element method.

SUPPLEMENTAL READING

Boyd (2001)
Canuto et al. (1988)
Chandrasekhar (1961)
Durran (1998)
Kundu & Cohen (2008)
Peyret (2002)

EXERCISES

1. *A 2D convection simulation using a poloidal decomposition*
   Convert a strictly 2D model of convection in a box that employs a vorticity-streamfunction representation to one that uses a poloidal decomposition (Eqs. 10.45).

2. *Critical Rayleigh number for convection in a 2.5D fluid box*
   Consider convection in a 2.5D box with impermeable and stress-free boundaries. Show that a linear stability analysis that expands T_n, $\omega_{y,n}$, and ψ_n in $\sin(m\pi z)$ as in Eqs. 3.7 and $v_{y,n}$ in $\cos(m\pi z)$ produces the same critical Rayleigh number as for strictly 2D convection (Eq. 3.8).

3. *Spherical harmonic recursion relations*
   Show that the spherical harmonics listed in Table 10.1 satisfy the relations 10.50.

4. *Spherical harmonic construction*
   Extend the list of spherical harmonics in Table 10.1 to $l = 4$ and $m = 0$ to 4 by using either Eq. 10.50a or Eq. 10.50b.

5. *Additional spherical harmonic recursion relations*
   Formulate the recursion relations for the following expressions: $\sin^2 \theta \, Y_l^m$, $\cos \theta \sin \theta \, \partial Y_l^m / \partial \theta$, $\partial / \partial \theta (\sin \theta \, Y_l^m)$, and $\cos \theta \sin^2 \theta \, Y_l^m$.

6. *Legendre transforms*
   Write two codes that simply test a Legendre transform to and from θ-space as described in Section 10.6.1. Have one do the summation as described in Eqs. 10.53 and 10.54 and have the other take advantage of the $(l - m)$ determined symmetries with respect to the equator.

7. *Poloidal magnetic field scalar and gauss coefficients*
   Confirm the relationship between the poloidal magnetic field scalar and gauss coefficients, Eqs. 10.68–10.71.

8. *Insulating magnetic boundary conditions and fields*
   Derive the electrically insulating magnetic boundary conditions on spherical boundaries and the expressions for the potential magnetic fields beyond the boundaries, Eqs. 10.84, 10.85, 10.87, and 10.88.

9. *Kinetic and magnetic energy densities in a spherical shell*
   Derive the expressions for the kinetic energy density, Eq. 10.92, and magnetic energy density, Eq. 10.93, within a spherical fluid shell.

10. *Streamfunction and poloidal-toroidal decompositions in 2.5D spherical geometry*
    Derive the relationships in 2.5D spherical geometry between the streamfunction and poloidal-toroidal formulations for mass flux (Eqs. 10.94 and 10.95) and for magnetic field (Eqs. 10.96 and 10.97).

11. *Axisymmetric meridional circulation*
    Assume a simple meridional circulation described by a single poloidal mass flux scalar, W_1^0. Let W_1^0 increase from zero at the bottom boundary, peak with a positive value at mid-depth, and decrease to zero at the top boundary. Examine the sense of this flow circulation in the meridian plane by checking the signs of v_r and v_θ at various locations. Confirm that this makes $\psi \geq 0$ everywhere, i.e., that the circulation in the meridian plane is northward near the bottom boundary and southward near the top boundary.

12. *Testing an FFT*
    Write a test program that compares the results from an FFT (fast Fourier transform) to those from a standard discrete Fourier transform, which is described in Eqs. 10.35 and 10.37. Consider one z_k-level and one time t. Choose random values for the Fourier coefficients, f_m.

13. *Axial dipole field with insulating boundary conditions*
    Show that the magnetic field defined by Eq. 10.98 satisfies the insulating boundary conditions, Eqs. 10.84 and 10.87, and that the current density vanishes below r_{bot} and above r_{top}. Derive the expression for the current density

inside the shell. Write a graphics code that plots the magnetic field lines for this field for given values of the radii and radial magnetic field intensities at the polar bottom and top boundaries.

COMPUTATIONAL PROJECTS

1. *Sponge layers in a 2D box*
   Test the effects of viscous and Rayleigh sponge layers in a simulation of convection within a 2D box and compare how large δ needs to be for a given value of A to produce comparable damping. Likewise, compare these two methods by varying A for a given δ.

2. *Convection with and without a mean zonal flow in a 2D box*
   Produce a simulation like the one illustrated in Fig. 10.1 and another with $\omega_{m=0} = 0$ and $\psi_{m=0} = 0$ for all z and t. Compare these two cases via movies of $T(x, z)$ and $\psi(x, z)$ and plots of their horizontally averaged profiles (as functions of z) of T and v_x.

3. *Convection with traveling-wave thermal boundary conditions in a 2D box*
   Produce a simulation like the one illustrated in Fig. 10.1 but with no mean zonal flow. Replace the bottom thermal boundary condition with a temperature that has the form of a traveling wave by selecting one horizontal mode, $m_o > 0$, frequency, ω_o, and real amplitude, A_o, and at $z = 0$ setting

   $$T_{m_o}(0, t) = A_o e^{-i\omega_o t}$$

   with

   $$T_{m \neq m_o}(0, t) = 0.$$

   This makes the time-dependent temperature at the bottom boundary

   $$T(x, 0, t) = 2A_o \, \text{Re}\left(e^{i(2\pi m_o x/L - \omega_o t)}\right)$$
   $$= 2A_o \, \cos(2\pi m_o x/L - \omega_o t),$$

   which is a horizontal wave with wavelength $L/(2\pi m_o)$ propagating in the positive x-direction (for $\omega_o > 0$) at a phase velocity of $(\omega_o L)/(2\pi m_o)$. Pick a frequency (ω_o) such that $2\pi/\omega_o$ is large relative to a typical convective turnover time (i.e., the depth of the domain divided by the average fluid velocity in the z-direction). Also try this type of time-dependent temperature condition on both the bottom and top boundaries using the same values of m_o, ω_o, and A_o, or using the same m_o and A_o but setting ω_o on the top boundary to $-\omega_o$ on the bottom boundary. Also try this with a sum of several modes m_o, each with different values of ω_o and A_o.

4. *Instability determined by the initial perturbation in a 2D box*
   Produce a simulation like the one illustrated in Fig. 10.1, but just before the shear instability begins (Fig. 10.1b) add a small external perturbation (to $T_m(z)$ or $\omega_m(z)$ for some m) in a way that will initiate the instability. Then rerun this case with the perturbation of the opposite sign.

5. *Internal gravity waves with no significant mean zonal flow in a 2D box*
   Produce a simulation with the parameters for the case illustrated in Fig. 10.1, but with the top and bottom boundary conditions on T_0 switched, i.e., simulate internal gravity waves. Let the initial condition be $T_0 = z$ and $T_1 = 0.5 \sin \pi z$.

6. *Convection with periodic flow through all boundaries of a 2D box*
   Sometimes one wishes to simulate convection only within a small part of a much larger fluid region. This can be done by imposing periodic boundary conditions on all four boundaries of a 2D box, i.e., allowing fluid flow through not only the side boundaries but also the bottom and top boundaries, which are no longer impermeable and stress-free. In this case a convenient method is an expansion of all variables in full Fourier series in both the x and z directions. Construct such a model.

7. *Prescribed fluid flow rates through boundaries of a 2D box*
   Simulate convection within a 2D box with a constant and equal total fluid flow rate through the side boundaries. That is, set the integral in z of v_x over each of the side boundaries to the same constant. Likewise, impose a constant and equal total mass flow rate through the bottom and top boundaries.

8. *Convection in a 2.5D box*
   Construct a nonlinear Boussinesq convection model for a 2.5D fluid box with all boundaries being impermeable and stress-free. Use a finite-difference method in z and Fourier expansions in x with a spectral-transform method to compute nonlinear terms. Describe the evolution of $v_y(x, z, t)$.

9. *Convection in a 2D annulus*
   Construct a nonlinear Boussinesq convection model for a 2D fluid annulus with impermeable and stress-free top and bottom boundaries. Use a Fourier spectral method in longitude and either a finite-difference method or a Chebyshev spectral method in radius. Compute nonlinear terms using a spectral-transform method. Solve the vorticity-streamfunction equations using either the direct matrix solution method (Section 10.3.2) or the influence matrix solution method (Section 10.3.3).

10. *Internal gravity waves in a 2D annulus*
    Do the *Internal gravity waves excited by a continuous central source* project described at the end of Chapter 6 in a 2D annulus geometry using the method described in *Convection in a 2D annulus*.

11. *Convection in a 2.5D spherical shell*
    Construct a nonlinear Boussinesq convection model for a 2.5D (axisymmetric) spherical shell with impermeable and stress-free bottom and top boundaries. Use spherical harmonic expansions in colatitude, Chebyshev polynomial expansions in radius, and a spectral-transform method to compute nonlinear terms.

PART 3

Additional Physics

Chapter Eleven

Magnetic Field

Magnetoconvection is the term usually used to describe thermal convection of an electrically conducting fluid within a background magnetic field maintained by some external mechanism (e.g., Roberts, 1967; Weiss, 1981a; Glatzmaier, 2005a; Rempel et al., 2009). For example, much of the solar magnetic field generated deep within the sun extends to the surface, where it is swept by the convection into intense small-scale magnetic flux concentrations called "sunspots," which inhibit outgoing heat flux and so appear darker than the surrounding photosphere. Understanding this mechanism was the motivation for much of the magnetoconvection research that Nigel Weiss and colleagues pioneered in the 1980s (Weiss, 1981a). Matthias Rempel and colleagues recently produced one of the most sophisticated and realistic 3D magnetoconvection simulations of a sunspot (Rempel et al., 2009). Magnetoconvection is also being studied as the mechanism that induces fields within the subsurface oceans of icy satellites orbiting within the background magnetic field of their parent giant planets, like Europa around Jupiter or Titan around Saturn. In these examples, the flow of electrically conducting fluid induces new magnetic field by doing work distorting the original field. However, in 2D the induced field would decay away if the background field were removed. A self-sustaining magnetohydrodynamic (MHD) *dynamo*, on the other hand, is a mechanism that maintains the magnetic field without any background field or external mechanism other than heat being transferred through the convecting domain. Stellar and planetary dynamos are of this type. Self-consistent simulations of MHD dynamos necessarily produce 3D magnetic fields (Cowling, 1934; Section 13.6). However, considerable insight can be gained by first doing 2D magnetoconvection and magneto-gravity wave studies.

We begin this chapter with a review of the magnetohydrodynamic equations. Then we explain how to make such a 2D model by adding an externally maintained background magnetic field to the Boussinesq model in a box with impermeable side boundaries (Chapter 4) or to any of the variations to that model described in this book. First we consider the case of a uniform vertical background field, which, for example, might be a simple model of the outer convective layer in the polar region of a planetary core or stellar interior that has a dipole-dominated global magnetic field. Then we consider the case of a uniform horizontal background field, which could represent the magnetohydrodynamics of the equatorial region of such a body. We also discuss how one could simulate a case of a uniform background field that is tilted relative to both the vertical and horizontal axes.

11.1 MAGNETOHYDRODYNAMICS

The magnetic induction equation describes how new magnetic field is continually generated by convection of electrically conducting fluid, which on average balances the removal of the field by diffusion. This equation is derived from the pre-Maxwell equations, Maxwell's set of electromagnetic equations modified by the *MHD approximation* (e.g., Roberts, 1967; Davidson, 2001). That is, when fluid velocities (v) are small relative to the speed of light (c) the displacement current in Maxwell's equations can be neglected. This approximation is certainly valid for the problems suggested in this book since we already assume these velocities are small relative to the speed of sound. Even supersonic flows usually have velocities much less than the speed of light.

We assume, as is usually done for the MHD of stellar and planetary interiors, a linear isotropic relationship between the (applied) magnetic field, $\mathbf{H}$, and the net magnetic flux density, $\mathbf{B} = \mu\mathbf{H}$, and between the (applied) electric field, $\mathbf{E}$, and the net electric displacement field, $\mathbf{D} = \epsilon\mathbf{E}$, where μ is magnetic permeability and ϵ is the permittivity (or dielectric constant) of the fluid. We also take μ and ϵ to be their "free-space" values because of the high temperature of liquid iron in terrestrial cores or of ionized hydrogen in stellar and giant planet interiors. This high temperature precludes any permanent magnetism; that is, the magnetic field would decay away if it were not being regenerated. The magnetic permeability of free-space, μ_o, is $4\pi \times 10^{-7}$ henry/m and the permittivity, ϵ_o, equals $1/(c^2\mu_o)$, where c is the speed of light in a vacuum. In addition, we neglect the plasma effects that would exist in low-density magnetospheres due to electrons, ions, and neutral particles flowing at different velocities; representing these effects would require a more complicated Eq. 11.3. We also assume the electrical conductivity is isotropic. The relevant equations, in SI units, are then

$$\nabla \times \mathbf{B} = \mu\mathbf{J}, \tag{11.1}$$

$$\nabla \times \mathbf{E} = -\frac{\partial \mathbf{B}}{\partial t}, \tag{11.2}$$

$$\mathbf{J} = \sigma\left(\mathbf{E} + \mathbf{v} \times \mathbf{B}\right), \tag{11.3}$$

where σ is electrical conductivity of the fluid.

Ampère's Law (Eq. 11.1) states that an electric current density, $\mathbf{J}$, generates a $\mathbf{B}$ field, which is usually just called the magnetic field. Maxwell's correction says that a magnetic field is also generated by a time-varying electric field, i.e., the displacement current density. However, this contribution is of order $(v/c)^2$ relative to the contribution $\mathbf{J}$ makes and therefore is neglected within the MHD approximation. Taking the divergence of Eq. 11.1 shows that this approximation makes the current density divergence-free:

$$\nabla \cdot \mathbf{J} = 0. \tag{11.4}$$

Faraday's Law (Eq. 11.2) states that a time-varying magnetic field induces an electric field, $\mathbf{E}$. (Note, the MHD approximation on Ampère's Law begins by scaling Faraday's Law to show that $|\mathbf{E}|/|\mathbf{B}|$ should be of order v and by noting that

$\epsilon\mu = 1/c^2$.) Taking the divergence of Faraday's Law gives Gauss's Law,

$$\nabla \cdot \mathbf{B} = 0 , \tag{11.5}$$

assuming the initial $\nabla \cdot \mathbf{B}$ vanishes. That is, there are no magnetic monopoles. Consequently, a vector potential, $\mathbf{A}$, can be defined such that

$$\mathbf{B} \equiv \nabla \times \mathbf{A} , \tag{11.6}$$

which ensures that $\mathbf{B}$ is divergence-free (Eq. 11.5).

$\mathbf{B}$, $\mathbf{E}$, $\mathbf{J}$, and $\mathbf{v}$ in Eqs. 11.1–11.3 are measured in the lab frame of reference. Within the MHD approximation (i.e., when dropping terms of order $(v/c)^2$), $\mathbf{B}$ and $\mathbf{J}$ are Galilean-invariant (frame-independent). However, the electric field measured in the frame moving with the fluid is $(\mathbf{E} + \mathbf{v} \times \mathbf{B})$, the electric field in the lab frame plus the "electromotive force". The electric current density is driven by the electric field the fluid experiences, i.e., *Ohm's Law*, Eq. 11.3.

The frame-dependence of the electric field also makes the electric charge density, q, frame-dependent since Gauss's Law for the electric field is

$$q = \epsilon \nabla \cdot \mathbf{E} . \tag{11.7}$$

Although conservation of charge is technically $\partial q / \partial t = -\nabla \cdot \mathbf{J}$, the amplitude of $\partial q / \partial t$ is of order $(v/c)^2$ smaller than the individual components of $\nabla \cdot \mathbf{J}$. Therefore, within the MHD approximation, Eq. 11.4 represents conservation of electric charge in all frames of reference. Equations 11.3, 11.4, and 11.7 then show that when observed in the lab frame of reference the charge density within a fluid moving with velocity $\mathbf{v}$ is

$$q = -\epsilon \nabla \cdot (\mathbf{v} \times \mathbf{B}) ,$$

whereas when the observer is moving with the fluid (i.e., $\mathbf{v} = 0$) the charge density vanishes.

The electric field (in the lab frame) is the sum of a static field, determined by the gradient of the electrostatic scalar potential, Φ_e, and an induced field determined by the time derivative of the vector potential, $\mathbf{A}$:

$$\mathbf{E} = -\nabla \Phi_e - \frac{\partial \mathbf{A}}{\partial t} . \tag{11.8}$$

Therefore,

$$\nabla \cdot \mathbf{E} = -\nabla^2 \Phi_e \tag{11.9}$$

(when choosing the Coulomb gauge, $\nabla \cdot \mathbf{A} = 0$), which by Eq. 11.7 relates the charge density to the electrostatic potential, and

$$\nabla \times \mathbf{E} = -\frac{\partial \nabla \times \mathbf{A}}{\partial t} , \tag{11.10}$$

which by Eq. 11.6 is Faraday's Law, Eq. 11.2. Note also that, according to Eqs. 11.1 and 11.6,

$$\mu \mathbf{J} = \nabla \times \nabla \times \mathbf{A}$$
$$= -\nabla^2 \mathbf{A} , \tag{11.11}$$

again assuming the Coulomb gauge.

The secondary role played by the electric field relative to the magnetic field within the MHD approximation also extends to the electromagnetic force and energy. The electric force density, $q\mathbf{E}$, is of order $(v/c)^2$ smaller than the magnetic force density, $\mathbf{J} \times \mathbf{B}$; so within the MHD approximation the Lorentz force density is taken to be just the magnetic force density. With Eqs. 11.1 and 11.5 (and the constant μ), this Lorentz force density can be written as

$$\mathbf{J} \times \mathbf{B} = -\nabla \left(\frac{B^2}{2\mu} \right) + \nabla \cdot \left(\frac{\mathbf{BB}}{\mu} \right). \tag{11.12}$$

This illustrates how the magnetic force density is the sum of a gradient of magnetic pressure, $B^2/2\mu$, and the divergence of the Maxwell stress tensor (or magnetic tension), $\mathbf{BB}/\mu$. Likewise, since the electric energy density, $\epsilon E^2/2$, is order $(v/c)^2$ smaller than the magnetic energy density, $B^2/2\mu$, the electromagnetic energy density, within the MHD approximation, is taken to be just $B^2/2\mu$.

With the electromagnetic force and energy being only a function of $\mathbf{B}$, it would be convenient to also have an equation for the evolution of $\mathbf{B}$ only in terms of $\mathbf{B}$ and the fluid flow. This is easily accomplished by eliminating the electric field and current density via Eqs. 11.1–11.3 to obtain the *magnetic induction equation*:

$$\frac{\partial \mathbf{B}}{\partial t} = \nabla \times (\mathbf{v} \times \mathbf{B}) - \nabla \times (\eta \nabla \times \mathbf{B}), \tag{11.13}$$

where the magnetic diffusivity, η, is $1/(\mu\sigma)$. When η is constant, Eq. 11.13 simplifies, with the help of Eq. 11.5, to

$$\frac{\partial \mathbf{B}}{\partial t} = \nabla \times (\mathbf{v} \times \mathbf{B}) + \eta \nabla^2 \mathbf{B}. \tag{11.14}$$

Note that having solved for $\mathbf{v}$ and $\mathbf{B}$, one can use Eq. 11.1 to compute $\mathbf{J}$ and then Eq. 11.3 to get $\mathbf{E}$.

Like viscous (ν) and thermal (κ) diffusivities, η has units of length squared per time. The magnetic diffusion time (the time for magnetic field to decay to $1/e$ of its original intensity) for a length scale of D is roughly D^2/η. In the Earth's liquid iron-rich core η is much larger than κ and so the dominant magnetic structures are likely larger than the dominant thermal structures. The opposite is true within the sun's convection zone because κ there is mainly due to radiative transfer, which transfers heat much more effectively than molecular conduction. Therefore, as mentioned, the dominant length scale of the magnetic field observed on the solar surface is small compared to a global scale.

Equation 11.14 shows how the magnetic field is induced by the flow of the electrically conducting fluid across an existing field. The resulting distorted field responds with a Lorentz force density (Eq. 11.12) on the fluid, which is added to the momentum equation 1.2.

By taking the dot product of Eq. 11.14 with $\mathbf{B}/\mu$ and using Eqs. 11.1–11.5 and several vector identities one obtains the equation for the rate of change of the magnetic energy density:

$$\frac{\partial}{\partial t} \left(\frac{B^2}{2\mu} \right) = -\nabla \cdot \left(\frac{\mathbf{E} \times \mathbf{B}}{\mu} \right) - \frac{|\mathbf{J}|^2}{\sigma} - \mathbf{v} \cdot (\mathbf{J} \times \mathbf{B}). \tag{11.15}$$

The first term on the right side is the convergence of the Poynting flux, $(\mathbf{E} \times \mathbf{B})/\mu$, the flux of electromagnetic energy. The second term on the right is minus the ohmic heating, which is added to the internal energy density equation (1.6) as part of Q. The third term on the right is minus the rate work is done by the Lorentz forces, which is added to the left side of the mechanical energy density equation (1.9). These two terms therefore cancel out when Eq. 11.15 is added to Eq. 1.10 to get the total energy density equation that now includes magnetic energy (although not a prescribed internal heating such as radiogenic):

$$\frac{\partial}{\partial t}\left(\rho e + \frac{1}{2}\rho v^2 + \rho\Phi + \frac{B^2}{2\mu}\right) = -\nabla\cdot\left[\left(\rho e + \frac{1}{2}\rho v^2 + \rho\Phi + p\right)\mathbf{v}\right] -$$
$$\nabla\cdot\left[-k\nabla T - \mathbf{v}\cdot\sigma + \frac{\mathbf{E}\times\mathbf{B}}{\mu}\right]. \qquad (11.16)$$

For stress-free impermeable boundaries the rate of change of the volume-integrated total energy density equals minus the surface integral of the thermal heat flux and Poynting flux out of the fluid domain.

Now, a few words about units. In SI units, $\mathbf{B}$ is in tesla, $\mathbf{J}$ is in amperes per meter squared, $\mathbf{E}$ in volts per meter, μ in henries per meter or newtons per ampere squared, and σ is in siemens per meter. See the appendix in Jackson (1998) for a comprehensive discussion of other systems of units. If CGS units were chosen for Eqs. 11.1–11.5 and if magnetic field were measured in gauss (10^{-4} tesla), the free-space μ_o (which is usually what μ is taken to be if the temperature is high) would be 4π; this is without any modifications to Eqs. 11.1–11.5.

11.2 MAGNETOCONVECTION WITH A VERTICAL BACKGROUND FIELD

Consider the nonlinear Boussinesq model of convection in a box with impermeable boundaries on all four sides as described in Chapter 4. To keep this model relatively simple, we take η, μ, and σ to be constants and force the two-dimensional constraints on the field: $B_y = 0$ and $\partial\mathbf{B}/\partial y = 0$. For an initial condition we set $\mathbf{B} = \mathbf{B}_o = B_o\hat{z}$, a uniform vertical magnetic field within the fluid domain. This $\mathbf{B}_o$ is an externally imposed *background* field that exists at all times. At time zero there is not yet an induced magnetic field.

We want to maintain the boundaries tangentially stress-free, so magnetic stress, $B_x B_z/\mu$ (Eq. 11.12), in addition to viscous stress, needs to vanish on them. For the top and bottom boundaries, $B_x B_z/\mu$ is the magnetic force in the x-direction per surface area normal to the z-direction; and for the side boundaries $B_x B_z/\mu$ is the magnetic force in the z-direction per surface area normal to the x-direction. In addition, we want magnetic field to permeate through the top and bottom boundaries. These conditions are accomplished by forcing the horizontal component of the field, B_x, to vanish on all four boundaries. We also want to maintain the net magnetic flux (per length in the y-direction) through the top and bottom boundaries at the initial (background) value,

$$\int_0^L B_z dx = B_o L .$$

(11.17)

Note, because the field does not permeate through the side boundaries and because of Eq. 11.5, Eq. 11.17 holds for all values of z within the box of fluid. This "externally maintained" boundary condition prevents the 2D field from decaying away.

Having specified the initial and boundary conditions on the magnetic field, we now describe a way to solve Eqs. 11.5 and 11.14 for the evolution of the magnetic field. As we have been using the vector streamfunction, $\psi \hat{y}$, to ensure that the fluid velocity is divergence-free, here we use a magnetic vector potential in the y-direction, $\mathbf{A} = A\hat{y}$, to ensure, by Eq. 11.6, that Eq. 11.5 is satisfied. That is,

$$\mathbf{B} = \nabla \times \mathbf{A} = -\frac{\partial A}{\partial z}\hat{x} + \frac{\partial A}{\partial x}\hat{z}$$

and therefore

$$B_x = -\frac{\partial A}{\partial z} \quad \text{and} \quad B_z = \frac{\partial A}{\partial x} .$$

(11.18a,b)

Also, since the current density, $\mathbf{J}$, is related to $\mathbf{A}$ via Eq. 11.11, and since $\mathbf{A}$ is in the y-direction, $\mathbf{J} = J\hat{y}$. That is, like the streamfunction and vorticity, the vector potential and current density are only in the y-direction but depend only on x, z, and t. Also, contours of A are magnetic field lines, which are everywhere tangent to $\mathbf{B}$ and have a density proportional to the local amplitude of $\mathbf{B}$ for the same reasons contours of ψ are streamlines of the flow as discussed in Section 2.1.

However, instead of taking the curl of the magnetic induction equation, as we did for the momentum equation to get a vorticity equation, here we write Eq. 11.14 in terms of the vector potential:

$$\frac{\partial \nabla \times \mathbf{A}}{\partial t} = \nabla \times (\mathbf{v} \times (\nabla \times \mathbf{A})) + \eta \nabla^2 \nabla \times \mathbf{A} .$$

(11.19)

Then, using the standard vector identity for $\nabla(\mathbf{A} \cdot \mathbf{v})$ and the 2D constraints, it is easily seen that in Eq. 11.19 $\mathbf{v} \times (\nabla \times \mathbf{A}) = -(\mathbf{v} \cdot \nabla)A$. Also, since the time derivative and the Laplacian both commute with the curl, we can "uncurl" Eq. 11.19 to get the familiar advection/diffusion equation for the y-component, A, of the magnetic vector potential:

$$\frac{\partial A}{\partial t} = -(\mathbf{v} \cdot \nabla)A + \eta \nabla^2 A .$$

(11.20)

Note that the integrating factor, the y-derivative of a electrostatic scalar potential, vanishes because of the 2D constraints. Therefore, by Eqs. 11.7, 11.8, and 11.9, the electrostatic potential makes no contribution to the electric field or to the electric charge density. Also note that the advection term in Eq. 11.20 equals $(\mathbf{v} \times \mathbf{B})_y$ and the diffusion term equals $-J/\sigma$.

Now, by using Eqs. 11.18, the integral boundary condition at the bottom and top boundaries, Eq. 11.17, equals the vector potential at $x = L$ minus that at $x = 0$. Therefore, for convenience, we set

$$A = 0 \text{ at } x = 0 \quad \text{and} \quad A = B_o L \text{ at } x = L .$$

(11.21)

In addition, to force B_x to vanish on all four boundaries we want

$$\frac{\partial A}{\partial z} = 0 \qquad (11.22)$$

on all boundaries, which means that Eq. 11.21 needs to hold for all z. The vanishing of B_x on all four boundaries also means that there is no Poynting flux, $(\mathbf{E} \times \mathbf{B})/\mu$, through the boundaries.

Now we need to decide how to solve these equations. We could employ a fully finite difference method (Section 9.3) or a fully spectral method (Section 9.4). However, here we again choose to Fourier expand in the x-direction and use finite differences in the z-direction (Part 1). Therefore, to satisfy these boundary conditions on A we define the vector potential as

$$A(x, z, t) = B_o x + \sum_{n=1}^{N_n} A_n(z, t) \, \sin(n\pi x/L) \qquad (11.23)$$

and force

$$\frac{\partial A_n}{\partial z} = 0 \text{ at } z = 0 \text{ and } D. \qquad (11.24)$$

(Note, in this chapter we revert to using n as the symbol for the horizontal mode number, as is done in Part 1.) The spectral magnetic induction equation is therefore

$$\frac{\partial A_n}{\partial t} = -[(\mathbf{v} \cdot \nabla)A]_n + \eta \left(\frac{\partial^2 A_n}{\partial z^2} - \left(\frac{n\pi}{L} \right)^2 A_n \right). \qquad (11.25)$$

This equation describes the evolution of the field by induction and diffusion.

Now consider the back reaction of the field on the flow via the Lorentz force in the momentum equation. As is described in Part 1, we curl the momentum equation to get the vorticity equation. We discussed, using Eqs. 2.1-2.4, how the curl of the nonlinear advection term is

$$-\nabla \times ((\mathbf{v} \cdot \nabla)\mathbf{v}) = -\nabla \times (\boldsymbol{\omega} \times \mathbf{v})$$
$$= -(\mathbf{v} \cdot \nabla)\omega \hat{\mathbf{y}}$$

since $\partial/\partial y = 0$, $v_y = 0$, $\nabla \cdot \mathbf{v} = 0$, and $\boldsymbol{\omega} = \nabla \times \mathbf{v}$. Similarly, since $\partial/\partial y = 0$, $B_y = 0$, $\nabla \cdot \mathbf{B} = 0$ and $\mu \mathbf{J} = \nabla \times \mathbf{B}$, the curl of the Lorentz force per mass,

$$\frac{1}{\rho_o} \nabla \times (\mathbf{J} \times \mathbf{B}) = \frac{1}{\rho_o} (\mathbf{B} \cdot \nabla) J \hat{\mathbf{y}}, \qquad (11.26)$$

needs to be added to the vorticity equation.

If the model being modified here is *nondimensional*, scale $\mathbf{B}$ by the constant B_o, $\mathbf{A}$ by $B_o D$, and multiply Eq. 11.14 by $D^2/\kappa B_o$ and Eq. 11.20 (and 11.25) by $D/\kappa B_o$. This results in nondimensional versions of the magnetic induction equation with the only difference in appearance being that η is replaced by $1/q$ in those equations, where

$$q \equiv \kappa/\eta \qquad (11.27)$$

is the *Roberts number* (after Paul H. Roberts). In addition, the nondimensional version of Eq. 11.21 requires $A = a$ at $x = a$ (the aspect ratio). Likewise, the nondimensional spectral vorticity equation (Eq. 2.11), but now with the Lorentz "torque" (Eq. 11.26) included, is

$$\frac{\partial \omega_n}{\partial t} = -\left[(\mathbf{v}\cdot\nabla)\omega\right]_n + \mathrm{RaPr}\left(\frac{n\pi}{a}\right)T_n + \mathrm{Pr}\left(\frac{\partial^2 \omega_n}{\partial z^2} - \left(\frac{n\pi}{a}\right)^2 \omega_n\right)$$

$$+\frac{\mathrm{Q}\,\mathrm{Pr}}{\mathrm{q}}\left[(\mathbf{B}\cdot\nabla)J\right]_n \,, \tag{11.28}$$

where

$$Q \equiv \frac{\sigma B_o^2 D^2}{\rho_o \nu} \tag{11.29}$$

is the *Chandrasekhar number*. For the remainder of this section we assume this nondimensional model.

With A expanded according to Eq. 11.23 and employing Eqs. 11.18, we have expansions for the two components of the (nondimensional) magnetic field,

$$B_x(x, z, t) = -\sum_{n=1}^{N_n} \frac{\partial A_n}{\partial z}\,\sin(n\pi x/a)\,, \tag{11.30}$$

$$B_z(x, z, t) = 1 + \sum_{n=1}^{N_n}\left(\frac{n\pi}{a}\right)A_n\,\cos(n\pi x/a)\,, \tag{11.31}$$

and, using Eq. 11.11, the nondimensional current density is

$$J(x, z, t) = \sum_{n=1}^{N_n} J_n(z, t)\,\sin(n\pi x/a)\,, \tag{11.32}$$

where

$$J_n = -\left(\frac{\partial^2 A_n}{\partial z^2} - \left(\frac{n\pi}{a}\right)^2 A_n\right)\,. \tag{11.33}$$

By Eq. 11.32, the current density vanishes on the side boundaries.

Notice the similarities in the equations for $\mathbf{B}$ and $\mathbf{v}$, and for A_n and ψ_n, and for J_n and ω_n. There are differences, however, in the way these equations are solved. At each time step, A_n is first updated using the $\partial A_n/\partial t$ equation (11.25) and then J_n is easily updated with Eq. 11.33 by simply taking spatial derivatives of A, whereas ω_n is updated using the $\partial \omega_n/\partial t$ equation (11.28) and then ψ_n is updated by solving the Poisson equation (2.12) via the tridiagonal solver and applying two boundary conditions on ψ_n. This difference is due to our choice of a horizontally integrated boundary condition on B_z, Eq. 11.17, instead of a local boundary condition.

Having discussed the magnetic boundary conditions, we also need to mention the initial condition on the vector potential. Since we want just the uniform background field in the z-direction initially, we set $A_n = 0$ everywhere for all n at $t = 0$, which makes the nondimensional $B_x = 0$ and $B_z = 1$ everywhere at $t = 0$.

The subsequent evolution of the system of equations now has additional diffusive and advective constraints on the nondimensional time step, Δt. Magnetic diffusion requires

$$\Delta t < (\Delta z)^2 q / 4$$

and the CFL condition requires

$$\Delta t < \Delta z / |V_{A_z}|,$$

where the *nondimensional* Alfvén velocity, the velocity a transverse disturbance in the field propagates along the field, is

$$\mathbf{V}_A = \left(\frac{Q \Pr}{q} \right)^{1/2} \mathbf{B}. \tag{11.34}$$

In dimensional units,

$$\mathbf{V}_A = \frac{\mathbf{B}}{(\rho_o \mu)^{1/2}}. \tag{11.35}$$

Note, in CGS units with $\mu = 4\pi$, ρ in gm/cm^3, and $\mathbf{B}$ in gauss, $\mathbf{V}_A$ from Eq. 11.35 would be in cm/s. The Boussinesq (and anelastic) approximations also require that $\mathbf{V}_A$ remains small relative to the local speed of sound, which effectively limits the amplitude of the magnetic field.

Next we consider in detail the addition of the nonlinear terms involving the magnetic field: the Lorentz torque in the vorticity equation (11.28) and the advection of the vector potential in the magnetic induction equation (11.25). Here we describe these for the Galerkin method (Chapter 4); but of course they could be computed via the spectral-transform method (Section 10.4).

The Lorentz torque in Eq. 11.28 is (using Eqs. 11.18)

$$\frac{Q \Pr}{q} [(\mathbf{B} \cdot \nabla) J]_n = \frac{Q \Pr}{q} \left[-\frac{\partial A}{\partial z} \frac{\partial J}{\partial x} + \frac{\partial A}{\partial x} \frac{\partial J}{\partial z} \right]_n. \tag{11.36}$$

Since A and J are both expanded in $\sin(n\pi x / a)$, as are ψ, and ω, and since A_n and J_n are related to each other (Eq. 11.33) in the same way ψ_n and ω_n are related (Eq. 3.5), the Galerkin formula for calculating the contribution to Eq. 11.36 due to the *induced* magnetic field is, other than the minus sign and the (Q Pr/q) factor, identical to Eq. 4.4 with ψ_n replaced with A_n and ω_n replaced with J_n for $n > 0$. In practice, one can simply add this magnetic contribution to the nonlinear advection of vorticity using the coding strategy described in Section 4.3.2. However, there is one more contribution that needs to be included: that due to the interaction of the induced current density with the externally maintained constant-background magnetic field, which is the "1" in Eq. 11.31 or, equivalently, the "$B_o x$" in Eq. 11.23. This linear part of the (nondimensional) Lorentz torque is simply $(Q \Pr/q) \partial J_n / \partial z$, which can be added in with the other linear terms in the vorticity equation.

The Galerkin formula for calculating the advection of the *induced* vector potential in Eq. 11.25 is also identical to Eq. 4.4 but now with just $\omega_{n'}$ replaced with $A_{n'}$ for $n > 0$. In practice, this nonlinear term can be added to the linear diffusion part of the time derivative of the vector potential just as the nonlinear terms in the time derivatives of the vorticity and temperature are added to their respective linear terms before updating in time. However, as for the Lorentz force, this advection of the vector potential has an additional contribution due to the background vector potential, the "$B_o x$" in Eq. 11.23. The (nondimensional) gradient of this is "1" in the x-direction; so the additional (linear) contribution to this advection term is simply $-(v_x)_n$, which is $\partial \psi_n / \partial z$.

The magnetic induction equation has an additional complication. Unlike T_n, ω_n, and ψ_n, which all vanish on the bottom and top boundaries, $\partial A_n / \partial z$, not A_n, needs to vanish on these boundaries. Therefore, Eq. 11.25 needs to be solved on the bottom and top boundaries to update A_n there while forcing Eq. 11.24. This requires knowing, on these boundaries, $\partial^2 A_n / \partial z^2$ for the magnetic diffusion term and $\partial \psi_n / \partial z$ for the advection term. We could employ "noncentered" finite-difference methods that approximate these boundary derivatives; but these are typically less accurate. Instead, there is a clever way to use centered finite-difference methods by defining ghost points one grid level beyond the boundaries based on the boundary condition.

Consider the magnetic diffusion term on the bottom boundary, which involves the second derivative with respect to z of A_n calculated according to Eq. 2.16 for ($k = 1$):

$$\left(\frac{\partial^2 A_n}{\partial z^2} \right)_1 = \frac{(A_n)_2 - 2(A_n)_1 + (A_n)_0}{(\Delta z)^2} .$$

The ghost point is $k = 0$; but, since

$$\left(\frac{\partial A_n}{\partial z} \right)_1 = \frac{(A_n)_2 - (A_n)_0}{2\Delta z} = 0 ,$$

the value at the ghost point, $(A_n)_0$, must always equal $(A_n)_2$. Therefore,

$$\left(\frac{\partial^2 A_n}{\partial z^2} \right)_1 = \frac{2((A_n)_2 - (A_n)_1)}{(\Delta z)^2} . \tag{11.37}$$

Likewise, at the top boundary ($k = N_z$)

$$\left(\frac{\partial^2 A_n}{\partial z^2} \right)_{N_z} = \frac{2((A_n)_{N_z-1} - (A_n)_{N_z})}{(\Delta z)^2} . \tag{11.38}$$

Now consider the advection term on the bottom boundary. This involves $(-v_z \partial A / \partial z)$, which vanishes there since both v_z and $\partial A / \partial z$ vanish there, and $(-v_x \partial A / \partial x)$, which does not vanish. The challenge here is calculating $-(v_x)_n$ on the bottom boundary, which is

$$\left(\frac{\partial \psi_n}{\partial z} \right)_1 = \frac{(\psi_n)_2 - (\psi_n)_0}{2\Delta z} .$$

We know from Part 1 that the impermeable and stress-free boundary conditions force both ψ_n and $\partial^2 \psi_n / \partial z^2$ to vanish on the bottom boundary. Therefore, by Eq. 2.16 $(\psi_n)_0$ must always equal $-(\psi_n)_2$, which makes

$$\left(\frac{\partial \psi_n}{\partial z} \right)_1 = \frac{(\psi_n)_2}{\Delta z} . \tag{11.39}$$

Likewise, at the top boundary

$$\left(\frac{\partial \psi_n}{\partial z}\right)_{N_z} = -\frac{(\psi_n)_{N_z-1}}{\Delta z}. \tag{11.40}$$

In summary, the vector potential, A_n, is updated by solving Eq. 11.25 on all z-levels, including the top and bottom boundaries where the finite difference representations of the z derivatives in the linear diffusion term and in the nonlinear advection term are calculated as described in Eqs. 11.37, 11.38, 11.39, and 11.40.

Before leaving this section, recall that we have assumed impermeable side boundaries. If instead we want permeable periodic side boundaries (Section 10.2), the vector potential, A, would be x (the background field) plus a full (complex) Fourier series. If, in addition, we allow for an x-independent x-component of B, we would also solve the $n = 0$ part of the magnetic induction equation for $A_0(z, t)$.

Finally, ohmic heating, $|\mathbf{J}|^2/\sigma$ (also called Joule heating), is an additional non-linear term in Eq. 1.6. However, like viscous heating, we neglect it here for simplification. See Eqs. 12.31 and 12.32 if including them is desired. Otherwise, the energy equation remains unchanged. Total energy, however, will not be conserved exactly when viscous and ohmic heating are neglected; but the system adjusts by having slightly less heat flow out through the top boundary, on average, compared to the heat flow in through the bottom boundary.

11.3 LINEAR ANALYSES: MAGNETIC

Before discussing nonlinear numerical simulations we check what can be learned about the stability and structure of magnetoconvection and the dispersion relation for magneto-gravity waves from analytical analyses without the nonlinear terms. It would also be wise to test whether the magnetic modifications discussed in the previous section that apply to the linear system of equations work before adding the nonlinear modifications.

11.3.1 Linear Stability: Magnetic

First consider the nondimensional linear equations and boundary conditions, Eqs. 3.3–3.6, for an unstable background temperature gradient but now with the *linear* part of the curl of the Lorentz force per mass, $(Q\,Pr/q)\partial J_n/\partial z$, added to the right side of the vorticity equation, (3.4). In addition, we have the relationship between J_n and A_n, Eq. 11.33; the (nondimensional) linear magnetic induction equation,

$$\frac{\partial A_n}{\partial t} = \frac{\partial \psi_n}{\partial z} + \frac{1}{q}\left(\frac{\partial^2 A_n}{\partial z^2} - \left(\frac{n\pi}{a}\right)^2 A_n\right); \tag{11.41}$$

and the (nondimensional) boundary conditions,

$$\frac{\partial A_n}{\partial z} = 0 \text{ at } z = 0 \text{ and } 1. \tag{11.42}$$

As discussed in Section 3.4, T_n, ω_n, and ψ_n are expanded in $\sin(m\pi z)$ to satisfy their top and bottom boundary conditions. To satisfy the boundary conditions on A_n, Eq. 11.42, let

$$A_n(z, t) = \sum_{m=1}^{N_m} A_{nm}(t)\, \cos(m\pi z),$$

where again the coefficients depend on the x-mode number n, the z-mode number m, and time t. Then, as is done in Section 3.4, substitute the expressions, now including A_n and J_n, into the linear equations (including the linear part of the Lorentz torque) with all time derivatives set to zero. The resulting critical Rayleigh number is

$$\mathrm{Ra}_{crit}(n, m) = \left(\frac{\pi}{a}\right)^4 \frac{\left(n^2 + (am)^2\right)^3}{n^2} + \left(\frac{\pi m}{n}\right)^2 \left(n^2 + (am)^2\right) Q. \qquad (11.43)$$

Again, cells that extend from the bottom to the top ($m = 1$) are more unstable. However, as Q increases, i.e., as either the intensity of the background magnetic field (B_o) increases or the electrical conductivity of the fluid (σ) increases, the Rayleigh number needed at the onset thermal convection increases. That is, the greater the Lorentz force, which tries to prevent fluid flow across magnetic field lines, the more thermal driving is required to maintain a given level of convection.

One can also check how the magnetic field affects the preferred spatial scale of the resulting convection by setting the derivative with respect to n of Eq. 11.43 to zero to find the critical horizontal mode number n_{crit}. The result, for $m = 1$, is

$$2\left(\frac{n_{crit}}{a}\right)^6 + 3\left(\frac{n_{crit}}{a}\right)^4 = \frac{Q}{\pi^2} + 1. \qquad (11.44)$$

For no magnetic field, $Q = 0$, this reduces to Eq. 3.9. As Q increases n_{crit} increases; that is, the larger the background magnetic field the smaller the preferred horizontal length scale because the Lorentz forces inhibit large horizontal flows across the vertical magnetic field.

Equation 11.44 can be solved for n_{crit} by checking what the left side of this equation would be for the $Q = 0$ value of n_{crit}, i.e., the nearest integer to $a/\sqrt{2}$, and then checking the left side for increasingly higher integer values of n_{crit} until the left side exceeds the right side, for the given value of Q. The actual n_{crit} will then be the integer that most closely satisfies Eq. 11.44. Also, based on this equation,

$$n_{crit} \to a\left(\frac{Q}{2\pi^2}\right)^{1/6} \quad \text{and} \quad \mathrm{Ra}_{crit} \to \pi^2 Q \quad \text{as} \quad Q \to \infty.$$

These analytic predictions for the critical Rayleigh (Eq. 11.43) and mode number (Eq. 11.44) can be used to check the *linear* magnetic modifications to the convection code. One could start with the original convection code and add an option to run as a linear code by skipping over the nonlinear terms. Then, as discussed in Chapter 3, one could check if the linear magnetic code produces the critical Rayleigh number, for a given a and Q, as predicted by Eqs. 11.43 and 11.44. As mentioned, this would be a good way to test the linear magnetic modifications before proceeding to the nonlinear modifications.

11.3.2 Internal Gravity Waves: Magnetic

The linear magnetic code could also be tested by simulating internal magneto-gravity waves in a stable temperature stratification. As is done in Section 6.1, drop the nonlinear terms and the diffusion terms. The resulting set of *dimensional* equations is

$$\frac{\partial T}{\partial t} = -\frac{\partial \psi}{\partial x}\frac{d\overline{T}}{dz}, \tag{11.45}$$

$$\frac{\partial \omega}{\partial t} = -g_o\alpha\frac{\partial T}{\partial x} + \frac{B_o}{\rho_o}\frac{\partial J}{\partial z}, \tag{11.46}$$

$$\frac{\partial A}{\partial t} = \frac{\partial \psi}{\partial z}B_o, \tag{11.47}$$

$$\omega = -\left(\frac{\partial^2}{\partial x^2} + \frac{\partial^2}{\partial z^2}\right)\psi, \tag{11.48}$$

$$J = -\frac{1}{\mu}\left(\frac{\partial^2}{\partial x^2} + \frac{\partial^2}{\partial z^2}\right)A. \tag{11.49}$$

By taking the time derivative of Eq. 11.46 and substituting in the other equations, this set of five equations and five unknowns can be reduced to one equation in terms of, for example, only ψ:

$$\left(\frac{\partial^2}{\partial x^2} + \frac{\partial^2}{\partial z^2}\right)\frac{\partial^2 \psi}{\partial t^2} = -N^2\frac{\partial^2 \psi}{\partial x^2} + V_A^2\left(\frac{\partial^2}{\partial x^2} + \frac{\partial^2}{\partial z^2}\right)\frac{\partial^2 \psi}{\partial z^2}, \tag{11.50}$$

where N is the Brunt-Väisälä frequency (Eq. 6.5) and V_A is the Alfvén speed (Eq. 11.35) based on the background field B_o. The resulting dispersion relation, relating the wave frequency, $\bar{\omega}$, to the wave vector, **k**, of a planewave solution (Eq. 6.6), is

$$\bar{\omega} = \left(N^2\frac{k_x^2}{k^2} + V_A^2 k_z^2\right)^{1/2}. \tag{11.51}$$

This shows that the frequency of an internal magneto-gravity wave increases with both stratification, N, and magnetic field intensity, V_A. However, the more horizontal the direction of the phase propagation the more the wave behaves like an internal gravity wave because the restoring forces on the transverse fluid motions are more in the direction of gravity. The more vertical the phase propagation the more the wave behaves like an Alfvén wave because the restoring forces on the transverse fluid motions are more in the direction perpendicular to the vertical magnetic field lines.

The linear magnetic code could then be tested by comparing simulated wave frequencies with those predicted in Eq. 11.51 for a set of wave vectors, k_x and k_z, a given background thermal structure, N, and magnetic field, V_A.

11.4 NONLINEAR SIMULATIONS: MAGNETIC

Now let's consider nonlinear simulations of magnetoconvection in a box with impermeable side boundaries. Choose a Ra significantly greater than the overall critical Rayleigh number for a given aspect ratio, a, and now also for a given Chandrasekhar number, Q. The larger the prescribed value of Q the more intense the background field and therefore the more difficult it will be for the system to convect.

A convenient way to illustrate the magnetic field is to plot magnetic field lines. As mentioned in Section 11.2 for this 2D problem, contours of A represent instantaneous magnetic field lines, which like the constant-ψ streamlines are not physical but only a graphical illustration of the local direction and intensity of the field. Note that magnetic field lines permeate through the top and bottom boundaries; whereas, because of the impermeable velocity boundary conditions, the fluid streamlines do not.

Consider the case illustrated on the right in Fig. 4.2 but with the addition of a relatively weak vertical background magnetic field ($Q = 10^2$) and with a magnetic diffusivity equal to the thermal diffusivity ($q = 1$). A snapshot of this case is displayed in the top row of Fig. 11.1. The kinetic energy is so much greater than the magnetic energy that the changes in the temperature and fluid flow profiles due to the presence of the field are extremely small. However, the magnetic field profile is significantly changed from its initial uniform profile, now having most of the field swept into the upflows and downflows, where it is nearly parallel to the flow, minimizing the inhibiting Lorentz forces. Nigel Weiss and colleagues pioneered the study of this "magnetic flux expulsion" effect (e.g., Weiss, 1966; Moore et al., 1973; Galloway et al., 1978; Weiss, 1981a,b) using similar 2D magnetoconvection models. Recall that the nonmagnetic case (illustrated on the right in Fig. 4.2) continues to be steady and to maintain the two-cell pattern even after five thermal diffusion times. The magnetic case here appears to be steady initially; but after about one thermal diffusion time the central downwelling is seen to slowly drift to one side and eventually replace the upwelling there, converting the pattern to a single cell like that on the left side of Fig. 4.2.

Now consider the same case but with a more intense background magnetic field by setting Q to 10^4. The Lorentz forces are now about 100 times larger; so the field acts like a stiff elastic medium, strongly resisting flows locally perpendicular to it. The resulting solution is nicely time-dependent, a combination of thermal convection and Alfvén waves. A snapshot of this solution is displayed in the middle row of Fig. 11.1. There is still a small tendency for the expulsion of magnetic flux from regions of strong horizontal flow and the concentration of it within regions of strong vertical flow; but since the pattern is so time-dependent and the field is more intense this process is not as efficient, as seen for the case illustrated in the top row of Fig. 11.1. Increasing Q further makes the motion change from convection to an oscillation. A value of Q greater than that predicted by Eq. 11.43 with the critical Rayleigh number and aspect ratio set to the Ra and a for this example, 10^6 and 3, respectively, m set to 1, and n determined by Eq. 11.44 will cause the convection to decay away.

Temperature   **Magnetic field lines**

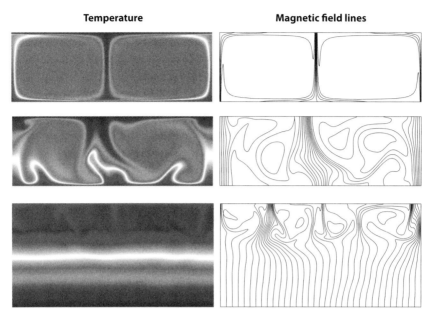

Figure 11.1 Snapshots of three magnetoconvection solutions, illustrated with profiles of temperature and vector potential (i.e., magnetic field lines) plotted on the horizontal (x) and vertical (z) grid (see *Color Plate* 4a for a color version of this figure). The snapshot illustrated in the top row is the case on the right in Fig. 4.2 with the addition of a relatively weak vertical background magnetic field ($Q = 10^2$ and q = 1). Initially it appears to be in steady state, like the nonmagnetic version; but after about one thermal diffusion time the pattern switches to a single cell like that on the left side of Fig. 4.2. The snapshot illustrated in the middle row is the same case but with a much more intense background field ($Q = 10^4$ and q = 1). This case is quite time-dependent. The snapshot illustrated in the bottom row is the case on the right in Fig. 6.2 with a relatively intense background field ($Q = 10^4$, q = 1). It too is time-dependent as it was without the field. For temperature, red represents hot buoyant fluid and blue cold heavy fluid. The horizontal-mean temperature is included in these images. Solid contours of the vector potential are magnetic field lines, entering the bottom boundary and exiting the top.

Finally, consider the case illustrated on the right in Fig. 6.2, now with a relatively intense vertical background field ($Q = 10^4$, q = 1). A snapshot of this case is displayed in the bottom row of Fig. 11.1. The Lorentz forces reduce the kinetic energy in the convection zone as the flow does work on the field there. Magnetogravity waves are continually excited at the interface and propagate through the lower stable region. However, the amplitudes of these waves are not sufficient to significantly distort the uniform field profile there.

11.5 MAGNETOCONVECTION WITH A HORIZONTAL BACKGROUND FIELD

Consider a magnetoconvection problem like that described in Section 11.2 but now with an imposed uniform *horizontal* background magnetic field $\mathbf{B}_o = B_o\hat{x}$. Again, we want no magnetic (or viscous) stress on the boundaries. However, now we want the field to permeate through the side boundaries and none to permeate through the top and bottom boundaries. To satisfy these constraints, we force B_z, and therefore $\partial A/\partial x$, to vanish on all four boundaries. Here again, A is the amplitude of the vector potential, which is in the y-direction.

We also want to maintain a net magnetic flux through the side boundaries equal to its initial (background) value:

$$\int_0^D B_x dz = B_o D .\qquad(11.52)$$

This integral constraint will hold for all values of x within the fluid because of Eq. 11.5 and of the top and bottom boundary conditions on B_z. Using Eqs. 11.18 and 11.52, the vector potential at $z = 0$ minus that at $z = D$ equals $B_o D$, so we can choose

$$A = 0 \text{ at } z = 0 \quad \text{and} \quad A = -B_o D \text{ at } z = D$$

for all x and t. Therefore, since we want $\partial A/\partial x$ to vanish on all four boundaries, the dimensional vector potential for this problem is

$$A(x, z, t) = -B_o z + \sum_{n=1}^{N_n} A_n(z, t) \, \cos(n\pi x/L)\qquad(11.53)$$

with

$$A_n = 0 \text{ at } z = 0 \text{ and } D .$$

The $-B_o z$ in Eq. 11.53 is the uniform, background, x-directed, magnetic field; in nondimensional format it is $-z$.

The vector potential version of the magnetic induction equation is the same as it is for the case of a vertical background field, Eq. 11.25. Note also that for this case of a horizontal background field the top and bottom boundary conditions on $A_{n>0}$ are simpler than those for the vertical background field case. That is, the equation for A_n (Eq. 11.25) does not need to be solved on the top and bottom boundaries and no ghost points are needed since the $A_{n>0}$ vanish on these boundaries.

Using Eqs. 11.18 and 11.53, the two components of the *nondimensional* magnetic field are

$$B_x(x, z, t) = 1 - \sum_{n=1}^{N_n} \frac{\partial A_n}{\partial z} \, \cos(n\pi x/a) ,$$

$$B_z(x, z, t) = -\sum_{n=1}^{N_n} \left(\frac{n\pi}{a}\right) A_n \, \sin(n\pi x/a) ,$$

and, by Eq. 11.11, the nondimensional current density, which is still only in the y-direction, is

$$J(x, z, t) = \sum_{n=1}^{N_n} J_n(z, t) \, \cos(n\pi x/a) \,,$$

where

$$J_n = -\left(\frac{\partial^2 A_n}{\partial z^2} - \left(\frac{n\pi}{a}\right)^2 A_n\right) \,,$$

the same as in Eq. 11.33. Note, for this case, the current density does not vanish on the side boundaries.

Now consider the nonlinear terms. We discuss their computation using the Galerkin method (Chapter 4), but of course they could instead be calculated using a spectral-transform method (Section 10.4). Consider first the Lorentz torque, Eq. 11.36, in the vorticity equation 11.28. Since A and J are both expanded in cosines, this nonlinear term involves products of sines and cosines as it does for the vertical background field. However, now the sines appear because of the derivatives with respect to x. Using Eq. 4.1c, the coefficients of $\sin(n\pi x/a)$ for the Lorentz torque in the vorticity equation are

$$\frac{Q \, \mathrm{Pr}}{q} [(\mathbf{B} \cdot \nabla) J]_n = -\frac{Q \, \mathrm{Pr}}{q} \frac{n\pi}{a} J_n - \frac{Q \, \mathrm{Pr}}{q} \frac{\pi}{2a} \sum_{n'=1}^{N_n} \sum_{n''=1}^{N_n}$$
$$\left[\left(-n' \frac{\partial A_{n''}}{\partial z} J_{n'} + n'' A_{n''} \frac{\partial J_{n'}}{\partial z} \right) \delta_{n''+n',n} \right.$$
$$\left. + \left(n' \frac{\partial A_{n''}}{\partial z} J_{n'} + n'' A_{n''} \frac{\partial J_{n'}}{\partial z} \right) (\delta_{n''-n',n} - \delta_{n'-n'',n}) \right] .$$

$$(11.54)$$

Note that the linear part of this (nondimensional) Lorentz torque, due to the induced electric current through the horizontal background field, is

$$\frac{Q \, \mathrm{Pr}}{q} \left(\frac{\partial J}{\partial x}\right)_n = -\frac{Q \, \mathrm{Pr}}{q} \frac{n\pi}{a} J_n \,.$$

The nonlinear contribution due to the induced current through the induced magnetic field, i.e., the double summation over n' and n'', is the same as that in Eq. 4.4 when $\psi_{n''}$ is replaced with $A_{n''}$ and $\omega_{n'}$ is replaced with $J_{n'}$, except for the factor $(Q \, \mathrm{Pr}/q)$ and the plus sign in front of the third line of Eq. 11.54. One can also compare this nonlinear contribution in Eq. 11.54 to that for the case of a vertical background field in Section 11.2, where the only difference is the sign in front of the second line in Eq. 11.54.

Next consider the advection of the induced vector potential in Eq. 11.25. We are now looking for coefficients of $\cos(n\pi/a)$ since the vector potential is now expanded in these. Therefore, for $n > 0$,

$$- [(\mathbf{v} \cdot \nabla) A]_n = \frac{n\pi}{a} \psi_n - \frac{\pi}{2a} \sum_{n'=1}^{N_n} \sum_{n''=1}^{N_n}$$

$$\left[\left(-n' \frac{\partial \psi_{n''}}{\partial z} A_{n'} + n'' \psi_{n''} \frac{\partial A_{n'}}{\partial z} \right) \delta_{n''+n',n} \right.$$

$$\left. + \left(n' \frac{\partial \psi_{n''}}{\partial z} A_{n'} + n'' \psi_{n''} \frac{\partial A_{n'}}{\partial z} \right) (\delta_{n''-n',n} + \delta_{n'-n'',n}) \right].$$

$$(11.55)$$

Note that the linear part of this (nondimensional) advection term, due to the advection of the horizontal background field, is

$$- \left(v_z \frac{\partial (-z)}{\partial z} \right)_n = \frac{n\pi}{a} \psi_n \,.$$

The nonlinear part of the advection term, i.e., the double summation, is the same as that in Eq. 4.6 when $T_{n'}$ is replaced with $A_{n'}$. One can also compare this nonlinear contribution in Eq. 11.55 to that for the case of a vertical background field in Section 11.2, where the only difference is the sign in front of the $\delta_{n''-n',n}$ term in the third line of Eq. 11.55.

The vorticity and magnetic induction equations are solved as they are for the vertical background field case except that, as mentioned above, there is no need to solve the induction equation on the top and bottom boundaries because now the A_n are forced to vanish on these boundaries.

Snapshots are shown in Fig. 11.2 for scenarios that are the same as those in Fig. 11.1 except that here there is a horizontal, instead of a vertical, background field. The weak horizontal magnetic field in the scenario illustrated in the top row initially gets deformed by a central downwelling and side upwellings, as depicted in the temperature snapshot in the top row of Fig. 11.1. The field grows more than order of magnitude until the Lorentz force becomes significant enough to destabilize the flow pattern, first causing it to evolve erratically and then allowing it to settle into a slightly time-dependent, single-cell pattern as seen in the top row of Fig. 11.2. Note that this statistically steady pattern maintains such a large induced magnetic field that the much smaller horizontal background field is not visible with the chosen contour interval for $\mathbf{A}$.

The middle scenario is for a horizontal background field one hundred times larger. This case is relatively time-dependent with both the flow and the field continually changing but always trying to be aligned as much as possible to avoid Lorentz forces.

The scenario in the bottom row is also time-dependent, and much more so in the upper convection zone than in the lower stable region. Notice how magnetic field lines near the interface connect both regions. This situation may be relevant to the global magnetic field well below the solar surface in the equatorial region, with the x-direction representing colatitude. On the other hand, the situation illustrated in

Temperature   **Magnetic field lines**

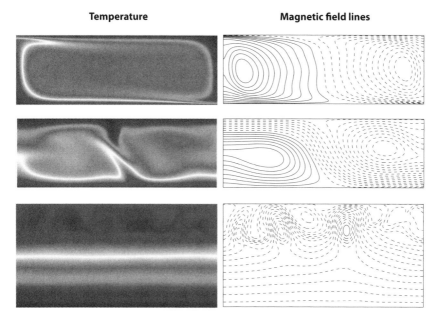

Figure 11.2 Snapshots of three magnetoconvection solutions, illustrated with profiles of temperature and vector potential (i.e., magnetic field lines) plotted on the horizontal (x) and vertical (z) grid (see *Color Plate* 4b for a color version of this figure). The cases represented in the top, middle, and bottom rows are the same as those in Fig. 11.1 except that here there is a horizontal background field. Note the dashed lines represent negative vector potentials, because of the arbitrary choice we made for the top and bottom boundary values of **A**.

Fig. 11.1 might be more like the solar field in the polar region, especially when the global field is dominantly dipolar.

It is important to remember, though, that these simulations are forced to be 2D, with both the field and the flow directed only in the x, z-plane. In 3D, however, the flow would prefer to form convection rolls with axes parallel to the constant-background horizontal field and solid-body flow around each axis, in which case magnetic induction and Lorentz forces vanish. However, for more complicated 3D flow and field patterns, like a sunspot, there is magnetic flux expulsion.

11.6 MAGNETOCONVECTION WITH AN ARBITRARY BACKGROUND FIELD

After studying magnetoconvection with an imposed uniform *vertical* background magnetic field in Section 11.2 and with a *horizontal* background field in Section 11.5 we are tempted to consider magnetoconvection within our cartesian box with a uniform background magnetic field in an arbitrary direction within the x, z-plane:

$$\mathbf{B}_o = B_{V,o}\hat{z} + B_{H,o}\hat{x} .$$

Since the magnetic induction equation is linear with respect to the magnetic field, we can define the total magnetic field as

$$\mathbf{B} = \mathbf{B}_V + \mathbf{B}_H = \nabla \times \mathbf{A} = \nabla \times (A_V + A_H)\hat{y}$$

and divide the vector potential version of the magnetic induction equation into vertical and horizontal parts:

$$\frac{\partial A_V}{\partial t} = -(\mathbf{v} \cdot \nabla) A_V + \eta \nabla^2 A_V ,$$

$$\frac{\partial A_H}{\partial t} = -(\mathbf{v} \cdot \nabla) A_H + \eta \nabla^2 A_H .$$

We might also adopt the boundary conditions from the purely vertical and horizontal background field cases:

$$B_{V,x} = -\frac{\partial A_V}{\partial z} = 0 ,$$

$$B_{H,z} = \frac{\partial A_H}{\partial x} = 0 ,$$

and

$$\int_0^L B_{V,z} dx = B_{V,o} L ,$$

$$\int_0^D B_{H,x} dz = B_{H,o} D .$$

However, there is a problem. We no longer satisfy our tangentially stress-free boundary conditions on our cartesian box, i.e., $B_x B_z / \mu$ does not vanish on the boundaries as it does for the purely vertical or purely horizontal background field cases. Allowing these stresses would require changing the boundary conditions on the velocity.

Proceeding anyway, we might continue to assume that all four boundaries are impermeable and solve the vertical and horizontal parts of the vector potential as we do in Sections 11.2 and 11.5, respectively. That is,

$$A_V(x, z, t) = B_{V,o}x + \sum_{n=1}^{N_n} A_{V,n}(z, t) \sin(n\pi x/L) ,$$

$$A_H(x, z, t) = -B_{H,o}z + \sum_{n=1}^{N_n} A_{H,n}(z, t) \cos(n\pi x/L) ,$$

with

$$\frac{\partial A_{V,n}}{\partial z} = 0 \ \text{at} \ z = 0 \ \text{and} \ D,$$

$$A_{H,n} = 0 \ \text{at} \ z = 0 \ \text{and} \ D.$$

Note, to make these nondimensional, scale both $\mathbf{B}_V$ and $\mathbf{B}_H$ with, for example, $B_{V,o}$.

However, the next problem is that the Lorentz torque (Eq. 11.26) in the vorticity equation *is* nonlinear with respect to the magnetic field. Therefore,

$$[(\mathbf{B} \cdot \mathbf{\nabla}) J]_n = [(\mathbf{B}_V \cdot \mathbf{\nabla}) J_V]_n + [(\mathbf{B}_H \cdot \mathbf{\nabla}) J_H]_n + [(\mathbf{B}_V \cdot \mathbf{\nabla}) J_H]_n + [(\mathbf{B}_H \cdot \mathbf{\nabla}) J_V]_n.$$

The first two terms on the right could be calculated as they are for the purely vertical and purely horizontal background field cases, respectively. The problem is that the third and fourth ("cross") terms involve sines squared and cosines squared, which, according to Eqs. 4.1a,b, are the sums of cosines. Since the vorticity for this case of impermeable side boundaries is expanded in $\sin(n\pi x/L)$, these $\cos(n\pi x/L)$ contributions to the Lorentz term in the vorticity equation (11.28) are incompatible. Again, the reason is that there would be stresses on the boundaries, which would require different boundary conditions on the velocity. Imposing permeable periodic boundary conditions (Section 10.2) on all four boundaries is one way that would work.

As a final remark about arbitrary background magnetic fields, we remind the reader that so far we have considered strictly 2D magnetoconvection in a cartesian box. Several very interesting studies (e.g., Spiegel & Weiss, 1982; Hughes & Weiss, 1995) have been done for magnetoconvection in a 2.5D box (Section 10.5) for which the background magnetic field is horizontal but in the y-direction, normal to the plane of the fluid flow; however, all variables are independent of y. These studies focus on double-diffusive instabilities and evolution (Chapter 7) due to buoyancy being partly thermal and partly magnetic (Section 12.4.1) with magnetic diffusivity η being much less than thermal diffusivity κ. A modified Boussinesq approximation, "magneto-Boussinesq," is employed in these studies of a very slightly density-stratified perfect gas. This approximation uses an anelastic version of $\mathbf{\nabla} \cdot \mathbf{v}$ (Eq. 12.15) in the magnetic induction equation and, like the anelastic approximation, retains the effects of the pressure perturbation in the equation of state and energy equation because what is assumed to be small is the *sum* of the perturbation gas and magnetic pressures.

SUPPLEMENTAL READING

Davidson (2001)
Gubbins & Herrero-Bervera (2007)
Jackson (1998)
Olson & Schubert (2007)
Roberts (1967)

EXERCISES

1. *Neglecting the displacement current density in Ampère's Law*
   Scale Faraday's Law to show that $|\mathbf{E}|/|\mathbf{B}|$ should be of order of the fluid velocity v and then show that the displacement current density in Maxwell's equations should be of order $(v/c)^2$.

2. *Derivation of the magnetic induction equation*
   Derive the magnetic induction equation (11.13) by eliminating the electric field and current density via Eqs. 11.1–11.3.

3. *Elsässer variables*
   Starting with the magnetic induction equation and the Boussinesq equations for mass and momentum conservation for the case of no buoyancy (or rotation) and assuming constant viscous and magnetic diffusivities, formulate equations for $\partial Z^+/\partial t$ and $\partial Z^-/\partial t$, where the

   $$\partial Z^\pm \equiv \mathbf{v} \pm \mathbf{V}_A ,$$

   $\mathbf{v}$ and $\mathbf{V}_A$ being, as usual, the fluid velocity and Alfvén velocity vectors, respectively. Let q be the sum of the fluid and magnetic pressures divided by density and $v^\pm$ be defined as $(v \pm \eta)/2$. Write the two equations only in terms of $Z^\pm$, q, and $v^\pm$.

4. *The advection term in the vector potential equation*
   Using the standard vector identity for $\nabla(\mathbf{A} \cdot \mathbf{v})$ and the 2D constraints, show that in the magnetic induction equation (11.19) $\mathbf{v} \times (\nabla \times \mathbf{A}) = -(\mathbf{v} \cdot \nabla)\mathbf{A}$.

5. *Critical Rayleigh number and horizontal mode number for convection in a box with a vertical background magnetic field*
   Derive the expression for the critical Rayleigh number, $\mathrm{Ra}_{crit}(n, m)$, (Eq. 11.43), and for the critical horizontal mode number, n_{crit} (Eq. 11.44), for magnetoconvection in a box with a vertical background field.

6. *Magneto-gravity wave for a vertical background field*
   Derive the magneto-gravity wave equation for a vertical background field (Eq. 11.50) by manipulating Eqs. 11.45–11.49. Also derive the phase and group velocities (Section 6.1) for this case.

7. *Magneto-gravity wave dispersion relation with diffusion*
   Derive the magneto-gravity wave dispersion relation for a vertical background field, like Eq. 11.51, but with viscous, thermal, and magnetic diffusion in Eqs. 11.45–11.47 for constant values of v, κ, and η.

8. *Magnetic energy flow through boundaries and Lorentz force on boundaries*
   Show that the total flow of magnetic energy through the impermeable boundaries (i.e., the integrated Poynting flux) vanishes for a vertical background field. What are the Lorentz forces on these boundaries? What magnetic energy flows and Lorentz forces exist on the boundaries for a horizontal background field?

9. *Critical Rayleigh number and horizontal mode number for convection in a box with a horizontal background field*
   As can be seen in Eqs. 11.54 and 11.55, the linear part of the Lorentz term in the vorticity equation is $(-Q \, \mathrm{Pr} \, n\pi \, J_n/aq)$ and the linear part of the advection

term in the induction equation is $(n\pi \psi_n/a)$. Since $A_n = 0$ at $z = 0$ and 1, set

$$A_n(z, t) = \sum_{m=1}^{N_m} A_{nm}(t) \, \sin(m\pi z).$$

Using the method described in Sections 3.4 and 11.3, show that the critical Rayleigh number for this horizontal background field case is

$$\text{Ra}_{crit}(n, m) = \left(\frac{\pi}{a}\right)^4 \frac{\left(n^2 + (am)^2\right)^3}{n^2} + \left(\frac{\pi}{a}\right)^2 \left(n^2 + (am)^2\right) Q. \quad (11.56)$$

Compare this to Eq. 11.43. Also derive the equation that determines the critical horizontal mode number (n_{crit}) for this case.

10. *Dispersion relation and phase and group velocities for magneto-gravity waves with a horizontal background field.*
    Using the method described in Sections 6.1 and 11.3, show for a horizontal background field that, when neglecting the nonlinear and diffusion terms, the dispersion relation, relating the wave frequency, $\bar{\omega}$, to the wave vector, $\mathbf{k}$, of a plane wave solution (Eq. 6.6), is

$$\bar{\omega} = \left(N^2 \frac{k_x^2}{k^2} + V_A^2 k_x^2\right)^{1/2}. \quad (11.57)$$

Compare this to Eq. 11.51. Notice how for this case the gravitational and magnetic restoring forces both act in the vertical direction. Also derive the phase and group velocities (Section 6.1) for this case.

11. *Magnetic energy per mode*
    As described for kinetic energy in Section 5.3, show that the nondimensional magnetic energy per mode n (per unit length in the y-direction), for either a vertical or horizontal background field, is

$$\text{ME}_n(z, t) = \frac{Q \, \text{Pr}}{q} \frac{a}{4} \left[\left(\frac{\partial A_n}{\partial z}\right)^2 + \left(\frac{n\pi}{a} A_n\right)^2\right].$$

This can be compared to the energy in the background field, which is just QPr/q.

COMPUTATIONAL PROJECTS

1. *Testing a linear code by finding the Ra_{crit} and n_{crit} for magnetoconvection in a box with a vertical background field*
   Find the Ra_{crit} and n_{crit} using a linear code for magnetoconvection in a box with a vertical background field and compare to the analytic predictions for several values of Q.

2. *Magneto-gravity wave dispersion relations: simulated compared to analytic*
   Using a linear diffusive magneto-gravity wave code for a vertical background field, compare simulated wave frequencies with those predicted in Eq. 11.51 for a set of wave vectors, k_x and k_z, a given background thermal structure,

N, and magnetic field, V_A. Likewise, compare the simulated frequencies with those predicted by the diffusive dispersion relation obtained in the *Magneto-gravity wave dispersion relation with diffusion* exercise.

3. *Nonlinear Simulations, Vertical background field: Poynting flux, current density, and electric field*

   Produce simulations of magnetoconvection in a box with a vertical background field, like those illustrated in Fig. 11.1. Calculate and plot the Poynting flux, the current density, and the electric field for these cases.

4. *Nonlinear simulations, vertical background field: simulated dispersion relation*

   For simulations of magnetoconvection and magneto-gravity waves in a box with a vertical background field, like those illustrated in Fig. 11.1, measure the dominant frequency and wavenumber of the waves in the stable region and compare to that predicted by Eq. 11.51.

5. *Testing a linear code by finding the Ra_{crit} and n_{crit} for magnetoconvection in a box with a horizontal background field*

   Find the Ra_{crit} and n_{crit} using a linear code for magnetoconvection in a box with a horizontal background field and compare to the analytic predictions for several different values of Q.

6. *Kinetic and magnetic energy vs. time and energy spectra*

   Simulate one of the time-dependent scenarios illustrated in this chapter and plot the total (z-integrated) $\mathrm{KE}(t)$ and $\mathrm{ME}(t)$ vs. time t (not including the background magnetic energy). Also plot, at a given time step, the spectra of total kinetic energy and total magnetic energy vs. mode n:

$$\int_0^1 \mathrm{KE}_n(z,t)\, dz \quad \text{and} \quad \int_0^1 \mathrm{ME}_n(z,t)\, dz \,.$$

7. *Semi-implicit magnetic induction equation*

   As described in Chapter 8 for the nonmagnetic case, modify the magnetoconvection model, for either the vertical or horizontal background field, by treating the diffusion term implicitly in the magnetic induction equation 11.25. Care needs to be taken in the construction of the implicit matrix operator for the vertical background magnetic field case because the vector potential is updated on the top and bottom boundaries.

8. *A time-dependent background field*

   Make B_o proportional to $\sin \omega_o t$ so the background magnetic field (either vertical or horizontal) has a time-dependent amplitude and direction. Pick a frequency (ω_o) such that $2\pi/\omega_o$ is large relative to a typical convective turnover time (i.e., the depth of the domain divided by the average fluid velocity in the z-direction).

9. *A nonuniform background field*

   Modify your code by replacing a uniform vertical background field with one that varies linearly in x, i.e., with $B_o x/L\hat{z}$. Likewise, replace a uniform horizontal background field with one that varies linearly in z, i.e., with $B_o z/D\hat{x}$.

Chapter Twelve

Density Stratification

For studies of convection and/or gravity waves in atmospheres and interiors of stars and planets that span several density scale heights it is important to account for the effects of large variations in density with depth, i.e., *density stratification*. (Here we usually mean *continuously* stratified, i.e., no discontinuities in density with depth. An exception would be a localized phase transition within a planetary interior.) As mentioned in Section 1.2, when the modeled domain spans a density scale height or more the Boussinesq approximation (which we have employed up to this point) is not valid. A fully compressible model, however, would need to resolve sound waves, which, because the CFL condition (Section 4.3.4), would require much smaller numerical time steps compared to the CFL condition based on the fluid velocity. If these fast-propagating small-amplitude waves are not important to the problem of interest and the fluid flows are very subsonic, it would be beneficial to remove them and be able to use a much larger numerical time step. One way to do this would be to treat the nonlinear advection terms implicitly; however, since these terms couple all the modes this would be very expensive (in terms of computer memory and time). A better way, which we describe in this chapter, would be to use the "anelastic approximation." Here anelastic models for 2D cartesian box and 2D cylindrical annulus geometries are described, using entropy and pressure as working thermodynamic variables or using temperature and pressure, for both convectively unstable and stable regions.

An alternative approximation may be a better choice for some problems. For example, there is the liquid anelastic approximation (e.g., Jarvis & McKenzie, 1980; Anufriev et al., 2005), which sits between the Boussinesq and anelastic approximations and can be used for density-stratified liquids when the product of the thermal expansion coefficient and the fluid's average background temperature is much less than unity. This constraint is satisfied for mantle convection and arguably also for convection in the Earth's liquid core. In this approximation the background density varies with depth, as in the anelastic approximation, but the density perturbation is assumed to be a function of the temperature (and possibly a compositional) perturbation but independent of the pressure perturbation, as in the Boussinesq approximation. In addition, viscous (and ohmic) heating are included in the energy equation, as they are for the anelastic approximation.

Another alternative, called the "low Mach number" approximation or the "pseudo-incompressible" approximation (e.g., Majda & Sethian, 1985; Durran, 1989; Cook & Riley, 1996; Bell et al., 2004; Almgren et al., 2006a; Almgren et al., 2006b; Lin et al., 2006; Lessani & Papalexandris, 2006; Plourde et al., 2008; Almgren et al., 2008; Zingale et al., 2009), sits between the anelastic approximation and a fully compressible model and has been used for simulations of combustion and of stellar convection. As would be imagined from its name, it assumes the fluid velocity is small relative to the local speed of sound; so, like the anelastic approximation, it has the advantage of filtering out sound waves. Also like the anelastic approximation, the pressure perturbation is assumed small relative to the background pressure. However, unlike the anelastic approximation, the density, entropy, and temperature perturbations are not assumed small relative to their respective background values. This could be useful for simulations of gravity waves in a strongly stable thermal stratification or thermal convection in a viscous mantle. In addition, the equation of state and mass conservation are formulated in ways that differ from the anelastic method. For example, the density perturbation is obtained from the mass conservation equation instead of the equation of state and the equation of state involves the background pressure instead of the pressure perturbation.

Here we focus on the anelastic approximation. The first global 3D anelastic simulations of a stellar dynamo were published in the early 1980s (Glatzmaier, 1984); that model was then modified to simulate mantle convection (Glatzmaier, 1988), the geodynamo (Glatzmaier & Roberts, 1996a), and giant planet dynamos (Glatzmaier, 2005b).

12.1 ANELASTIC APPROXIMATION

The anelastic approximation to the equations of motion (Batchelor, 1953; Ogura & Phillips, 1962; Braginsky, 1964; Gough, 1969; Gilman & Glatzmaier, 1981; Ginet & Sudan, 1987; Lantz, 1992; Braginsky & Roberts, 1995; Lantz & Fan, 1999; Braginsky & Roberts, 2007; Jones, 2007; Berkoff et al., 2010; Jones et al., 2011) was formulated to avoid the severe CFL condition due to sound waves while including the effects of density stratification. Slightly different approaches were taken to derive the anelastic equations and the equations evolved from a nondissipative plane-parallel model to models in spherical geometry with viscous dissipation and magnetic fields. Like the Boussinesq approximation, the anelastic approximation filters out sound waves by neglecting the $\partial \rho / \partial t$ term in the mass conservation equation (1.1). This is justified via a formal scale analysis (e.g., Gough, 1969; Gilman & Glatzmaier, 1981) based on the assumption that the *Mach number* (M = fluid velocity / sound speed) is much less than unity and the assumption that the thermodynamic perturbations have amplitudes of order M^2 relative to their respective horizontally averaged values. These assumptions are usually well satisfied in the interiors of planets and stars. However, since the speed of sound is proportional the square root of the temperature (Eq. 12.8), vigorous convective flows in some atmospheres (where the temperature is relatively low) can approach or exceed the local sound speed; in these cases a fully compressible set of equations needs to be solved. On the other hand, the density in the interiors of our solar system terrestrial planets (their mantles and cores) typically varies by no more than about 20% from the bottom to the top boundaries; the Boussinesq approximation has usually been used to simulate these problems and is arguably valid. "Super-Earths," which are now being discovered around other stars, with masses on the order of ten times the

Earth's mass, likely have much larger variations of density through their interiors; the Boussinesq approximation would not be valid for these. The Boussinesq approximation is also not valid for global models that simulate the entire convecting region or even just a portion of the outer region of a star or giant planet, over which density varies by several orders of magnitude. An anelastic or fully compressible model is needed for such problems to adequately represent the huge expansion rising fluid parcels experience and contraction sinking parcels experience, which significantly affect the flow pattern and how it transports of heat, angular momentum, and kinetic energy.

12.1.1 Equation of State

There are several slightly different ways to formulate the anelastic equations; however, all are based on the assumptions that the fluid velocities are very subsonic and the thermodynamic perturbations are small relative to their horizontal means. Here we describe the thermodynamic variables as sums of their prescribed "reference state" values ($\bar{\rho}$, $\overline{T}$, and $\bar{p}$) and the perturbations (ρ, T, and p) relative to these reference state values such that $|\rho|/\bar{\rho}$, $|T|/\overline{T}$, and $|p|/\bar{p}$ are small. The horizontally average (mean) state, which is a function of the vertical coordinate and time, is the sum of the horizontally independent time-independent reference state and the horizontally averaged time-dependent thermodynamic perturbations. An anelastic simulation will not numerically "blow up" if these anelastic assumptions are violated; therefore, the validity of an anelastic simulation always needs to be checked by comparing the resulting fluid velocities with the local sound speed and comparing the resulting thermodynamic perturbations with their respective reference state values.

The reference state thermodynamic variables are functions of only the vertical coordinate (z or r) and are in hydrostatic equilibrium:

$$\frac{d\bar{p}}{dz} = -\bar{g}\bar{\rho} \,. \tag{12.1}$$

We choose these variables to have an isentropic, or nearly isentropic, stratification in the *convection zone*,

$$\frac{d\overline{S}}{dz} \approx 0 \,, \tag{12.2}$$

where $\overline{S}$ is the specific entropy of the reference state. Note, since local thermodynamic equilibrium (LTE) is usually a very good approximation, processes are assumed reversible; therefore, we use the terms isentropic (i.e., constant entropy) and adiabatic (i.e., no heat transfer) interchangeably when referring to the temperature and density stratifications of an atmosphere in hydrostatic equilibrium in comparison to the change in temperature and density that an ideal test parcel would experience when vertically displaced within the atmosphere (Section 1.1.1). Unless the planet or star is exploding or imploding, it is in hydrostatic equilibrium to first order; convection tends to keep the thermal stratification very nearly isentropic (isothermal if Boussinesq, as illustrated in Fig. 4.2). A subadiabatic thermal stratification is chosen for a stably stratified region.

Here we assume a *perfect gas* equation of state, which is a good approximation for the atmosphere and interior of a star and the outer region and atmosphere of a giant planet:

$$(\bar{p} + p) = R(\bar{\rho} + \rho)(\overline{T} + T),$$

where R is the gas constant divided by the mean mass per particle (in atomic mass units) for the gas, which is assumed a constant; that is, we assume the gas has a constant chemical mixture and ionization state. Note, $R = c_p - c_v$ and $\gamma = c_p/c_v$, where c_v and c_p are the specific heat capacities at constant volume and pressure, respectively. We also assume these are constants. The equation of state for the reference state is therefore

$$\bar{p} = R\bar{\rho}\overline{T} \tag{12.3}$$

and the linearized equation of state for the perturbations is

$$\frac{p}{\bar{p}} = \frac{\rho}{\bar{\rho}} + \frac{T}{\overline{T}}. \tag{12.4}$$

The perfect gas approximation assumes intermolecular forces are negligible and particle collisions are elastic. An *ideal gas* is like a perfect gas but allows molecular, atomic, and nuclear reactions, which change the number density of the gas particles (ions and free electrons); therefore, c_p, c_v, and R vary in space and time. Here, however, we assume a perfect gas.

The specific entropy for a perfect gas is defined, to within an arbitrary constant, S_o, as

$$(\overline{S} + S) = c_v \ln(\bar{p} + p) - c_p \ln(\bar{\rho} + \rho) + S_o.$$

Therefore,

$$\overline{S} = c_v \ln \bar{p} - c_p \ln \bar{\rho} + S_o \tag{12.5}$$

and the linearized version for the perturbations is

$$S = c_v \frac{p}{\bar{p}} - c_p \frac{\rho}{\bar{\rho}}. \tag{12.6}$$

Rearranging Eq. 12.6 gives the density perturbation in terms of the entropy and pressure perturbations, which is what is needed for the buoyancy term:

$$\rho = \frac{\bar{\rho}}{\gamma \bar{p}} p - \frac{\bar{\rho}}{c_p} S = \overline{\left(\frac{\partial \rho}{\partial p}\right)_S} p + \overline{\left(\frac{\partial \rho}{\partial S}\right)_p} S. \tag{12.7}$$

As mentioned, one of the anelastic assumptions is that the fluid velocity is very subsonic; to check this one needs to know the local sound speed. A sound wave is very nearly an adiabatic and reversible process because of the low amplitude and high frequency of the associated pressure, density, and temperature perturbations. Therefore, the isentropic (i.e., constant entropy) sound speed is a very good approximation for the actual sound speed. The reference state isentropic sound speed, c_s, which is usually just called the "adiabatic sound speed," can easily be derived for a perfect gas using Eqs. 12.3 and 12.5:

$$c_s \equiv \overline{\left(\frac{\partial p}{\partial \rho}\right)_S^{1/2}} = (\gamma R\overline{T})^{1/2} .$$ (12.8)

Note, c_s varies with position according to the square root of the reference state temperature. In the interior of the sun, for example, fluid velocities are very subsonic. However, in the photosphere fluid velocities are somewhat greater, due to the lower density, and the sound speed is significantly less, due to the lower temperature; so there the anelastic approximation is not valid.

As also mentioned (Eq. 12.2), we usually set $d\overline{S}/dz = 0$ in the convection zone. That is, the reference state temperature gradient is adiabatic and can be easily calculated by starting with Eq. 12.5, using Eq. 12.3 to write $\bar{\rho}$ in terms of $\bar{p}$ and $\overline{T}$, taking the z-derivative of this expression for $\overline{S}$, and using Eq. 12.1. This gives the reference state entropy gradient,

$$\frac{d\overline{S}}{dz} = \frac{c_p}{\overline{T}}\left[\frac{d\overline{T}}{dz} + \frac{\bar{g}}{c_p}\right],$$ (12.9)

which when set to zero gives the adiabatic temperature gradient for a perfect gas in hydrostatic equilibrium:

$$\left(\frac{d\overline{T}}{dz}\right)_{AD} = -\frac{\bar{g}}{c_p} .$$ (12.10)

Substituting this back into Eq. 12.9 shows that the reference state entropy gradient, in general for a hydrostatic perfect gas, is a measure of difference between the reference state and adiabatic temperature gradients:

$$\frac{d\overline{S}}{dz} = \frac{c_p}{\overline{T}}\left[\frac{d\overline{T}}{dz} - \left(\frac{d\overline{T}}{dz}\right)_{AD}\right] = c_p\left[\frac{d\ln\overline{T}}{dz} - \left(\frac{d\ln\overline{T}}{dz}\right)_{AD}\right].$$ (12.11)

In addition, manipulation of Eqs. 12.1, 12.3, and 12.5 shows that the difference between the reference state and adiabatic temperature gradients for a hydrostatic perfect gas can be written as

$$\left[\frac{d\overline{T}}{dz} - \left(\frac{d\overline{T}}{dz}\right)_{AD}\right] = \frac{1}{\gamma}\left[\frac{d\overline{T}}{dz} - (\gamma - 1)\overline{T}\frac{d\ln\bar{\rho}}{dz}\right],$$ (12.12)

which will be useful.

In a convectively *stable* region the actual temperature gradient is less steep (i.e., less negative) than the adiabatic temperature gradient and therefore $d\overline{S}/dz > 0$. In these regions fluid motions take the form of internal gravity waves (Chapter 6); and the Brunt-Väisälä frequency (Section 12.4.2) is proportional to the square root of the local entropy gradient:

$$N = \left(\frac{\bar{g}}{c_p}\frac{d\overline{S}}{dz}\right)^{1/2} .$$ (12.13)

Recall, when comparing Eqs. 6.5 and 12.13, that the temperature perturbation in a Boussinesq model should be interpreted as the temperature relative to the adiabat. Also, since buoyancy could occur due to compositional gradients in addition to temperature gradients, a more general expression is

$$N = \left(\bar{g} \left[\left(\frac{d \ln \bar{\rho}}{dz} \right)_{AD} - \frac{d \ln \bar{\rho}}{dz} \right] \right)^{1/2} .$$

Recall, however, that even when $N > 0$, double-diffusive convection can occur if there is a heavy constituent within the fluid that diffuses much more slowly than temperature (Chapter 7).

12.1.2 Mass Conservation

A formal scale analysis (e.g., Gilman & Glatzmaier, 1981) chooses a length scale, D, which could be the depth of the convection zone, and a fluid velocity scale that is a Mach number, M, times a typical $(\bar{p}/\bar{\rho})^{1/2}$, i.e., M times a characteristic speed of sound. The time scale is therefore the ratio of the length and velocity scales. Also, in anticipation of having to use turbulent (eddy) diffusivities instead of actual molecular values, the thermal and viscous diffusivities scale as the product of the velocity and length scales and so are order M times the sound speed times D. Then, assuming the pressure *perturbation* (p) scales like the Reynolds stress ($\bar{\rho}\mathbf{v}\mathbf{v}$), p and the other thermodynamic *perturbations* (because of Eq. 12.4) are of order M^2 times their respective reference state values.

As a result, the anelastic form of *mass conservation* (Eq. 1.1) to a scaled order of M is

$$\nabla \cdot \bar{\rho}\mathbf{v} = 0 . \tag{12.14}$$

Like the Boussinesq version (Eq. 1.11), it is linear (because $\bar{\rho}$ is known); but now the mass flux, $\bar{\rho}\mathbf{v}$, is divergence-free, not just the velocity. That is,

$$\nabla \cdot \mathbf{v} = -h_\rho v_z , \tag{12.15}$$

where

$$h_\rho \equiv \frac{1}{\bar{\rho}} \frac{d\bar{\rho}}{dz} = \frac{d \ln \bar{\rho}}{dz} \tag{12.16}$$

is the negative inverse of the local reference state density scale height.

The neglect of $\partial \rho / \partial t$ in Eq. 12.14, which is arguably smaller by a factor of M^2 than the individual components of $\nabla \cdot \bar{\rho}\mathbf{v}$, is what eliminates the pressure wave equation, and therefore sound waves, by severing the direct connection between $\partial^2 \rho / \partial t^2$ and $\nabla^2 p$, which would otherwise exist when equating the time derivative of Eq. 1.1 and the divergence of Eq. 1.2. Consequently, as in a Boussinesq model, all pressure perturbations that occur during a time step are immediately known everywhere within the domain and have adjusted, instead of requiring a finite amount of time for them to travel (via sound waves) throughout the domain as would occur in a fully compressible model. Also, as in a Boussinesq model, the density perturbations are obtained each time step via an equation of state (Eq. 12.7) instead of via mass conservation, as they would be in a fully compressible model.

12.1.3 Momentum Conservation with Entropy as a Variable

The anelastic form of *momentum conservation* (Eq. 1.2), to a scaled order of M^2 after subtracting out the reference state hydrostatic equilibrium (Eq. 12.1), is

$$\bar{\rho}\frac{\partial \mathbf{v}}{\partial t} = -\bar{\rho}(\mathbf{v}\cdot\boldsymbol{\nabla})\mathbf{v} - \boldsymbol{\nabla} p - \bar{\rho}\boldsymbol{\nabla}\Phi - \rho\boldsymbol{\nabla}\overline{\Phi}$$
$$+ \boldsymbol{\nabla}\cdot\left[2\bar{\rho}\bar{\nu}\left(e_{ij} - \frac{1}{3}(\boldsymbol{\nabla}\cdot\mathbf{v})\,\delta_{i,j}\right)\right]. \tag{12.17}$$

Recall that (Eq. 1.5) the viscous force density, the last term on the right of Eq. 12.17, would reduce to

$$\bar{\rho}\bar{\nu}\left(\nabla^2\mathbf{v} + \frac{1}{3}\boldsymbol{\nabla}(\boldsymbol{\nabla}\cdot\mathbf{v})\right)$$

if the dynamic viscosity $\bar{\rho}\bar{\nu}$ were constant in space.

The gravitational potential energy per mass is $(\overline{\Phi}+\Phi)$, where the reference state gravitational acceleration is

$$\bar{g} = -\bar{g}\hat{z} = -\boldsymbol{\nabla}\overline{\Phi} \tag{12.18}$$

and Φ is the perturbation in the gravitational potential energy per mass. This perturbation, Φ, is due to density perturbations throughout the domain and could be calculated by solving

$$\nabla^2\Phi = 4\pi G\rho \tag{12.19}$$

after updating the density perturbation and setting $\nabla^2\Phi$ to zero at the boundaries, assuming no external density perturbations.

The "self-gravity" term $-\bar{\rho}\boldsymbol{\nabla}\Phi$ in Eq. 12.17 is seldom included in thermal convection simulations because it is usually quite small when most of the mass of the body is below the simulated domain and the density perturbations are small. However, when the simulated domain extends down close to the center of the star or planet or when it includes the center (where $\bar{g}$ vanishes) this term can be relatively significant and should not be neglected. It is included here, which, as is shown next, requires no extra work unless one needs to calculate the actual pressure perturbation.

Consider the working thermodynamic variables to be the perturbations in the pressure, p, and entropy, S. Then an equation of state (Eq. 12.7) is needed to write the density perturbation, ρ, in Eq. 12.17 in terms of p and S. Now, unlike a Boussinesq model, p appears in both the pressure gradient term and the buoyancy term. To avoid having to solve for this pressure perturbation (as we have avoided solving for it in our Boussinesq models) Lantz (1992) and Braginsky & Roberts (1995) independently showed, with no additional approximation other than those already made for the anelastic equations with an *isentropic* reference state, that the sum of the pressure gradient and gravity terms in Eq. 12.17,

$$-\boldsymbol{\nabla} p - \bar{\rho}\boldsymbol{\nabla}\Phi - \rho\boldsymbol{\nabla}\overline{\Phi}, \tag{12.20}$$

can be simplified. Write the first two terms in Eq. 12.20 as

$$-\nabla p - \bar{\rho}\nabla\Phi = -\bar{\rho}\nabla\left(\frac{p}{\bar{\rho}} + \Phi\right) - \frac{p}{\bar{\rho}}\nabla\bar{\rho}. \qquad (12.21)$$

Then, assuming a reference state that is isentropic (Eq. 12.2) and in hydrostatic equilibrium (Eq. 12.1), write the gradient of $\bar{\rho}$ as

$$\nabla\bar{\rho} = \frac{d\bar{\rho}}{dz}\hat{z} = \overline{\left(\frac{\partial\rho}{\partial p}\right)_s}\frac{d\bar{p}}{dz}\hat{z} = -\overline{\left(\frac{\partial\rho}{\partial p}\right)_s}\bar{\rho}\bar{g}\hat{z}. \qquad (12.22)$$

Substituting Eq. 12.22 into the last term on the right in Eq. 12.21 gives

$$-\nabla p - \bar{\rho}\nabla\Phi = -\bar{\rho}\nabla\left(\frac{p}{\bar{\rho}} + \Phi\right) + p\overline{\left(\frac{\partial\rho}{\partial p}\right)_s}\bar{g}\hat{z}. \qquad (12.23)$$

Next, using the reference state gravitational acceleration (Eqs. 12.18) and our linear perturbation equation of state (Eqs. 12.7), write the other term in Eq. 12.20 as

$$-\rho\nabla\overline{\Phi} = -\rho\bar{g}\hat{z} = -\left[\overline{\left(\frac{\partial\rho}{\partial p}\right)_s}p + \overline{\left(\frac{\partial\rho}{\partial S}\right)_p}S\right]\bar{g}\hat{z}. \qquad (12.24)$$

Finally, adding Eqs. 12.23 and 12.24 and defining the "reduced pressure" as

$$P \equiv \frac{p}{\bar{\rho}} + \Phi \qquad (12.25)$$

gives

$$-\nabla p - \bar{\rho}\nabla\Phi - \rho\nabla\overline{\Phi} = -\bar{\rho}\nabla P - \overline{\left(\frac{\partial\rho}{\partial S}\right)_p}\bar{g}S\hat{z}. \qquad (12.26)$$

The critical result here is that the part of the buoyancy term due to the pressure perturbation has been absorbed by the pressure gradient term, leaving just the (nonpotential) buoyancy force due to the entropy perturbation, which Braginsky & Roberts (1995) call the "co-density":

$$C \equiv \frac{1}{\bar{\rho}}\overline{\left(\frac{\partial\rho}{\partial S}\right)_p}S = -\frac{S}{c_p}.$$

Substituting the *Lantz-Braginsky-Roberts* simplification (Eq. 12.26) into the anelastic momentum equation (12.17), assuming a constant $\bar{\rho}\bar{\nu}$, and dividing the equation by $\bar{\rho}$ transforms the anelastic momentum equation into

$$\frac{\partial\mathbf{v}}{\partial t} = -(\mathbf{v}\cdot\nabla)\mathbf{v} - \nabla P - \overline{\left(\frac{\partial\rho}{\partial S}\right)_p}\frac{\bar{g}}{\bar{\rho}}S\hat{z} + \bar{\nu}\left(\nabla^2\mathbf{v} + \frac{1}{3}\nabla(\nabla\cdot\mathbf{v})\right), \qquad (12.27)$$

which makes it look very similar to the Boussinesq momentum equation (1.13) when p/ρ_o in that equation is replaced with P and αT is replaced with S/c_p. Note, for a perfect gas, the reference state thermal expansion coefficient, α, is $1/\bar{T}$ and

$$\overline{\left(\frac{\partial\rho}{\partial S}\right)_p} = -\frac{\bar{\rho}}{c_p}.$$

The Lantz-Braginsky-Roberts formulation (Eq. 12.26) was derived assuming an isentropic reference state. If instead one wanted to simulate a domain (or part of a domain) that is convectively stable, Eq. 12.22 would need to be modified to account for $d\overline{S}/dz > 0$ (Rogers & Glatzmaier, 2005a). The reference state density gradient would now be a function of both the pressure and entropy gradients:

$$
\nabla\bar{\rho} = \left[-\overline{\left(\frac{\partial\rho}{\partial p}\right)_S}\bar{\rho}\bar{g} + \overline{\left(\frac{\partial\rho}{\partial S}\right)_p}\frac{d\overline{S}}{dz}\right]\hat{z}. \tag{12.28}
$$

Using Eq. 12.28 in place of 12.22 gives a more general anelastic momentum equation (again assuming a constant $\bar{\rho}\bar{v}$):

$$
\frac{\partial\mathbf{v}}{\partial t} = -(\mathbf{v}\cdot\nabla)\mathbf{v} - \nabla P + \frac{1}{c_p}\left(\bar{g}S + \frac{1}{\bar{\rho}}\frac{d\overline{S}}{dz}p\right)\hat{z} + \bar{v}\left(\nabla^2\mathbf{v} + \frac{1}{3}\nabla(\nabla\cdot\mathbf{v})\right). \tag{12.29}
$$

However, if the reference state were nearly adiabatic, i.e., $d\overline{S}/dz \approx M^2 c_p/D$, the $1/(c_p\bar{\rho})d\overline{S}/dz\ p$ part of the buoyancy term in Eq. 12.29 could be dropped within the anelastic approximation. Moreover, as we show in Section 12.4.2 and was first pointed out by Brown et al. (2012), this additional pressure term could cause problems with energy conservation within the anelastic approximation; therefore, dropping this term, even where the reference state is not nearly adiabatic, would probably be a wise choice.

The assumption of a constant dynamic viscosity, $\bar{\rho}\bar{v}$, in Eqs. 1.5, 12.27, and 12.29 is not necessary. When this assumption is made, $\bar{v}$ varies in radius inversely with $\bar{\rho}$. Besides the convenience of a somewhat simpler viscous force density, there is a physical argument for making this assumption when the viscous diffusivity, $\bar{v}$, is taken to be a parameterization of subgrid-scale turbulent eddies. The mixing done by these eddies may be more effective in the upper less-dense part of a density-stratified domain since the smallest resolved eddies there tend to have more kinetic energy than they do at greater depths. That is, the amplitude of the fluid velocity tends to increase with decreasing density and the size of the dominant eddies tends to decrease with the density scale height, (i.e., also with radius). If, however, one prescribes a depth-dependent dynamic viscosity, the viscous term should include derivatives of $\bar{\rho}\bar{v}$ (Eqs. 1.2, 1.3). This is not a concern for our Boussinesq models, for which both $\bar{\rho}$ and $\bar{v}$ are constant in space. Note that the anelastic momentum equation also has a viscous term proportional to $\nabla\cdot\mathbf{v}$, which vanishes for the Boussinesq approximation.

12.1.4 Internal Energy Conservation with Entropy as a Variable

Now consider the anelastic form of *internal energy conservation* (Eq. 1.6), correct to order M^5. Note that

$$
(\bar{\rho}+\rho)(\overline{T}+T)\frac{d}{dt}(\overline{S}+S) \approx \bar{\rho}\overline{T}\frac{dS}{dt} + \bar{\rho}\overline{T}v_z\frac{d\overline{S}}{dz} + (\rho\overline{T}+\bar{\rho}T)v_z\frac{d\overline{S}}{dz}, \tag{12.30}
$$

where we continue to assume $\partial\overline{S}/\partial t = 0$. If $d\overline{S}/dz$ were set to zero in the *convection zone*, only the first term on the right side of this expression, which has a scaled amplitude of order M^3, would survive there. The second term would also survive if $d\overline{S}/dz$ were prescribed as negative in the convection zone with a scaled amplitude of order M^2. However, then the amplitude of the third term on the right should be of order M^2 smaller than that of the second term; and so we neglect this third term.

On the other hand, $d\overline{S}/dz$ can be quite large (and positive) in a *convectively stable region* below or above a stellar convection zone compared to its amplitude in the convection zone because of the efficiency of radiative heat transport. Although gravity waves, via the associated thermal advection and diffusion, can alter the mean temperature gradient, the effect is usually significant only near the interface between the convection zone and the stable region where they are excited by overshooting convective plumes. One exception, however, is the role of gravity waves in double-diffusive convection (e.g., Stellmach et al., 2011).

If $d\overline{S}/dz$ were as large as c_p/D the Brunt-Väisälä frequency (Eq. 12.13) would be of order $(\bar{g}/D)^{1/2}$, roughly the inverse of the free-fall time through the depth D. The corresponding maximum gravity wave speed (Eq. 6.9 and 6.10) would then be of order $(\bar{g}D)^{1/2}$ (for a horizontally directed phase velocity with a wavelength of $2\pi D$). For comparison, the sound speed (using Eqs. 12.1 and 12.16) is

$$\overline{\left(\frac{\partial p}{\partial \rho}\right)}^{-1/2} = \left(\frac{d\bar{p}}{dz}\bigg/\frac{d\bar{\rho}}{dz}\right)^{1/2} = \left(\frac{\bar{g}}{|h_\rho|}\right)^{1/2}.$$

Therefore, if $d\overline{S}/dz$ were of order c_p/D and if the local density scale height, $1/|h_\rho|$, were not significantly greater than the depth of the stable region, D, the maximum gravity wave speed could be of order of the sound speed; in such a case the anelastic approximation would not be valid (Lantz & Fan, 1999).

The amplitude of $d\overline{S}/dz$ could therefore be limited for anelastic simulations; certainly the resulting wave velocities in a stable region (in addition to fluid velocities in the convection zone) need to be compared to the local sound speed. Internal gravity waves, however, tend to have frequencies smaller than N (due to the penetration angle of overshooting convective plumes) and wavelengths much smaller than D (because of the small horizontal scale of these plumes) and therefore have much smaller wave velocities than the maximum estimated here. In addition, the density scale height within a stable region far below the surface is typically much larger than the radius of the star. Therefore, $d\overline{S}/dz$ could conceivably be of order c_p/D deep within a stable region without the gravity wave speed or the fluid velocity approaching the sound speed. However, the third term on the right in Eq. 12.30 would still be of order M^2 smaller than the second term and therefore should still be neglected.

Note, here we have been considering the amplitude of the *reference state* entropy gradient within a stable region. The anelastic approximation also assumes that the *entropy perturbation*, S, is of order $c_p M^2$ (Eq. 12.6); this could be violated in a thermal boundary layer within the *convection zone*, where $d\overline{S}/dz$ is set to zero but $|\partial S/\partial z|$ is large. One could also define $d\overline{S}/dz$ to be zero within the *stable region* and attribute the change in entropy across the stable region to the entropy perturbation, which, if comparable to c_p, would violate the above mentioned anelastic assumption. The bottom line is that the amplitudes of the resulting fluid velocities and thermodynamic perturbations need to be checked for each anelastic simulation to determine the validity of the anelastic approximation for the given simulation.

The anelastic version of the thermal diffusion part of the internal energy conservation equation (1.6) also needs some discussion. This heating term represents the convergence of diffusive heat flux, which is partly the usual radiative (or molecular) heat flux $(-c_p \bar{\kappa}_R \bar{\rho} \nabla(\overline{T} + T))$ driven by a temperature gradient, and an additional heat flux $(-\bar{\kappa} \bar{\rho} \overline{T} \nabla(\overline{S} + S))$ driven by an entropy gradient. This second contribution is commonly added (e.g., Glatzmaier, 1984) because for global simulations of thermal convection in stars and planets one cannot afford the spatial resolution needed to resolve all length scales down to the viscous dissipation scale. Instead, a common practice is to truncate the numerical resolution at a much larger scale and dissipate the energy that cascades down to this scale by prescribing an enhanced diffusivity, $\bar{\kappa}$. This "turbulent" or "eddy" diffusivity is a crude representation of the transport and cascade of energy by the unresolved "subgrid-scale" convective eddies. Since the actual eddies are driven by an entropy gradient (similar to the idea behind *mixing length* theory in stellar structure models), it is reasonable to make their parameterized diffusive heat flux be proportional to the local entropy gradient. The convergence of this turbulent diffusive heat flux tries to make the entropy constant in space, as does turbulent convection. On the other hand, the convergence of the usual diffusive heat flux (proportional to the temperature gradient) tries to make the temperature constant in space. (Note, the viscous diffusivity, $\bar{v}$, likewise represents a turbulent transport of momentum by subgrid-scale eddies.)

It could be argued that the T part of this thermal diffusion can be neglected since its magnitude is so much less than $\overline{T}$; however, technically this should be checked since the spatial derivatives of $\overline{T}$ and T are being compared here. Keeping the T part would require calculating the pressure perturbation each time step to write T in terms of S and p. To simplify our model, we will drop the terms involving temperature perturbation, T.

The anelastic internal energy conservation equation also includes viscous heating and, if magnetic, also ohmic heating:

$$Q = 2\bar{\rho}\bar{v}\left(e_{ij}e_{ij} - 1/3(\nabla\cdot\mathbf{v})^2\right) + |\mathbf{J}|^2/\bar{\sigma}\,, \tag{12.31}$$

where $e_{i,j}$ is the rate of strain tensor (Eq. 1.4) and $\mathbf{J}$ is the electric current density (Section 11.1). The viscous heating part of Q can also be written, for our 2D cartesian box, as

$$2\bar{\rho}\bar{v}\left[\left(\frac{\partial v_x}{\partial x}\right)^2 + \left(\frac{\partial v_z}{\partial z}\right)^2 + \frac{1}{2}\left(\frac{\partial v_x}{\partial z} + \frac{\partial v_z}{\partial x}\right)^2 - \frac{h_\rho^2}{3}v_z^2\right]\,. \tag{12.32}$$

The heating terms in Eq. 12.31 can be computed each time step with the other nonlinear terms. In addition, there may be a reference state heating term representing, for example, nuclear burning or radioactive decay or gravitational contraction or the convergence (heating) or divergence (cooling) of radiative heat flux (Section 6.2.2). We combine all purely reference state heating or cooling terms into one prescribed function:

$$\overline{Q} = \nabla\cdot(\bar{\kappa}\bar{\rho}\overline{T}\nabla\overline{S} + c_p\bar{\kappa}_R\bar{\rho}\nabla\overline{T}) + \cdots\,. \tag{12.33}$$

Therefore, our anelastic internal energy equation is

$$\bar{\rho}\overline{T}\frac{\partial S}{\partial t} = -\bar{\rho}\overline{T}\mathbf{v}\cdot\nabla S - \bar{\rho}\overline{T}v_z\frac{d\overline{S}}{dz} + \nabla\cdot\left(\bar{\kappa}\bar{\rho}\overline{T}\nabla S\right) + Q + \overline{Q}. \tag{12.34}$$

Now consider again a convectively *stable* (subadiabatic) reference state. Such a region can be prescribed by defining $\overline{S}$ as an increasing function of z. If the domain also has a convective region, both $\overline{S}$ and $d\overline{S}/dz$ need to be continuous at the interface between the convective region, where $d\overline{S}/dz$ is zero or slightly negative, and the subadiabatic region, where $d\overline{S}/dz > 0$. The entropy perturbation, $S(x, z, t)$, is relative to this $\overline{S}(z)$ in both regions and the magnitude of this perturbation must be much less than c_p (of order $M^2 c_p$) for the anelastic approximation to be valid because of Eq. 12.6 (Lantz & Fan, 1999; Berkoff et al., 2010).

Another consideration is that within a convectively stable (subadiabatic) region the eddy diffusive heat flux ($-\bar{\kappa}\bar{\rho}\overline{T}\,d\overline{S}/dz$) is directed downward. However, the degree of turbulence there would likely be significantly less than it is in a convection zone and so $\bar{\kappa}$ should be significantly less in a stable region. For a stable region below a convection zone, $\bar{\kappa}$ should be prescribed to decrease by orders of magnitude within a short distance below the interface between the stable and unstable zones. The large-scale (resolved) convective heat flux at the top of the stable region is also on average downward directed because convective plumes overshoot into the stable region nearly adiabatically. This makes them become hotter than the surrounding subadiabatic environment; therefore, when the overshooting plumes deposit their heat in the upper part of the stable region they steepen the temperature gradient there toward an adiabatic profile. Ideally the z-dependence of $\bar{\kappa}$ should evolve as a function of the vigor and extent of the resolved overshooting convection. In any case, however, the reference state radiative (or molecular) heat flux ($-c_p\bar{\kappa}_R\bar{\rho}d\overline{T}/dz$) in a stable region, which is upward directed, needs to be large enough to more than cancel the downward eddy and convective heat fluxes so a net upward heat flux is maintained.

12.1.5 Temperature as a Variable

Instead of choosing the entropy perturbation, S, to be one of the working thermodynamic variables, one could choose, say, the temperature perturbation, T. To do this we start with the conservation of internal energy equation (1.6) but now work with the internal energy and pressure work instead of the entropy. For a perfect gas, the rate of change of the internal energy density is (again writing each thermodynamic variable as the sum of its reference state value and its perturbation)

$$(\bar{\rho} + \rho)\frac{d}{dt}(\bar{e} + e) = (\bar{\rho} + \rho)c_v\frac{d}{dt}(\overline{T} + T) \tag{12.35}$$

and for an anelastic fluid the rate pressure does work per volume is (using Eq. 12.15)

$$(\bar{p} + p)\nabla\cdot\mathbf{v} = -(\bar{p} + p)h_\rho v_z. \tag{12.36}$$

Also, since we are usually simulating a turbulent fluid and using an eddy diffusivity, we again describe the diffusive heat flux as being partly the traditional conductive heat flux driven by the reference state temperature gradient and partly a turbulent subgrid-scale heat flux driven by the entropy gradient, i.e., the difference between the temperature gradient and the adiabatic temperature gradient. Therefore, we replace

$$-\bar{k}\nabla(\overline{T} + T)$$

with

$$-c_p\bar{\kappa}_R\bar{\rho}\frac{d\overline{T}}{dz}\hat{z} - c_p\bar{\kappa}\bar{\rho}\left(\nabla(\overline{T} + T) - \left(\frac{d\overline{T}}{dz}\right)_{AD}\hat{z}\right),$$

where again $\bar{\kappa}_R$ is the radiative (or molecular) thermal diffusivity and $\bar{\kappa}$ is the (much larger) turbulent thermal diffusivity. This makes the thermal diffusion term be

$$\nabla\cdot(c_p\bar{\kappa}\bar{\rho}\nabla T) + \nabla\cdot\left(c_p\bar{\kappa}\bar{\rho}\left[\frac{d\overline{T}}{dz} - \left(\frac{d\overline{T}}{dz}\right)_{AD}\right]\hat{z}\right) + \nabla\cdot\left(c_p\bar{\kappa}_R\bar{\rho}\frac{d\overline{T}}{dz}\hat{z}\right) . \quad (12.37)$$

The second term in expression 12.37 either vanishes or is scaled to order M^3 within the (adiabatic) convection zone. In a convectively stable region this term represents the convergence of the downward turbulent heat flux $(-c_p\bar{\kappa}\bar{\rho}[d\overline{T}/dz - (d\overline{T}/dz)_{AD}]\hat{z})$ as does the convergence of $(-\bar{\kappa}\bar{\rho}\overline{T}d\overline{S}/dz\,\hat{z})$ in Eq. 12.34 (see also Eq. 12.11). Setting expressions 12.35 plus 12.36 equal to expressions 12.37 plus 12.31 gives the energy equation according to Eq. 1.6. When expanding the terms in this energy equation, using Eqs. 12.9–12.12 and noting that for a perfect gas in hydrostatic equilibrium

$$(h_\rho\bar{p})/(c_v\bar{\rho}) = (\gamma - 1)h_\rho\overline{T} ,$$

two of the terms that independently are of order M are of order M^3 when combined:

$$\left(c_v\frac{d\overline{T}}{dz}\bar{\rho} - h_\rho\bar{p}\right)v_z = c_v\bar{\rho}\left[\frac{d\overline{T}}{dz} - (\gamma - 1)h_\rho\overline{T}\right]v_z$$

$$= c_p\bar{\rho}\left[\frac{d\overline{T}}{dz} - \left(\frac{d\overline{T}}{dz}\right)_{AD}\right]v_z = \bar{\rho}\overline{T}\frac{d\overline{S}}{dz}v_z .$$

Also, using Eqs. 12.3 and 12.4, the terms with the perturbation density and pressure are

$$\left(c_v\frac{d\overline{T}}{dz}\rho - h_\rho p\right)v_z = -c_v\bar{\rho}\left((\gamma - 1)h_\rho T - \gamma\left[\frac{d\overline{T}}{dz} - \left(\frac{d\overline{T}}{dz}\right)_{AD}\right]\frac{\rho}{\bar{\rho}}\right)v_z .$$

The term proportional to $\rho/\bar{\rho}$ is order M^2 smaller than the others and so is dropped.

Using these anelastic perfect-gas expressions in the conservation of internal energy equation gives us an equation for updating the temperature perturbation:

$$\frac{\partial T}{\partial t} = -\mathbf{v}\cdot\nabla T + (\gamma - 1)h_\rho Tv_z - \gamma\left[\frac{d\overline{T}}{dz} - \left(\frac{d\overline{T}}{dz}\right)_{AD}\right]v_z$$

$$+ \frac{1}{c_v\bar{\rho}}\nabla\cdot(c_p\bar{\kappa}\bar{\rho}\nabla T) + \frac{\overline{Q} + Q}{c_v\bar{\rho}} . \quad (12.38)$$

The viscous and ohmic heating, Q, is defined in Eq. 12.31 and, as we do for the entropy version of the energy equation (Eq. 12.34), we combine all the purely reference state terms into

$$\overline{Q} = \nabla \cdot \left(c_p \bar{\kappa} \bar{\rho} \left[\frac{d\overline{T}}{dz} - \left(\frac{d\overline{T}}{dz} \right)_{AD} \right] \hat{z} \right) + \nabla \cdot \left(c_p \bar{\kappa}_R \bar{\rho} \frac{d\overline{T}}{dz} \right) + \cdots .$$

Also note that, using Eq. 12.15, the first two advection terms on the right in Eq. 12.38 can be written as

$$-\mathbf{v} \cdot \nabla T + (\gamma - 1) h_\rho T v_z = -\nabla \cdot (T\mathbf{v}) + (\gamma - 2) h_\rho T v_z ,$$

which is usually more accurate for numerical calculations.

Recall that $[d\overline{T}/dz - (d\overline{T}/dz)_{AD}]$ is set either to zero or to something slightly negative in the convection zone. As discussed above, in a convectively stable region it is positive and could have a larger magnitude; but, where it is large, v_z and $\bar{\kappa}$ are small.

Switching the working thermodynamic variable from S to T also requires new thermodynamic derivatives for the perturbation equation of state. The density perturbation in terms of T and p, using Eq. 12.4, is now

$$\rho = \frac{\bar{\rho}}{\bar{p}} p - \frac{\bar{\rho}}{\overline{T}} T = \overline{\left(\frac{\partial \rho}{\partial p} \right)_T} p + \overline{\left(\frac{\partial \rho}{\partial T} \right)_p} T \qquad (12.39)$$

and the entropy perturbation using Eq. 12.6 is

$$S = -\frac{R}{\bar{p}} p + \frac{c_p}{\overline{T}} T = \overline{\left(\frac{\partial S}{\partial p} \right)_T} p + \overline{\left(\frac{\partial S}{\partial T} \right)_p} T . \qquad (12.40)$$

In addition, if using the Lantz-Braginsky-Roberts formulation, modified for a reference state that is not necessarily adiabatic (Rogers & Glatzmaier, 2005a), the buoyancy term in Eq. 12.29 written in terms of T and p (using Eq. 12.11) would be

$$\frac{1}{c_p} \left(\bar{g} S + \frac{1}{\bar{\rho}} \frac{d\overline{S}}{dz} p \right) = \left[\left(\frac{T}{\overline{T}} - \frac{\gamma - 1}{\gamma} \frac{p}{\bar{p}} \right) \bar{g} + R \left[\frac{d\overline{T}}{dz} - \left(\frac{d\overline{T}}{dz} \right)_{AD} \right] \frac{p}{\bar{p}} \right]$$

$$= \left(T - \overline{\left(\frac{\partial T}{\partial p} \right)} p \right) \frac{\bar{g}}{\overline{T}} = -C\bar{g} , \qquad (12.41)$$

where again C is the co-density and

$$\overline{\left(\frac{\partial T}{\partial p} \right)} \equiv \frac{d\overline{T}}{dz} \left(\frac{d\bar{p}}{dz} \right)^{-1} = -\frac{d\overline{T}}{dz} \frac{1}{\bar{\rho}\bar{g}} , \qquad (12.42)$$

which equals $(\gamma - 1)\overline{T}/(\gamma \bar{p})$ in a convection zone with an adiabatic reference state. Therefore, the anelastic momentum equation (for constant $\bar{\rho}\bar{\nu}$) is

$$\frac{\partial \mathbf{v}}{\partial t} = -(\mathbf{v} \cdot \nabla)\mathbf{v} - \nabla P + \left(T - \overline{\left(\frac{\partial T}{\partial p} \right)} p \right) \frac{\bar{g}}{\overline{T}} \hat{z} + \bar{\nu} \left(\nabla^2 \mathbf{v} + \frac{1}{3} \nabla(\nabla \cdot \mathbf{v}) \right) . \qquad (12.43)$$

However, as mentioned for Eq. 12.29 and discussed in Section 12.4.2, dropping the extra contribution to buoyancy due to the pressure perturbation in Eq. 12.43, i.e., the $\overline{(\partial T/\partial p)}\ p$ term, improves energy conservation (Brown et al., 2012).

Other modifications, in addition to adding ohmic heating (Eq. 12.31), would be needed if one were simulating magnetoconvection (Chapter 11). The Lorentz force per mass, $\mathbf{J} \times \mathbf{B}/\bar{\rho}$, would need to be added to the right sides of Eqs. 12.29 and 12.43. Ahead, where we take the curl of the momentum equation to obtain a vorticity equation, the dependence of $\bar{\rho}$ on z or r produces an additional magnetic term that does not exist in the Boussinesq magnetoconvection models. In addition, the magnetic diffusivity, $\bar{\eta}$, which we assumed to be constant in our Boussinesq models, may need to depend on z or r. For example, the magnetic diffusivity, which is inversely proportional to the electrical conductivity (Section 11.1), decreases by several orders of magnetic with depth in the outer region of a gaseous giant planet (e.g., Jupiter or Saturn) as pressure ionization increases the number density of free electrons. In such a study one would need to account for this significant radial dependence of $\bar{\eta}$ in the magnetic diffusion term of Eq. 11.13 and likewise the radial dependence of the electrical conductivity, $\bar{\sigma}$, in the ohmic heating term of Eq. 12.31.

In summary, within the anelastic approximation, the equation of state is described by Eqs. 12.4 and 12.6 and mass conservation by Eq. 12.14. Internal energy conservation is Eq. 12.34 if using S as a variable and 12.38 if using T. Momentum conservation is Eq. 12.29 if using S as a variable and 12.43 if using T; however, as suggested, one could drop the extra pressure perturbation contribution to codensity in these two equations to improve energy conservation when employing the Lantz-Braginsky-Roberts formulation of the anelastic approximation. This section considered fluid in a cartesian box; but of course the equations apply as well for an annulus or sphere with a central gravity.

12.2 REFERENCE STATE: POLYTROPES

Before describing modifications to the numerical method that are required, we describe possible choices for a reference state. A convenient choice is a polytrope, i.e., one for which pressure and density satisfy hydrostatic equilibrium (Eq. 12.1) and pressure is proportional to density to the power of $(n + 1)/n$:

$$\bar{p} = p_o \left(\frac{\bar{\rho}}{\rho_o} \right)^{(n+1)/n} , \tag{12.44}$$

where here n is the polytropic index (not the horizontal mode number). The motivation for this is that the density profiles for several types of stellar and planetary interiors can be approximated with such a simple relationship when coupled with hydrostatic equilibrium (Clayton, 1984). For example, the simplest case, an $n = 0$ polytrope, is a constant-density (incompressible liquid) body (Eq. 12.46), which roughly approximates the mantles and cores of terrestrial planets. An $n = 3/2$ polytrope with the perfect gas equation of state has an adiabatic thermal stratification, which is appropriate for a convection zone. An $n = 3/2$ polytrope also approximates an electron degenerate nonrelativistic gas, i.e., an ionized gas that derives most of its pressure from free electrons that are in their lowest energy states (and is not a strong function of temperature); this, for example, is appropriate for the interiors of white dwarfs. An $n = 3$ polytrope approximates stellar interiors that are dominated by radiation pressure (e.g., the radiative interiors of main sequence stars like the sun) or by a relativistic degenerate electron gas (e.g., a white dwarf with mass just below the Chandrasekhar limit, 1.46 solar masses).

One-dimensional evolutionary models predict giant planets to be convecting throughout their fluid interiors or in some regions possibly undergoing *semiconvection* (Chapter 7). However, the fluid within a giant planet transforms slowly with depth from a gas in the outer regions to a liquid in the deep interior where it is partially electron degenerate. Therefore, although the interior of a giant planet may be nearly adiabatic throughout, an $n = 1$ polytrope (instead of $n = 3/2$) gives the best overall fit for the interior of a gas giant like Jupiter (Hubbard, 1975).

12.2.1 2D Fluid Box

Consider first a cartesian box with a constant gravitational acceleration, $-g_o \hat{z}$. The polytropic relationship between $\bar{p}$ and $\bar{\rho}$ (Eq. 12.44) is usually written as

$$\bar{p}(z) = p_o \Theta^{n+1}(z), \tag{12.45}$$

$$\bar{\rho}(z) = \rho_o \Theta^{n}(z), \tag{12.46}$$

where p_o and ρ_o are the reference state pressure and density at the bottom boundary ($z = 0$), respectively, and $\Theta(z)$ is the nondimensional polytropic function, which is unity at $z = 0$. Substituting Eqs. 12.45 and 12.46 into Eq. 12.1 gives

$$\Theta(z) = 1 - \frac{z}{z_o},$$

where z_o is

$$z_o = \frac{(n + 1) p_o}{g_o \rho_o}; \tag{12.47}$$

z_o is greater than the depth of the box, D. The constants, n and z_o determine the z-dependent density scale height, $-h_\rho^{-1}$ (Eq. 12.16), where for this polytrope

$$h_\rho = \frac{d \ln(\Theta^n)}{dz} = -\frac{n}{z_o - z}.$$

The number of density scale heights from the bottom boundary ($z = 0$) to the top boundary ($z = D$) is

$$N_\rho = -\int_0^D h_\rho dz = n \ln\left(\frac{z_o}{z_o - D}\right).$$

That is,

$$\frac{\bar{\rho}(0)}{\bar{\rho}(D)} = \left(\frac{z_o}{z_o - D}\right)^n = e^{N_\rho};$$

so as $z_o \to \infty$, $N_\rho \to 0$ (the Boussinesq limit) and as $z_o \to D$, $N_\rho \to \infty$ (the large stratification limit).

In addition, as in the previous section, we assume here a perfect gas (Eq. 12.3). Therefore,

$$\overline{T} = \frac{\overline{p}}{R\overline{\rho}} = T_o\Theta(z),$$

(12.48)

where

$$T_o = \frac{p_o}{R\rho_o}.$$

(12.49)

The reference state temperature gradient is then

$$\frac{d\overline{T}}{dz} = -\frac{T_o}{z_o}.$$

(12.50)

If T_o/z_o equals g_o/c_p, the reference state is adiabatic and, by Eqs. 12.47 and 12.49, $n = 1/(\gamma - 1)$, where $\gamma = c_p/c_v$. On the other hand, if T_o/z_o is less than g_o/c_p, $n > 1/(\gamma - 1)$ and the reference state is subadiabatic; or if T_o/z_o is greater than g_o/c_p, $n < 1/(\gamma - 1)$ and the reference state is superadiabatic. It follows from Eq. 12.9 that the reference state entropy gradient for this polytropic perfect gas is

$$\frac{d\overline{S}}{dz} = \frac{(n(\gamma - 1) - 1)c_p}{\gamma(z_o - z)} = -\frac{(n(\gamma - 1) - 1)c_p}{n\gamma}h_\rho,$$

(12.51)

which, if positive, becomes more subadiabatic with height z. Note that the value of the ratio of specific heats, γ, is determined by the composition of the gas and its ionization state. Knowing the value of γ for a given (perfect) gas tells us what value of the polytropic index, n, is required for an adiabatic stratification. For example, γ for a monatomic perfect gas is $5/3$; so, for such a gas, an $n = 3/2$ would be an adiabatic reference state and an $n > 3/2$ would be a subadiabatic reference state.

The negative of the inverse temperature scale height follows from Eqs. 12.48 and 12.50:

$$h_T \equiv \frac{d\ln\overline{T}}{dr} = -\frac{1}{z_o - z} = \frac{h_\rho}{n}.$$

Thermodynamic derivatives would also be needed to calculate the temperature perturbation given the pressure and entropy perturbations:

$$T = \overline{\left(\frac{\partial T}{\partial S}\right)_p}S + \overline{\left(\frac{\partial T}{\partial p}\right)_S}p.$$

(12.52)

For a perfect gas (Eq. 12.40)

$$\overline{\left(\frac{\partial T}{\partial S}\right)_p} = \frac{\overline{T}}{c_p}$$

and

$$\overline{\left(\frac{\partial T}{\partial p}\right)_S} = \frac{1}{c_p\overline{\rho}}.$$

Therefore, for this plane-layer polytrope, one can define the reference state by choosing, for example, n, N_ρ, D, p_o, ρ_o, c_p, c_v, $\overline{\nu}$, and $\overline{\kappa}$.

12.2.2 2D Fluid Annulus

Now consider the reference state for a density-stratified annulus or sphere. Substituting the (spherical) expression for the gravitational acceleration (Eq. 10.9) into the equation for hydrostatic equilibrium (Eq. 10.11), multiplying through by $r^2/\bar{\rho}$, and then differentiating both sides with respect to r gives

$$\frac{1}{r^2}\frac{d}{dr}\left(\frac{r^2}{\bar{\rho}}\frac{d\bar{p}}{dr}\right) = -4\pi G\bar{\rho}, \tag{12.53}$$

which is another version of hydrostatic equilibrium. Now assume a polytropic equation of state:

$$\bar{p} = K\bar{\rho}^{(n+1)/n}, \tag{12.54}$$

where

$$K = p_o\rho_o^{-(n+1)/n}$$

and p_o and ρ_o are now the values of $\bar{p}$ and $\bar{\rho}$ at the center ($r = 0$), respectively. Substituting Eq. 12.54 into Eq. 12.53 gives the spherically symmetric polytropic equation for $\bar{\rho}(r)$:

$$\frac{(n+1)K}{4\pi Gnr^2}\frac{d}{dr}\left(\frac{r^2}{\bar{\rho}^{(n-1)/n}}\frac{d\bar{\rho}}{dr}\right) = -\bar{\rho}. \tag{12.55}$$

Equation 12.55 is second order and so requires two boundary conditions; it is convenient to apply both at the center ($r = 0$). One condition is that $\bar{\rho}(0) = \rho_o$; the top boundary radius is then defined to be where $\bar{\rho}$ first goes to zero. The second condition is that the radial gradient of $\bar{\rho}$ needs to vanish at $r = 0$ because of the spherical symmetry constraint.

Again we write $\bar{\rho}$ and $\bar{p}$ in terms of a polytropic function, Θ, now a function of radius:

$$\bar{\rho}(r) = \rho_o\Theta^n(r), \tag{12.56}$$
$$\bar{p}(r) = p_o\Theta^{n+1}(r). \tag{12.57}$$

We also define a scaled radius

$$\xi \equiv \frac{r}{\alpha}, \tag{12.58}$$

where

$$\alpha \equiv \left(\frac{(n+1)K}{4\pi G\rho_o^{(n-1)/n}}\right)^{1/2} = \left(\frac{(n+1)p_o}{4\pi G\rho_o^2}\right)^{1/2}. \tag{12.59}$$

Substituting Eqs. 12.56, 12.58, and 12.59 into Eq. 12.55 gives the *Lane-Emden* equation:

$$\frac{d}{d\xi}\left(\xi^2\frac{d\Theta}{d\xi}\right) = -\Theta^n\xi^2. \tag{12.60}$$

The central boundary conditions are now

$$\Theta = 1 \quad \text{and} \quad \frac{d\Theta}{d\xi} = 0 \quad \text{at} \quad \xi = 0. \tag{12.61}$$

The top boundary, $\xi_{top} = r_{top}/\alpha$, is chosen to be some value less than the smallest value of ξ that makes Θ vanish; the closer ξ_{top} is to this value the greater the density stratification.

Analytic solutions exist for Eq. 12.60 with these boundary conditions for $n = 0$, 1, and 5. The polytropic function for $n = 0$ (the constant-density case) is

$$\Theta(\xi) = 1 - \frac{\xi^2}{6}$$

with Θ vanishing at $\xi = \sqrt{6}$. For $n = 1$ the function is

$$\Theta(\xi) = \frac{\sin \xi}{\xi} \tag{12.62}$$

with ξ vanishing at π. The $n = 5$ solution goes to zero at $\xi = \infty$ and is usually not appropriate for planets and stars. For other values of n, Eq. 12.60 needs to be solved numerically, for which a generalized Newton-Raphson iteration method on a finite-difference representation of Eq. 12.60 works well.

Having obtained an analytic or numerical solution for the polytropic function, Θ, the gravitational acceleration can be obtained via Eq. 10.9, using Eq. 12.60:

$$\begin{aligned}
\bar{g}(r) &= \frac{4\pi G}{r^2} \int_0^r \bar{\rho} r^2 dr \\
&= \frac{4\pi G \rho_o \alpha^3}{r^2} \int_0^\xi \Theta^n \xi^2 d\xi \\
&= 4\pi G \rho_o \alpha \left(-\frac{d\Theta}{d\xi} \right) .
\end{aligned} \tag{12.63}$$

To obtain the reference state temperature we again assume a perfect gas:

$$\bar{T} = \frac{\bar{p}}{R\bar{\rho}} = T_o \Theta , \tag{12.64}$$

where $T_o = p_o/R\rho_o$. The reference state temperature gradient is therefore

$$\frac{d\bar{T}}{dr} = \frac{T_o}{\alpha} \frac{d\Theta}{d\xi} \tag{12.65}$$

and, using Eqs. 12.59 and 12.63–12.65, the reference state entropy gradient, Eq. 12.9, is

$$\begin{aligned}
\frac{d\bar{S}}{dr} &= \frac{c_p}{\bar{T}} \left[\frac{d\bar{T}}{dr} + \frac{\bar{g}}{c_p} \right] \\
&= \frac{c_p (n(\gamma - 1) - 1)}{\alpha \gamma} \left(-\frac{d \ln \Theta}{d\xi} \right) .
\end{aligned} \tag{12.66}$$

Note, $(-d \ln \Theta/d\xi)$ vanishes at the center ($\xi = 0$) and is positive out to ξ_{top}. Therefore, as is the case for the cartesian-box polytrope (Eq. 12.51), the reference state is adiabatic if $n = 1/(\gamma - 1)$ and subadiabatic if $n > 1/(\gamma - 1)$.

We also need the inverse density scale height for a Lane-Emden polytrope, which is

$$h_\rho = \frac{d \ln \bar{\rho}}{dr} = \frac{n}{\alpha} \frac{d \ln \Theta}{d\xi} ;$$

therefore the entropy gradient for a hydrostatic perfect gas in an annulus or sphere, Eq. 12.66, can also be written as

$$\frac{d\bar{S}}{dr} = -\frac{c_p (n(\gamma - 1) - 1)}{n\gamma} h_\rho , \tag{12.67}$$

which is identical to what it is for a cartesian-box polytrope (Eq. 12.51). The number of density scale heights spanning the depth of the annulus is

$$N_\rho = -\int_{bot}^{top} h_\rho dr = n \ln \left(\frac{\Theta(r_{bot})}{\Theta(r_{top})} \right) .$$

That is,

$$\frac{\bar{\rho}(r_{bot})}{\bar{\rho}(r_{top})} = e^{N_\rho} .$$

In addition, the inverse temperature scale height (which is also negative) is

$$h_T = \frac{d \ln \bar{T}}{dr} = \frac{1}{\alpha} \frac{d \ln \Theta}{d\xi} = \frac{h_\rho}{n} .$$

12.2.3 2D Fluid Annulus: A Simpler Polytrope

As mentioned above, polytropes with n between 3/2 (fully convective) and 3 (fully radiative) are most relevant for the interiors of stars; however, only numerical solutions exist for these polytropic indices. If one is interested in simulating only the outer portion of a star or planet that represents a small portion of the total mass of the body, an approximation can be made that provides an analytic solution for any value of the polytropic index. Consider our annulus, which represents the equatorial plane. Let M_o represent the mass below the bottom radius, r_{bot}. Then the gravitational acceleration (Eq. 10.9) for $r \geq r_{bot}$ is

$$\bar{g}(r) = \frac{G}{r^2} \left(M_o + 4\pi \int_{r_{bot}}^{r} \bar{\rho} r^2 dr \right) . \tag{12.68}$$

If M_o were much greater than the mass within $r_{bot} \leq r \leq r_{top}$, the gravitational acceleration within the annulus could be approximated as

$$\bar{g}(r) = \frac{G M_o}{r^2} . \tag{12.69}$$

That is, the gravitational force due to the mass within the annulus is neglected. In this case the Lane-Emden equation does not need to be solved. Instead, hydrostatic equilibrium (Eq. 10.11) is now

$$\frac{d\bar{p}}{dr} = -\frac{G M_o}{r^2} \bar{\rho} \tag{12.70}$$

and, when the polytropic profiles (Eqs. 12.56 and 12.57) are substituted into Eq. 12.70 and $\rho_o \Theta^n(r_{bot})$ is the prescribed ρ_{bot}, we get a simple analytic expression for this (approximate) polytropic function:

$$\Theta(r) = \left(\frac{\rho_{bot}}{\rho_o}\right)^{1/n} - \frac{G M_o \rho_o}{(n+1) p_o}\left(\frac{1}{r_{bot}} - \frac{1}{r}\right), \tag{12.71}$$

which decreases with radius as $1/r$ for any polytropic index. Since Eq. 12.71 is defined for $r \geq r_{bot}$, it would be more convenient to use ρ_{bot}, p_{bot}, and T_{bot} as input parameters instead of the central values, ρ_o, p_o, and T_o. To do this, define a modified polytropic function,

$$\Theta_1(r) \equiv \frac{\Theta(r)}{\Theta(r_{bot})} = 1 - \frac{G M_o \rho_{bot}}{(n+1) p_{bot}}\left(\frac{1}{r_{bot}} - \frac{1}{r}\right), \tag{12.72}$$

which is unity at $r = r_{bot}$. Then Eqs. 12.56, 12.57, and 12.64 become

$$\bar{\rho}(r) = \rho_{bot} \Theta_1^n(r), \tag{12.73}$$

$$\bar{p}(r) = p_{bot} \Theta_1^{n+1}(r), \tag{12.74}$$

$$\overline{T}(r) = T_{bot} \Theta_1(r). \tag{12.75}$$

The number of density scale heights spanning the depth is now

$$N_\rho = -n \ln \Theta_1(r_{top}) \tag{12.76}$$

and the (negative) inverse scale height for density is now

$$h_\rho(r) = n \frac{d \ln \Theta_1}{dr}$$

and for temperature is

$$h_T(r) = \frac{d \ln \Theta_1}{dr}.$$

The corresponding reference state entropy gradient is

$$\frac{d\overline{S}}{dr} = -\frac{c_p (n(\gamma - 1) - 1)}{n(n + 1)(\gamma - 1)} h_\rho, \tag{12.77}$$

which has the same relationship (Eqs. 12.51 and 12.67) between the polytropic index, n, and the ratio of specific heats, γ, that determines if the atmosphere is adiabatic (i.e., isentropic), subadiabatic, or superadiabatic.

This simpler polytrope is used as the reference state for the anelastic benchmark described in Section 13.5.2.

12.2.4 Alternative Reference States

So far we have assumed one constant value of the polytropic index, n, for the entire domain. To be able to simulate a stellar interior with a convection zone above a stable radiative interior, like the sun, or a convective core below a stable radiative envelop, like a much more massive star, we would like to be able to use $n = 3/2$ for the convective zone and $n = 3$ for the radiative region with a smooth continuous transition between the two. This would require a modified numerical method for solving Eq. 12.60 (e.g., Rappaport et al., 1983).

Instead of using a polytropic model for the reference state, one could choose an arbitrary $\bar{\rho}(r)$ profile or a polynomial fit to the density profile produced with a 1D stellar or planetary evolution model. With this $\bar{\rho}(r)$, one would integrate Eq. 10.9 to get $\bar{g}(r)$ and then integrate the hydrostatic equilibrium equation, 10.11, to get $\bar{p}(r)$. The reference state temperature, $\overline{T}(r)$, and entropy, $\overline{S}(r)$, could also be obtained as polynomial fits to the output from a 1D internal structure model. One could instead assume a perfect gas equation of state (Eq. 12.3) to obtain the reference state temperature; however, since the density profile was arbitrarily chosen, this gas would not necessarily be an adiabatic perfect gas. Another choice would be to approximate the temperature profile by simply setting

$$\overline{T}(r) = T_{bot}\left(\frac{\bar{\rho}(r)}{\rho_{bot}}\right)^{\gamma_G} \tag{12.78}$$

if one had a reasonable estimate for a constant Grüneisen parameter,

$$\gamma_G = \overline{\left(\frac{\partial \ln T}{\partial \ln \rho}\right)}_S. \tag{12.79}$$

For example, this might be a good approximation for a convecting mantle or core of a terrestrial planet, which do not have significant density stratifications. The reference state thermodynamic derivatives could then be calculated as is done in Eq. 12.42 and h_T would be $\gamma_G h_\rho$. (Note, for an adiabatic perfect gas, $\gamma_G = \gamma - 1$, where $\gamma = c_p/c_v$.)

12.3 NUMERICAL METHOD: ANELASTIC

Having discussed the anelastic equations and possible reference states, we now describe the modifications to the numerical methods employed in our Boussinesq models that are needed to convert them to anelastic models. Here we consider both cartesian (x, z) box geometry (as we have for most of this book) and cylindrical (r, ϕ) annulus geometry (Section 10.3).

As with our cartesian Boussinesq models, we define a streamfunction, ψ, that constrains the fluid velocities to everywhere satisfy mass conservation. However, for our anelastic models we define ψ such that

$$\bar{\rho}\mathbf{v} = \nabla \times \psi \hat{\mathbf{y}} = -\frac{\partial \psi}{\partial z}\hat{\mathbf{x}} + \frac{\partial \psi}{\partial x}\hat{\mathbf{z}} \tag{12.80}$$

to satisfy the anelastic mass conservation constraint, $\nabla \cdot \bar{\rho}\mathbf{v} = 0$ (Eq. 12.14). Therefore, now

$$v_x = -\frac{1}{\bar{\rho}}\frac{\partial \psi}{\partial z} \quad \text{and} \quad v_z = \frac{1}{\bar{\rho}}\frac{\partial \psi}{\partial x} \tag{12.81a,b}$$

and, since vorticity is $\boldsymbol{\omega} = \nabla \times \mathbf{v} = \omega \hat{\mathbf{y}}$, the amplitude of vorticity in the y-direction is

$$\omega = -\frac{1}{\rho}\left[\nabla^2\psi - h_\rho\frac{\partial\psi}{\partial z}\right] = -\frac{1}{\rho}\left[\frac{\partial^2\psi}{\partial x^2} + \frac{\partial^2\psi}{\partial z^2} - h_\rho\frac{\partial\psi}{\partial z}\right]. \qquad (12.82)$$

For our anelastic models in cylindrical geometry (r, ϕ), we have

$$\bar{\rho}\mathbf{v} = \nabla\times\psi\hat{\mathbf{z}} = \frac{1}{r}\frac{\partial\psi}{\partial\phi}\hat{\mathbf{r}} - \frac{\partial\psi}{\partial r}\hat{\boldsymbol{\phi}}; \qquad (12.83)$$

so

$$v_r = \frac{1}{r\bar{\rho}}\frac{\partial\psi}{\partial\phi} \quad \text{and} \quad v_\phi = -\frac{1}{\bar{\rho}}\frac{\partial\psi}{\partial r} \qquad (12.84\text{a,b})$$

and

$$\omega = -\frac{1}{\bar{\rho}}\left[\nabla^2\psi - h_\rho\frac{\partial\psi}{\partial r}\right] = -\frac{1}{\bar{\rho}}\left[\frac{\partial^2\psi}{\partial r^2} + \left(\frac{1}{r} - h_\rho\right)\frac{\partial\psi}{\partial r} + \frac{1}{r^2}\frac{\partial^2\psi}{\partial\phi^2}\right]. \qquad (12.85)$$

Next consider the anelastic momentum conservation equation (12.29 or 12.43). As in our Boussinesq models, we take the curl of the momentum equation to obtain the vorticity equation (here first for the cartesian box geometry):

$$\frac{\partial\omega}{\partial t} = -\mathbf{v}\cdot\nabla\omega + h_\rho\omega v_z + \bar{g}\frac{\partial C}{\partial x} + \bar{\nu}\nabla^2\omega + \frac{d\bar{\nu}}{dz}\left(\frac{\partial\omega}{\partial z} - \frac{4h_\rho}{3\bar{\rho}}\frac{\partial^2\psi}{\partial x^2}\right). \qquad (12.86)$$

Here again C is co-density (Eq. 12.41), for which, as mentioned above, the extra pressure contribution (in either the entropy or temperature formulation) could be dropped to better conserve energy numerically. Also, Eq. 12.86 is for a constant dynamic viscosity, $\bar{\rho}\bar{\nu}$. If, on the other hand, $\bar{\nu}$ were an arbitrary function of z and $h_\nu \equiv 1/\bar{\nu}\, d\bar{\nu}/dz$, the full viscous term in the vorticity equation 12.86 would be

$$\bar{\nu}\left[\nabla^2\omega + h_\nu\left(\nabla^2 v_x - \frac{h_\rho}{3}\frac{\partial v_z}{\partial x}\right)\right.$$
$$+ \left(\frac{\partial}{\partial z}(h_\rho + h_\nu) + h_\nu(h_\rho + h_\nu)\right)\left(\frac{\partial v_x}{\partial z} + \frac{\partial v_z}{\partial x}\right)$$
$$\left.+ (h_\rho + h_\nu)\left(\frac{\partial^2 v_x}{\partial z^2} - \frac{\partial^2 v_z}{\partial x\partial z} - \frac{2h_\rho}{3}\frac{\partial v_z}{\partial x}\right)\right]. \qquad (12.87)$$

The additional terms, compared to the Boussinesq vorticity equation (2.4), are due to the density stratification, which makes $\nabla\cdot\mathbf{v} = -h_\rho v_z$, and due to the z-dependence of the viscous diffusivity, $\bar{\nu}$. Since we will assume $\bar{\nu}$ is inversely proportional to $\bar{\rho}$, $d\bar{\nu}/dz = -\bar{\nu}h_\rho$. To further simplify this model one could neglect the $d\bar{\nu}/dz$ term since it could be argued that this traditional molecular viscosity formulation is really not a correct model for turbulent "eddy" diffusion anyway. However, the additional term, $h_\rho\omega v_z$, coming from the curl of advection cannot be neglected. It represents the reduction of the amplitude of a fluid parcel's vorticity (angular velocity) as it rises (positive v_z) and therefore expands due to the decrease in the background density it experiences. Likewise, a sinking parcel contracts and therefore its vorticity increases in amplitude. This is a form of conservation of angular momentum applied to a local fluid parcel. Note also that this term can be combined with the advection term,

$$-\mathbf{v}\cdot\nabla\omega + h_\rho\omega v_z = -\left[v_x\frac{\partial\omega}{\partial x} + v_z\left(\frac{\partial\omega}{\partial z} - h_\rho\omega\right)\right] = -\nabla\cdot(\omega\mathbf{v}),$$

which is a somewhat more accurate way to numerically calculate these nonlinear terms.

For the cylindrical annulus geometry and for a constant dynamic viscosity, $\bar{\rho}\bar{\nu}$, the vorticity equation is

$$\frac{\partial\omega}{\partial t} = -\mathbf{v}\cdot\nabla\omega + h_\rho\omega v_r + \frac{\bar{g}}{r}\frac{\partial C}{\partial\phi} + \bar{\nu}\left(\frac{\partial^2\omega}{\partial r^2} + \frac{1}{r}\frac{\partial\omega}{\partial r} + \frac{1}{r^2}\frac{\partial^2\omega}{\partial\phi^2}\right)$$

$$+\frac{d\bar{\nu}}{dr}\left(\frac{\partial\omega}{\partial r} - \frac{4h_\rho}{3r^2\bar{\rho}}\frac{\partial^2\psi}{\partial\phi^2}\right). \tag{12.88}$$

Again, the two advection terms could be combined as $-\nabla\cdot(\omega\mathbf{v})$. Also, the $d\bar{\nu}/dr$ term could be neglected to simplify the model; however, it would be relatively easy to include, especially if ω, ψ, and S were solved simultaneously in a semi-implicit matrix operation (Section 8.2).

Note that the curl operation on the gradient of the reduced pressure in Eqs. 12.29 and 12.43 eliminates this term in Eqs. 12.86 and 12.88. However, the horizontal gradient of the pressure perturbation still exists, via the co-density (Eq. 12.41), unless one uses S, instead of T, for a working thermodynamic variable and prescribes an adiabatic reference state. That is, the buoyancy torque in Eq. 12.86 is

$$\bar{g}\frac{\partial C}{\partial x} = -\left(\frac{\bar{g}}{\bar{T}}\frac{\partial T}{\partial x} + \frac{1}{\bar{T}\bar{\rho}}\frac{d\bar{T}}{dz}\frac{\partial p}{\partial x}\right) = -\left(\frac{\bar{g}}{c_p}\frac{\partial S}{\partial x} + \frac{1}{c_p\bar{\rho}}\frac{d\bar{S}}{dz}\frac{\partial p}{\partial x}\right)$$

and in Eq. 12.88 the buoyancy torque is

$$\frac{\bar{g}}{r}\frac{\partial C}{\partial\phi} = -\left(\frac{\bar{g}}{\bar{T}r}\frac{\partial T}{\partial\phi} + \frac{1}{\bar{T}\bar{\rho}r}\frac{d\bar{T}}{dr}\frac{\partial p}{\partial\phi}\right) = -\left(\frac{\bar{g}}{c_p r}\frac{\partial S}{\partial\phi} + \frac{1}{c_p\bar{\rho}r}\frac{d\bar{S}}{dr}\frac{\partial p}{\partial\phi}\right).$$

One way to calculate the horizontal gradient of the pressure perturbation is to get it from the horizontal component of the momentum equation (Rogers & Glatzmaier, 2005a). If simulating flows in a box (Eq. 12.29), it would be

$$\frac{1}{\bar{\rho}}\frac{\partial p}{\partial x} = -\frac{\partial v_x}{\partial t} - [(\mathbf{v}\cdot\nabla)\mathbf{v}]_x + \bar{\nu}\left(\nabla^2 v_x - \frac{h_\rho}{3}\frac{\partial v_z}{\partial x}\right);$$

if in an annulus, it would be

$$\frac{1}{\bar{\rho}r}\frac{\partial p}{\partial\phi} = -\frac{\partial v_\phi}{\partial t} - [(\mathbf{v}\cdot\nabla)\mathbf{v}]_\phi + \bar{\nu}\left(\nabla^2 v_\phi - \frac{v_\phi}{r^2} + \left(\frac{2}{r^2} - \frac{h_\rho}{3r}\right)\frac{\partial v_r}{\partial\phi}\right).$$

The time derivatives in these two expressions would be calculated from the previous time step, $(v_{x,t} - v_{x,t-\Delta t})/\Delta t$. Also note, we have neglected the perturbation in the gravitational potential energy in these pressure perturbation terms, which should actually be included, and calculated via Eq. 12.19, if the bottom boundary is at or close to the body's center.

However, as mentioned above for Eqs. 12.29 and 12.43, discussed in Section 12.4.2, and suggested by Brown et al. (2012), the extra horizontal pressure gradient term in the buoyancy torque could be dropped because, within the anelastic approximation, it prevents exact energy conservation. Dropping this term simplifies the code since then there is no need to solve the horizontal component of the momentum equation.

Another modification to the vorticity equation would be needed if one were simulating magnetoconvection (Chapter 11). The curl of the Lorentz force (per mass),

$$\nabla \times \left(\frac{1}{\rho} \mathbf{J} \times \mathbf{B} \right) = \frac{1}{\rho} \nabla \times (\mathbf{J} \times \mathbf{B}) + \nabla(\frac{1}{\rho}) \times (\mathbf{J} \times \mathbf{B}) = \frac{1}{\rho} \left[(\mathbf{B} \cdot \nabla J) - h_\rho J B_z \right] \hat{y},$$

would appear on the right side in Eqs. 12.86 and likewise for Eq. 12.88, now with the additional term involving h_ρ.

Now consider what modifications are needed for the top and bottom boundary conditions. If we use the entropy perturbation as one of the working thermodynamic variables, the top and bottom boundaries would be more conveniently chosen to be isentropic, i.e., constant entropy, instead of isothermal. In this case, for example, we could set $S = \Delta S/2$ on the bottom boundary and $S = -\Delta S/2$ on the top boundary for a convectively unstable regime or vice versa for a stable regime. If, on the other hand, we use the temperature perturbation as a working variable, it would be more convenient to force isothermal top and bottom boundaries, i.e., constant temperature. In this case we could set $T = \Delta T/2$ on the bottom boundary and $T = -\Delta T/2$ on the top boundary, or vice versa for a stable region. Recall that T is relative to the vertically dependent $\overline{T}$, which has an adiabatic profile (or nearly one) in the convection zone.

This choice also affects the definition of the Rayleigh number. If we choose to use nondimensional variables, we could define the Rayleigh number as

$$\mathrm{Ra} = \frac{\bar{g}_{mid} \Delta S \, D^3}{c_p \bar{v}_{mid} \bar{\kappa}_{mid}} ; \tag{12.89a}$$

or, if using temperature,

$$\mathrm{Ra} = \frac{\bar{g}_{mid} \alpha \Delta T \, D^3}{\bar{v}_{mid} \bar{\kappa}_{mid}} , \tag{12.89b}$$

where the representative thermal expansion coefficient, α, could be set to $1/\overline{T}_{mid}$ for a perfect gas. As mentioned in Section 10.3, when simulating an annulus the gravitational acceleration will be a function of radius and now also the viscous and thermal diffusivities could be functions of radius, so we choose their mid-depth values for the definition of the Rayleigh number. Likewise, if $\bar{v}$ and $\bar{\kappa}$ have different radial dependencies, the Prandtl number could also be defined based on their mid-depth values.

The impermeable boundary condition for the box geometry (at $z = 0$ and D) is

$$v_z = 0 \quad \text{so} \quad \psi = 0 \quad \text{and} \quad \frac{\partial^2 \psi}{\partial x^2} = 0$$

and the stress-free condition is

$$\frac{\partial v_x}{\partial z} = 0 \quad \text{so} \quad \frac{\partial^2 \psi}{\partial z^2} - h_\rho \frac{\partial \psi}{\partial z} = 0 \quad \text{and} \quad \omega = 0.$$

For the annulus geometry the impermeable condition (at $r = r_{bot}$ and r_{top}) is

$$v_r = 0 \quad \text{so} \quad \psi = 0 \quad \text{and} \quad \frac{\partial^2 \psi}{\partial \phi^2} = 0$$

and the stress-free condition is

$$\frac{\partial}{\partial r} \left(\frac{v_\phi}{r} \right) = 0 \quad \text{so} \quad \frac{\partial^2 \psi}{\partial r^2} - \left(\frac{1}{r} + h_\rho \right) \frac{\partial \psi}{\partial r} = 0 \quad \text{and} \quad \omega = -\frac{2}{r\bar{\rho}} \frac{\partial \psi}{\partial r}.$$

Note that, like the Boussinesq annulus (Eq. 10.17), the vorticity on the impermeable stress-free boundaries of an anelastic annulus equals twice the local angular velocity:

$$\omega = \frac{2v_\phi}{r} \quad \text{at} \quad r_{bot} \quad \text{and} \quad r_{top}.$$

Also, like the Boussinesq annulus (Section 10.3), the anelastic annulus requires ω and ψ to be solved simultaneously or via the method in Section 10.3.3, placing all boundary conditions on ψ because ω does not vanish on the boundaries.

With these boundary conditions the anelastic equations can be solved using basically the same methods already described for the various Boussinesq models: spectral or finite-difference spatial discretization, explicit or semi-implicit time integration, impermeable or periodic side boundaries, cartesian box or cylindrical annulus geometry. There is also a choice of convection (unstable thermal stratification) or internal gravity waves (stable thermal stratification) or a combination; and there's a choice of using the entropy perturbation or the temperature perturbation as a working thermodynamic variable. Of course, the anelastic reference state density and temperature are functions of z or r and, if desired, also $\bar{g}$, $\overline{S}$, $\bar{v}$, and $\bar{\kappa}$ could be functions of z or r. Nonlinear viscous heating and ohmic heating should also be added to the internal energy equation to conserve total energy (Eq. 1.10). One also needs to choose to work with either dimensional variables as presented in this chapter or nondimensional variables as presented in Part 1.

The choice of using entropy or temperature as one of the working thermodynamic variables deserves some additional consideration. As shown, when using entropy with an adiabatic reference state and the Lantz-Braginsky-Roberts formulation, the pressure perturbation does not contribute directly to the buoyancy term in the momentum equation (unless one chooses not to drop the extra pressure term resulting from a nonadiabatic reference state). This could result in a more accurate solution because of the role of the pressure perturbation within the anelastic approximation. As mentioned above, filtering out sound waves effectively makes the sound speed infinite by requiring the pressure perturbation at each time step to be whatever it needs to be throughout the domain to make the divergence of the momentum equation 12.17 vanish everywhere and so satisfy the anelastic mass conservation equation 12.14. Consequently, this pressure perturbation could be relatively large. However, if this pressure were needed to formulate the buoyancy forces via the equation of state, it might, under certain extreme conditions, dominate the entropy or temperature contributions to buoyancy. This would be noticed if the diagnostic density (calculated via the equation of state) had a spatial pattern much more like the pressure perturbation than the entropy perturbation (or than the temperature perturbation if using temperature as a variable). Likewise, when using entropy as a variable, if the diagnostic temperature perturbation, calculated from Eq. 12.52, had the same sign and pattern of the pressure perturbation instead of the entropy perturbation, there would be reason for concern. That is, the numerical pressure may not be an accurate representation of the thermodynamic gas pressure.

12.4 LINEAR ANALYSES: ANELASTIC

12.4.1 Linear Stability: Anelastic

Before describing examples of nonlinear anelastic simulations we discuss some linear stability issues that are affected by density stratification. We have discussed the critical Rayleigh number for thermal convection in a box with (Eqs. 11.43 and 11.56) and without (Eq. 3.8) magnetic fields; however, these have been for the onset of Boussinesq convection, the equations which have constant coefficients. If we consider the linear anelastic equations and set the time derivatives to zero (as we do for the Boussinesq stability problems in Sections 3.4 and 11.3), we would have three coupled equations in terms of the z-dependent Fourier modes of S, ω, and ψ but now with coefficients that are arbitrary functions of z because $\bar{g}$, $\bar{\rho}$, $\bar{\kappa}$, and $\bar{\nu}$ are z-dependent. Therefore, solving for the critical Rayleigh number and mode number for a given set of linear equations with z-dependent coefficients usually requires a numerical solution. Alternatively, one could use a linear code with time derivatives to iterate on the Rayleigh number, as described in Section 3.3, to find the critical Rayleigh number, for which the numerical solution neither increases nor decreases exponentially with time.

Glatzmaier & Gilman (1981a,c) investigate the stability of anelastic thermal convection within a 3D spherical fluid shell by expanding in spherical harmonics and numerically solving the resulting eigenvalue problem using a generalized Newton-Raphson relaxation method on a finite-difference grid in radius. They describe how the structure of the most unstable cell at the onset of convection depends of the density stratification, rotation rate, boundary conditions, diffusivity profiles in radius, and on the shell depth. For example, they find that prescribing a constant viscous diffusivity results in larger fluid velocity amplitudes in the low-density upper region of the convection zone, whereas a constant dynamic viscosity (i.e., a viscous diffusivity inversely proportional to density) results in somewhat larger velocities in the lower region; these effects increase as the density stratification increases. They also find that increasing the density stratification increases the critical Rayleigh number and the critical horizontal mode number, i.e., makes the system more stable and reduces the size of the most unstable mode. A smaller horizontal size is preferred because, since the convective cell tends to be concentrated in the upper region for constant $\bar{\nu}$ and in the lower region for constant $\bar{\nu}\bar{\rho}$, the effective vertical extent of the cell is less than the total depth; and the most efficient convection tends to occur when the horizontal and vertical dimensions of the cell are similar.

Berkoff et al. (2010) investigate the effects of density stratification in a 3D fluid box on the onset of magnetoconvection in a thermally unstable stratification and on the magnetic buoyancy instability in a thermally stable stratification. Magnetic buoyancy occurs within a local concentration of magnetic field because the gas pressure and therefore gas density for a local isothermal environment need to be slightly less than they are outside the intense magnetic field since magnetic pressure, $B^2/2\mu$, contributes to the total pressure. They compare these linear instabilities for the anelastic approximation to those for a fully compressible model.

For magnetoconvection they prescribe a uniform vertical background magnetic field and find, as did Glatzmaier & Gilman (1981a,c) for the nonmagnetic spherical case, that the critical Rayleigh number and critical mode number increase with density stratification. They find that the linear anelastic predictions are very close to the linear fully compressible predictions when the reference state is adiabatic or nearly adiabatic as long as the magnetic field, which inhibits convection, does not become so large that the Alfvén velocity (which is proportional to the square of the magnetic field intensity) approaches the sound speed.

For the magnetic buoyancy problem Berkoff et al. (2010) prescribe a slightly subadiabatic thermal stratification with a horizontal magnetic field that increases linearly with depth. They find that the accuracy of the linear anelastic predictions decreases as the growth rate of the instability increases, which occurs as the intensity of the magnetic field increases. So, as for the magnetoconvection case, both the validity and accuracy of the magnetic anelastic approximation depend on the amplitude of the magnetic field.

12.4.2 Internal Gravity Waves: Anelastic

Consider now how a density stratification affects internal gravity waves in a subadiabatic region. As we do in Section 6.1 for the Boussinesq approximation, here we derive a dispersion relation that relates the frequency, $\bar{\omega}$, to the horizontal wavenumber, k_x, the vertical wavenumber, k_z, and the Brunt-Väisälä frequency, N (Eq. 12.13). However, now the reference state gravitational acceleration, $\bar{g}$, and the entropy gradient, $d\bar{S}/dz$, and therefore N, can be functions of the height, z.

The anelastic, linear, nondiffusive versions of the internal energy equation (12.34) and vorticity equation (12.86) are, respectively,

$$\frac{\partial S}{\partial t} = -v_z \frac{d\bar{S}}{dz} = -\frac{1}{\bar{\rho}} \frac{d\bar{S}}{dz} \frac{\partial \psi}{\partial x} \tag{12.90}$$

and

$$\frac{\partial \omega}{\partial t} = \bar{g} \frac{\partial C}{\partial x} = -\frac{1}{c_p} \left(\bar{g} \frac{\partial S}{\partial x} + \frac{1}{\bar{\rho}} \frac{d\bar{S}}{dz} \frac{\partial^2 \psi}{\partial z \partial t} \right). \tag{12.91}$$

In Eq. 12.90 we used the streamfunction expression for v_z (Eqs. 12.81) and in Eq. 12.91 we used the general expression for the co-density (12.41) and the

x-component of the momentum equation (12.17), neglecting viscosity and the perturbation gravitational potential:

$$\bar{\rho}\frac{\partial v_x}{\partial t} = -\frac{\partial p}{\partial x} = -\frac{\partial^2 \psi}{\partial z \partial t} .$$

However, as pointed out by Brown et al. (2012), this system of equations leads to a dispersion relation that makes the frequency, $\bar{\omega}$, complex; therefore, energy is not conserved because the imaginary part of this $\bar{\omega}$ would cause the wave amplitude to change exponentially with time even though there are no energy sources or sinks in this example. Their suggestion for fixing this, which we follow here, is to drop the extra pressure contribution to co-density in Eq. 12.91, which makes the vorticity equation

$$\frac{\partial \omega}{\partial t} = -\frac{\bar{g}}{c_p}\frac{\partial S}{\partial x} . \tag{12.92}$$

That is, we use the same vorticity equation in a stable (subadiabatic) reference state that we would use in an adiabatic reference state. Note, the usual part of the pressure contribution to buoyancy is still present within the gradient of the reduced pressure, P, when employing the Lantz-Braginsky-Roberts formulation of the anelastic equations.

Then, taking the time derivative of Eq. 12.92 and substituting in Eqs. 12.90 and 12.13 gives

$$\frac{\partial^2 \omega}{\partial t^2} = \frac{N^2}{\bar{\rho}}\frac{\partial^2 \psi}{\partial x^2} ; \tag{12.93}$$

taking the second derivative with respect to time of the streamfunction equation (12.82) gives

$$\frac{\partial^2 \omega}{\partial t^2} = -\frac{1}{\bar{\rho}}\left(\frac{\partial^2}{\partial x^2} + \frac{\partial^2}{\partial z^2} - h_\rho \frac{\partial}{\partial z}\right)\frac{\partial^2 \psi}{\partial t^2} . \tag{12.94}$$

By combining Eqs. 12.93 and 12.94 we have

$$-\left(\frac{\partial^2}{\partial x^2} + \frac{\partial^2}{\partial z^2} - h_\rho \frac{\partial}{\partial z}\right)\frac{\partial^2 \psi}{\partial t^2} = N^2 \frac{\partial^2 \psi}{\partial x^2} . \tag{12.95}$$

Now we again assume a planewave solution. However, here we need to account for how the density stratification affects wave kinetic energy. To simplify this example we assume h_ρ, the negative inverse density scale height, is a constant; so the reference state density is simply

$$\bar{\rho} = \rho_o e^{h_\rho z} .$$

Instead of using Eq. 6.6, here we need the velocity amplitude to depend on the background density so the integrated wave kinetic energy is constant in the vertical, i.e., a constant $\bar{\rho}v^2$. Therefore, the velocity amplitude needs to be

$$v \propto e^{-h_\rho z/2} .$$

That is, the wave velocity amplitude increases with height as the density decreases. Since the fluid velocity and streamfunction are related via Eqs. 12.81, which involve

$\bar{\rho}$, we set

$$\psi(x, z, t) = \psi_o \, e^{[i(k_x x + k_z z - \bar{\omega} t) + h_\rho z/2]} , \tag{12.96}$$

where k_x and k_z are real constants. One could consider this formulation as having a complex vertical wavenumber with the imaginary part being $-h_\rho/2$. Substituting Eq. 12.96 into Eq. 12.95 gives us the anelastic internal gravity wave dispersion relation:

$$\bar{\omega}^2 = \frac{N^2 k_x^2}{k_x^2 + k_z^2 + \frac{h_\rho^2}{4}} . \tag{12.97}$$

That is, the greater the density stratification the smaller the wave frequency for a given Brunt-Väisälä frequency and set of wavenumbers. Note, the Boussinesq dispersion relation (Eq. 6.7) is recovered when $h_\rho = 0$.

Equation 12.97 shows that this wave frequency is purely real; therefore, energy is conserved. If the extra pressure contribution to buoyancy were not dropped in Eq. 12.91, the frequency, $\bar{\omega}$, would have an imaginary part, which would cause the wave energy to change exponentially with time. This could easily affect the accuracy of the solution and, for low-viscosity cases, could lead to a numerical instability in a nonlinear simulation. Therefore, maintaining energy conservation within an anelastic simulation is probably worth any inaccuracy that develops due to the neglect of this pressure term.

Note that an alternative way to derive Eq. 12.97 (Brown et al., 2012) is to consider a parcel displacement vector, ξ, defined as

$$\mathbf{v} \equiv \frac{\partial \xi}{\partial t} ,$$

where

$$\xi(x, z, t) = \xi_\mathbf{0} \, e^{[i(k_x x + k_z z - \bar{\omega} t) - h_\rho z/2]} .$$

The vertical component of this displacement is related to our streamfunction via

$$\psi = -\frac{\bar{\rho} \bar{\omega}}{k_x} \xi_z .$$

Also note that although density stratification affects the frequency and amplitude of an internal gravity wave, a simple examination of anelastic mass conservation shows that these waves are transverse waves, as they are in a Boussinesq fluid. That is, replacing the velocity, $\mathbf{v}$, with the mass flux, $\bar{\rho}\mathbf{v}$, in Eq. 6.8 and using the anelastic streamfunction defined in Eqs. 12.81 shows that the fluid velocity is always perpendicular to the direction of the phase propagation.

12.5 NONLINEAR SIMULATIONS: ANELASTIC

12.5.1 2D Fluid Box Simulation

Now consider nonlinear anelastic simulations of thermal convection. As a first example, consider a 2D density-stratified fluid in a box with an aspect ratio of 2 and periodic side boundaries. The Rayleigh and Prandtl numbers are Ra $= 10^{10}$ and Pr $= 1$, respectively. An adiabatic reference state is prescribed using a polytropic index $n = 1.5$ (Section 12.2.1) with density at the bottom boundary being 148 times larger than that at the top boundary ($N_\rho = 5$). The thermal and viscous diffusivities are constants. The solution is obtained with a Fourier spectral method in the horizontal direction using 682 modes (2048 grid points) and a finite-difference method in the vertical direction on 2048 Chebyshev grid points (Section 9.1). The time integration employs a semi-implicit time integration scheme (Section 8.2) with the nonlinear terms computed via a spectral-transform method (Section 10.4). The code was run in parallel using MPI (Section 9.5.2) on 512 processors.

A snapshot of the entropy perturbation in this simulation is shown in Fig. 12.1. The large vertical flow structure at middle is a downwelling, which is warmer than the two cells that straddle it in the lower part of the box. The centers of these cells are closer to the bottom boundary because of mass conservation as sinking gas contracts and rising gas expands. Much of the circulating fluid never reaches the top boundary during a convective turnover time and much of the fluid that does reach the upper region does not have enough time to completely cool before being swept downward. The small cold (blue) sinking plume that originates within the top boundary layer contracts significantly as it sinks and mixes into the warmer fluid. This flow pattern is very different from the vertically symmetric patterns seen in Boussinesq convection.

The Reynolds number for this turbulent convection is Re $\approx 5 \times 10^4$. The flows, driven within very superadiabatic boundary layers, maintain a roughly isentropic mean thermal stratification within the bulk of the fluid. This is seen in Fig. 12.2, which shows the nonlinearly maintained time-averaged mean entropy profile for this case. The mean entropy gradient is actually slightly positive (subadiabatic) in the bulk of the fluid because the convective heat flux is so efficient, causing the upper region to heat up and the lower region to cool. This increases the steepness of the very negative (superadiabatic) entropy gradients in the shallow thermal boundary layers, which is needed to maintain sufficient diffusive heat flux through the boundaries. Note, the nondimensional entropy in this figure is one at the bottom boundary ($z = 0$) and zero at the top boundary ($z = 1$); the thermal boundary layers are so shallow they are difficult to see in the figure. The high-order Chebyshev grid is needed to resolve these boundary layers.

Also shown in Fig. 12.2 is the kinetic energy spectrum for this case (Fig. 12.1), where the log of kinetic energy is plotted vs. the log of the mode number. This volume-averaged kinetic energy is greatest for the large length scales (small horizontal mode numbers) and decreases through a turbulent inertial subrange as the scales decrease before reaching the viscous dissipation range at the smallest length scales where kinetic energy deceases more rapidly with increasing mode number. Kinetic energy in this plot drops 24 powers of 10 from mode 1 to 682.

12.5.2 2D Fluid Annulus Simulation

Next consider an example of anelastic convection in an annulus (Fig. 12.3) with a less significant density stratification, $N_\rho = 3$; so the density at the bottom

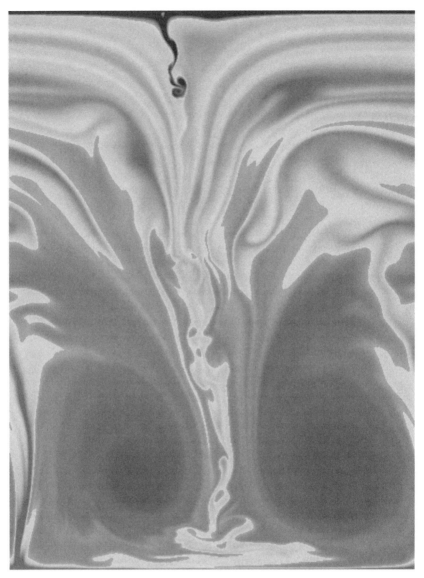

Figure 12.1 A snapshot of the entropy perturbation in an anelastic simulation of thermal convection in a 2D box with Ra $= 10^{10}$, Pr $= 1$, $N_\rho = 5$, and an aspect ratio of 0.75 (see *Color Plate* 5 for a color version of this figure). Reds represent hot fluid, yellows warm, and blues cold fluid relative to the volume-averaged mean entropy.

boundary is now 20 times greater than that at the top boundary (Fig. 12.4). Again we choose an adiabatic polytrope for the reference state with index $n = 1$, but now using a Lane-Emden polytrope (Section 12.2.2). Therefore, the reference state density and temperature both vary in radius according to Eq. 12.62. The thermal

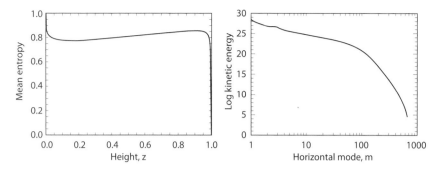

Figure 12.2 (Left) The horizontally averaged and time-averaged entropy perturbation for the case illustrated in Fig. 12.1. The entropy is scaled by ΔS and plotted as a function of height. (Right) The kinetic energy spectrum. The log of the (scaled) volume-averaged kinetic energy is plotted as a function of the log of the horizontal mode number.

and viscous diffusivities are both set to be inversely proportional to the reference state density. The bottom boundary radius is 20% of the top boundary radius and the Rayleigh and Prandtl numbers are Ra $= 1.3 \times 10^{10}$ and Pr $= 1$, respectively. The solution is obtained with a Chebyshev-Fourier spectral method using 1366 Fourier modes (4096 longitudinal grid points) in the horizontal direction and 1537 Chebyshev modes in the vertical direction (Section 9.4). The time integration is done using a semi-implicit time integration scheme that solves each Fourier mode of the entropy, vorticity, and streamfunction simultaneously (Section 8.2); however, to maintain stability without having to use a time step significantly smaller than the CFL condition, the linear terms are treated fully implicitly ($\alpha = 1$). The nonlinear terms are computed with a spectral-transform method (Section 10.4). The code was run in parallel using MPI (Section 9.5.2) on 512 processors.

Figure 12.3 is a snapshot of the entropy perturbation from this anelastic simulation. Unlike the snapshot in the Boussinesq simulation seen in Fig. 10.4, this one is taken well after the shear instabilities have destroyed the initial well-defined single-plume dipolar flow pattern. However, even after many convective turnovers (a million computational time steps), there tends to be no more than two main large-scale downflows, composed of many small-scale turbulent eddies, and no more than two main, but thin, upflow plumes. The most typical pattern is one large turbulent downflow (like the one centered at "12 o'clock" in Fig. 12.3) and one thin, distorted upflow (at "6 o'clock"), i.e., a turbulent dipolar flow. Unlike the Boussinesq case, for which the size of an eddy remains relatively constant as it sinks or rises, here boundary layer instabilities at the top boundary combine into a large downflow plume that contracts as it falls through the density stratification. Likewise, small boundary layer instabilities at the bottom boundary develop into an upflow plume that expands and deforms as it rises.

Another clear difference between this anelastic convection compared to Boussinesq convection is the mean entropy profile in radius, which is maintained by the

Figure 12.3 A snapshot of the entropy perturbation in an anelastic simulation of thermal
         convection within a 2D annulus after a million computational time steps with
         $N_\rho = 3$, Ra $= 1.3 \times 10^{10}$, and Pr $= 1$ (see *Color Plate* 6 for a color version of
         this figure). Reds represent hot fluid and blues represent cold fluid relative to
         the constant reference state entropy, with the crossover from blue to yellow at
         $S = S_{bot} + \Delta S/2$.

convergence of radial heat flux. Since $\bar{\rho}\bar{\kappa}$ is a constant in this simulation, the diffusive heat flux $(-\bar{\rho}\bar{\kappa}\overline{T}\,dS/dr)$ is proportional to $\overline{T}$, which is 20 times greater at the
bottom boundary than at the top. Therefore, the magnitude of the radial gradient
of the horizontally averaged (mean) entropy perturbation at the bottom is typically
20 times smaller than it is at the top boundary in order to maintain roughly the
same amount of heat flow through the top and bottom boundaries. Consequently,
since the depths of the two thermal boundary layers are comparable, the mean entropy drops less through the bottom thermal boundary layer compared to that at the
top. This makes the mean entropy throughout the bulk of the domain (Fig. 12.4),
which is kept nearly constant in radius by the nonlinear convective heat flux, closer
to the bottom entropy boundary value instead of midway between the bottom and
top values, as is the case for Boussinesq convection. This is also seen in larger
stratifications as in Fig. 12.2. Note, this is also apparent in Fig. 12.3, where the

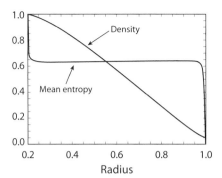

Figure 12.4 The prescribed reference state density and the evolved mean entropy pertur-
            bation maintained by the convection that is illustrated in Fig. 12.3. Both are
            normalized and plotted vs. the normalized radius.

color changes from blue to yellow wherever the normalized entropy perturbation
equals one half.

SUPPLEMENTAL READING

Clayton (1984)
Olson & Schubert (2007)

EXERCISES

1. *Adiabatic sound speed*
   Verify Eq. 12.8 for the adiabatic sound speed in a perfect gas.
2. *Viscous heating*
   Verify the viscous heating term for a 2D cartesian box (Eq. 12.32) starting
   from its expression in Eq. 12.31.
3. *Polytrope n = 1*
   Verify that the polytropic function for $n = 1$ (Section 12.2.2) is a solution to
   the Lane-Emden equation (12.60). Verify that it satisfies the central boundary
   conditions, Eq. 12.61, and that $\bar{g}$, h_ρ, and $d\overline{S}/dr$ for a perfect gas vanish at the
   center ($\xi = 0$). Derive the central density, pressure, and temperature in terms
   of the total mass (M_o) and outer radius (R_o) (where the polytropic function
   vanishes) for a giant planet model approximated as a polytrope of $n = 1$ with
   a perfect gas equation of state and a mean mass per particle of 0.5 atomic
   mass units.
4. *Viscous torque*
   Verify the viscous torque term in Eq. 12.87, for which $\bar{v}$ is an arbitrary function of z. Also verify the viscous torque term in Eq. 12.88, for which the
   dynamic viscosity, $\bar{\rho}\bar{v}$, is constant.

5. *Anelastic Nusselt number*

   The turbulent diffusive heat flux is $-\bar{\rho}\bar{\kappa}\overline{T}\partial S/\partial z$. Assume $\bar{\rho}\bar{\kappa}$ is constant and $\overline{T} = T_{bot}(1 - z/z_o)$. Formulate a reasonable Nusselt number (Section 5.3) for an anelastic fluid in a cartesian box.

6. *Anelastic vorticity*

   Verify the expressions for vorticity in Eqs. 12.82 and 12.85.

7. *Anelastic internal gravity wave dispersion relation.*

   Derive the anelastic dispersion relation (Eq. 12.97).

COMPUTATIONAL PROJECTS

1. *Anelastic critical Rayleigh number for a cartesian box*

   Using a linear anelastic code for thermal convection of a polytropic perfect gas in a box, iterate on the Rayleigh number, as described in Section 3.3, to find the critical Rayleigh number as a function of the horizontal mode number for a few different aspect ratios, Prandtl numbers, polytropic indices, and N_ρ. Let the $\bar{\rho}\bar{\nu}$ and $\bar{\rho}\bar{\kappa}\overline{T}$ be constants.

2. *Anelastic critical Rayleigh number for an annulus*

   Repeat the *Anelastic critical Rayleigh number for a cartesian box* exercise for an annulus with a few different ratios of top to bottom radii.

3. *Anelastic convection in an annulus*

   Do the simulation described in Section 12.5.2 but at a lower Rayleigh number and with less spatial resolution.

Chapter Thirteen

Rotation

The effects of rotation on convection and gravity waves can be significant. Certainly flows in the atmospheres, oceans, and liquid cores of terrestrial planets are dominated by the Coriolis forces, as are the interiors of giant planets and stars. The sum of centrifugal and gravitational forces can go to zero at the top boundary of a rapidly rotating star or accretion disk. Poincaré forces arise due to the time rate of change of the planetary rotation rate.

We begin this chapter with the derivation and discussion of parts of the inertial term in the momentum equation that exist due to the rotation of the body in the inertial frame, which are usually called "forces" when solved in a rotating frame of reference: the Coriolis, centrifugal, and Poincaré forces. We then describe the modifications needed to add these effects of rotation to our previous models of convection and gravity waves in 2D cartesian box and cylindrical annulus geometries, both of which now lie within a rotating equatorial plane. We also describe 2.5 D rotating models, which represent axisymmetric simulations with flows and fields varying within a meridian plane. Finally, we discuss 3D rotating spherical-shell MHD dynamo models.

13.1 CORIOLIS, CENTRIFUGAL, AND POINCARÉ FORCES

For a rotating star or planet, it is usually much more convenient to solve for the fluid flow and magnetic field in a frame of reference rotating relative to the inertial frame, i.e., relative to the universe or at least to the local neighborhood that has a negligible angular velocity compared to that of the rotating body of interest. Precessional and tidal torques due to neighboring bodies can cause the direction and/or amplitude of the body's rotation rate to change; in such a case the model's rotating frame of reference could be designed to be time-dependent. If, on the other hand, there are no external torques acting on the body, the frame's rotation rate is usually made to be constant and equal to the rate for which the total angular momentum of the body is, and remains, zero. The amplitude of the frame's rotation rate may also be designed to vary space; for example, when simulating a planetary disk, one may choose to use a radially dependent Keplerian rotating frame. However, here we assume that the rotation rate is constant in space and time. Since Newton's laws of motion apply to the inertial frame, we need to account for the accelerations due to the rotating frame and therefore need to mathematically describe the time rate of change of a vector in the rotating frame relative to its time rate of change in the inertial frame.

Consider a frame of reference rotating with angular velocity, $\boldsymbol{\Omega}$, relative to the inertial frame. As mentioned, $\boldsymbol{\Omega}$ is independent of space but its amplitude and direction could depend on time (although we will typically assume it is also time-independent). We use subscripts I and R to distinguish between time derivatives of a vector observed in the inertial and rotating frames, respectively. Note, time derivatives of scalars do not depend on the frame of reference and therefore do not have these subscripts. Let $\hat{\boldsymbol{x}}$, $\hat{\boldsymbol{y}}$, and $\hat{\boldsymbol{z}}$ be the cartesian unit vectors describing a coordinate system fixed in the rotating frame with its rotation axis (relative to the inertial frame) passing through the origin of this cartesian coordinate system.

Now consider an arbitrary vector, $\mathbf{A}$, observed in the rotating frame:

$$\mathbf{A} = A_x\hat{\boldsymbol{x}} + A_y\hat{\boldsymbol{y}} + A_z\hat{\boldsymbol{z}},$$

where A_x, A_y, and A_z are the three components of the vector. The full Lagrangian time derivative of this vector $\mathbf{A}$ in the inertial frame is

$$\left[\frac{d\mathbf{A}}{dt}\right]_I = \frac{dA_x}{dt}\hat{\boldsymbol{x}} + A_x\left[\frac{d\hat{\boldsymbol{x}}}{dt}\right]_I + \frac{dA_y}{dt}\hat{\boldsymbol{y}} + A_y\left[\frac{d\hat{\boldsymbol{y}}}{dt}\right]_I$$

$$+\frac{dA_z}{dt}\hat{\boldsymbol{z}} + A_z\left[\frac{d\hat{\boldsymbol{z}}}{dt}\right]_I$$

$$= \left[\frac{d\mathbf{A}}{dt}\right]_R + A_x(\boldsymbol{\Omega} \times \hat{\boldsymbol{x}}) + A_y(\boldsymbol{\Omega} \times \hat{\boldsymbol{y}}) + A_z(\boldsymbol{\Omega} \times \hat{\boldsymbol{z}}).$$

Here we make use of the fact that the time rate of change, as observed in the inertial frame, of the unit vector $\hat{\boldsymbol{x}}$, which is rotating at a rate $\boldsymbol{\Omega}$ relative to the inertial frame, is simply $(\boldsymbol{\Omega} \times \hat{\boldsymbol{x}})$ and likewise for the other two unit vectors. Therefore,

$$\left[\frac{d\mathbf{A}}{dt}\right]_I = \left[\frac{d\mathbf{A}}{dt}\right]_R + \boldsymbol{\Omega} \times \mathbf{A}, \qquad (13.1)$$

which is the mathematical relationship we need between time derivatives of a vector taken in the inertia and rotating frames.

Next let the arbitrary vector, $\mathbf{A}$, be the position vector, $\mathbf{r}$, of a fluid parcel relative to the origin of the rotating coordinate system, which as mentioned is on the axis of rotation. We could also define

$$\mathbf{r}_I = x_I\hat{\boldsymbol{x}}_I + y_I\hat{\boldsymbol{y}}_I + z_I\hat{\boldsymbol{z}}_I$$

as being the *same* position vector but measured in a cartesian coordinate system fixed within the inertial frame. Then, at any moment in time, although the individual components would not (usually) be equal, i.e., $x \neq x_I$, $y \neq y_I$, and $z \neq z_I$, the total vectors would be equal, i.e., $\mathbf{r} = \mathbf{r}_I$.

However, the time derivative of the position vector does depend on the frame in which it is measured. The time derivative of $\mathbf{r}$, according to Eq. 13.1, is

$$\left[\frac{d\mathbf{r}}{dt}\right]_I = \left[\frac{d\mathbf{r}}{dt}\right]_R + \boldsymbol{\Omega} \times \mathbf{r}. \qquad (13.2)$$

But the time derivative on the left of Eq. 13.2 is the fluid velocity as observed in the inertial frame, $\mathbf{v}_I$, and the time derivative on the right side is the fluid velocity as observed in the rotating frame, $\mathbf{v}_R$. That is,

$$\mathbf{v}_I = \mathbf{v}_R + \boldsymbol{\Omega} \times \mathbf{r}. \tag{13.3}$$

Newton's second law of motion, the basis for our momentum conservation equation, says the (*Net force per mass*) = (*Acceleration in the inertial frame*), which, using Eqs. 13.1 and 13.3, is

$$\left[\frac{d\mathbf{v}_I}{dt}\right]_I = \left[\frac{d}{dt}(\mathbf{v}_R + \boldsymbol{\Omega} \times \mathbf{r})\right]_I$$

$$= \left[\frac{d}{dt}(\mathbf{v}_R + \boldsymbol{\Omega} \times \mathbf{r})\right]_R + \boldsymbol{\Omega} \times (\mathbf{v}_R + \boldsymbol{\Omega} \times \mathbf{r}). \tag{13.4}$$

The first term on the right of Eq. 13.4 equals

$$\left[\frac{d\mathbf{v}_R}{dt}\right]_R + \left[\frac{d\boldsymbol{\Omega}}{dt}\right]_R \times \mathbf{r} + \boldsymbol{\Omega} \times \mathbf{v}_R$$

and the second term on the right is

$$\boldsymbol{\Omega} \times \mathbf{v}_R + \boldsymbol{\Omega} \times (\boldsymbol{\Omega} \times \mathbf{r}).$$

Therefore, after rearranging terms, an observer in the rotating frame measures the fluid acceleration as

$$\left[\frac{d\mathbf{v}_R}{dt}\right]_R = 2\mathbf{v}_R \times \boldsymbol{\Omega} + (\boldsymbol{\Omega} \times \mathbf{r}) \times \boldsymbol{\Omega} + \mathbf{r} \times \left[\frac{d\boldsymbol{\Omega}}{dt}\right]_R + \text{Net force per mass}. \tag{13.5}$$

The first three terms on the right of Eq. 13.5 are accelerations relative to the inertial frame but are called forces per mass in the rotating frame. The first one is the Coriolis force per mass, which is always perpendicular to the velocity and therefore does no work. That is, it does not change kinetic energy; it only changes the direction of the flow.

The second term on the right of Eq. 13.5 is the centrifugal force per mass, which is directed away from the rotation axis with an amplitude of $\Omega^2 r$, where r is the cylindrical radius. It can also be written as $-\nabla\Phi_c$, where the centrifugal potential energy per mass, Φ_c, is $-\Omega^2 r^2/2$. This makes $\nabla^2\Phi_c = -2\Omega^2$. Usually, the gravitational (Φ) and centrifugal (Φ_c) potentials are combined and called the "geopotential": $\Phi_g \equiv \Phi + \Phi_c$; and therefore $\nabla^2\Phi_g = 4\pi G\rho - 2\Omega^2$, where here ρ is the total mass density.

The third term on the right of Eq. 13.5 is the Poincaré force per mass. It occurs when the amplitude or direction of the body's angular velocity changes in time. This is usually due to external gravitational forces that cause precession or tides. Here we assume $\boldsymbol{\Omega}$ is a constant and so there is no Poincaré force.

Now we drop the subscript R and let the fluid velocity, $\mathbf{v}$ (and the magnetic field, $\mathbf{B}$), be that observed in the rotating frame of reference. The Eulerian time derivative of velocity is then

$$\frac{\partial\mathbf{v}}{\partial t} = -(\mathbf{v}\cdot\nabla)\mathbf{v} + 2\mathbf{v} \times \boldsymbol{\Omega} - \nabla P + \cdots, \tag{13.6}$$

where P is the reduced pressure (Eq. 12.25). Since the advection term can be written as

$$-(\mathbf{v}\cdot\nabla)\mathbf{v} = \left(\mathbf{v}\times\boldsymbol{\omega} - \tfrac{1}{2}\nabla\mathrm{v}^2\right),$$

where the *vector* vorticity $\boldsymbol{\omega}$ is $\nabla\times\mathbf{v}$,

$$\frac{\partial\mathbf{v}}{\partial t} = \mathbf{v}\times(\boldsymbol{\omega}+2\boldsymbol{\Omega}) - \nabla\left(P+\tfrac{1}{2}\mathrm{v}^2\right) + \cdots. \tag{13.7}$$

Note that $2\boldsymbol{\Omega} = \nabla\times(\boldsymbol{\Omega}\times\mathbf{r})$ is called the "planetary vorticity" since $(\boldsymbol{\Omega}\times\mathbf{r})$ is the rotation velocity observed from the inertial frame; therefore the term $(\boldsymbol{\omega}+2\boldsymbol{\Omega})$, which equals $\nabla\times(\mathbf{v}+\boldsymbol{\Omega}\times\mathbf{r})$, is the fluid vorticity at position $\mathbf{r}$ as measured in the inertial frame and $(\mathbf{v}+\boldsymbol{\Omega}\times\mathbf{r})$ is the fluid velocity in the inertial frame.

Our strategy has been to take the curl of the momentum equation (13.7) to get the vorticity equation. As previously seen, this removes the gradient term and leaves the advection of vorticity; but now there are two additional Coriolis terms:

$$\frac{\partial\boldsymbol{\omega}}{\partial t} = -(\mathbf{v}\cdot\nabla)\boldsymbol{\omega} - (\boldsymbol{\omega}+2\boldsymbol{\Omega})\nabla\cdot\mathbf{v} + ((\boldsymbol{\omega}+2\boldsymbol{\Omega})\cdot\nabla)\,\mathbf{v} + \cdots. \tag{13.8}$$

(Note, there would be two additional Coriolis terms if $\boldsymbol{\Omega}$ were not constant in space.) The second term on the right in Eq. 13.8 is called the *compressional torque* because the vorticity of a fluid parcel increases as it sinks through a density stratification and so contracts (i.e., $\nabla\cdot\mathbf{v}$ is negative) and likewise decreases as the parcel rises. This term obviously vanishes for a Boussinesq model. The third term is called the *stretching torque* because the vorticity of the parcel increases where there is a gradient of its velocity parallel to its total vorticity. This term vanishes for strictly 2D convection because both $\boldsymbol{\omega}$ and $\boldsymbol{\Omega}$ are perpendicular to the plane of the fluid flow.

If it can be argued that centrifugal, Poincaré, buoyancy, viscous, and nonlinear forces are all negligible compared to Coriolis and pressure gradient forces in Eq. 13.6 and if the fluid is incompressible and in steady state, the curl of the remaining Coriolis and pressure gradient terms would leave just

$$(2\boldsymbol{\Omega}\cdot\nabla)\mathbf{v} = 0, \tag{13.9}$$

the classic *Proudman-Taylor Theorem* (Proudman, 1916; Taylor, 1917). This states that in a *geostrophic flow* to first order there is no gradient parallel to the rotation axis for the fluid velocity. If the boundaries containing the fluid were impermeable (and preventing the identical nonzero fluid velocities on the northern and southern boundaries), this theorem further implies that the velocity parallel to the rotating axis also vanishes everywhere. Flows within a spherical shell would therefore be only within planes parallel to the equatorial plane. The resulting columnar flow patterns are called *Taylor columns* and are seen in both laboratory experiments and 3D numerical simulations for laminar flows dominated by Coriolis forces. Typically Taylor columns refer to nonaxisymmetric *convective* columns that encircle the rotation axis and have alternating signs of local vorticity. However, a cylindrical column axisymmetric longitudinal flow centered on the rotation axis within a sphere or spherical shell would also be called a Taylor column, one for which the fluid parcels maintain a constant latitude and radius. The axisymmetric part of a 3D flow that has this pattern would be called a zonal flow that is "constant on cylinders".

13.2 2D ROTATING EQUATORIAL BOX

Again we assume that a 2D cartesian box represents a small region within a star or planet. The problem discussed in this section is strictly two-dimensional, i.e., not only are there no gradients in the y-direction but there are also no flows or fields in the y-direction, as we assumed in previous chapters. In this case the rotation axis needs to be in the $\hat{y}$-direction; otherwise there would be unbalanced Coriolis forces in the y-direction. Therefore, we consider the plane of the fluid flow and magnetic field, i.e., the x, z-coordinate plane, to be the equatorial plane of the rotating body. We also assume the angular velocity of the rotating frame of reference, Ω, is constant in space and time and that the total angular momentum of the fluid box vanishes in this rotating frame. In addition, $\bar{g}(z)$ will now represent the sum of the gravitational acceleration and the centrifugal acceleration; and, since our model is a cartesian box with this net geopotential acceleration everywhere in the negative z-direction, we are assuming the size of the box is small relative to the distance to the rotation axis.

13.2.1 2D Box: Equations

Since we are considering a strictly 2D case in this section, the vorticity equation (in the y-direction), based on Eqs. 12.15, 12.86, and 13.8, assuming $\boldsymbol{\Omega} = \Omega \hat{y}$ and $\mathbf{g} = -\bar{g}\hat{z}$, is

$$\frac{\partial \omega}{\partial t} = -\mathbf{v}\cdot\nabla\omega + (\omega + 2\Omega)h_\rho v_z + \bar{g}\frac{\partial C}{\partial x} + \text{Viscous and magnetic terms},$$

where here ω is the amplitude of the y-component of vorticity. As mentioned in Section 12.3, the first two terms on the right can be combined, giving

$$\frac{\partial \omega}{\partial t} = -\nabla\cdot(\omega\mathbf{v}) + 2\Omega h_\rho v_z + \bar{g}\frac{\partial C}{\partial x} + \text{Viscous and magnetic terms}. \quad (13.10)$$

If the model uses nondimensional variables, as we do for most of Part 1, and employs a Fourier expansion with permeable and periodic side boundaries (Section 10.2), the vorticity equation would be

$$\frac{\partial \omega_m}{\partial t} = -[\nabla\cdot(\omega\mathbf{v})]_m + \text{Ek}^{-1}\text{Pr}\, h_\rho \left(\frac{2\pi i m}{a\bar{\rho}}\right)\psi_m + \cdots,$$

where m is the horizontal mode number and Ek is the *Ekman number*:

$$\text{Ek} = \frac{\bar{v}_{mid}}{2\Omega D^2}. \quad (13.11)$$

The Ekman number is a measure of the viscous effects relative to the Coriolis effects. Sometimes Ek is defined without the "2" in its denominator, in which case an additional factor of 2 would appear in equation. Also, sometimes the square root of the *Taylor number*, Ta, is used instead of Ek^{-1}, where

$$Ta = \frac{4\Omega^2 D^4}{\bar{v}_{mid}^2}.$$

The Coriolis term in Eq. 13.10 is linear and can be treated explicitly with the buoyancy term each numerical time step using Eqs. 12.81b to calculate v_z. Alternatively, if the complex Fourier coefficients of ω and ψ are updated simultaneously (Sections 10.3.2 and 10.3.3), this term could be treated implicitly.

It is obvious from Eq. 13.10 that for strictly 2D Boussinesq convection with the rotation axis normal to the plane of the flow, Coriolis forces have no effect on the flow. They exist but the perturbation pressure can always adjust to produce gradients that exactly cancel the Coriolis forces everywhere at all times. However, within a density-stratified fluid, local vorticity is generated as a fluid parcel rises or sinks.

13.2.2 2D Box: Simulations

Now let's examine some nonlinear simulations of anelastic rotating thermal convection in a density-stratified fluid within a cartesian box with permeable periodic side boundaries. To appreciate the effects of convective driving, rotation, and background magnetic field Fig. 13.1 illustrates four cases as snapshots of entropy perturbation (Glatzmaier, 2005a). All four cases employ an adiabatic (plane-layer) polytropic reference state (Section 12.2.1) with a polytropic index $n = 1$, which makes the ratio of specific heats $\gamma = 2$ and approximates the deep interior of a giant gas planet like Jupiter. However, we simulate here just a small region of the interior, setting the number of density scale heights spanning the box to $N_\rho = 0.2$, i.e., the density at the bottom is 22% greater than that at the top. The viscous, thermal, and magnetic diffusivities are equal (i.e., Pr = q = 1) and constant in radius for all four cases. All four cases have an aspect ratio of $a = 2$. The bottom and top boundaries are impermeable, stress-free, and constant entropy; the side boundaries are permeable and periodic (Section 10.2). Also, the mean horizontal flow and field are suppressed by maintaining $\omega_{m=0}$, $\psi_{m=0}$, $A_{m=0}$, and $J_{m=0}$ equal to zero at all z and t.

The equations are solved using a second-order semi-implicit scheme (Section 8.2). The relatively extreme values of the Rayleigh and Ekman numbers used in cases (b)–(d) require relatively high spatial resolution. We use 1333 complex Fourier modes, computing the nonlinear terms with a spectral-transform method (Section 10.4) on 4001 horizontal grid points. In the vertical direction we use a centered second-order finite-difference method on 2001 Chebyshev grid levels (Section 9.1). These cases were run on parallel computers using 128 processors.

Case (a) has a relatively small Rayleigh number (Eq. 12.89a), $Ra = 3 \times 10^6$; convection is driven a million times more strongly for the other cases with Ra $= 3 \times 10^{12}$ by using viscous and thermal diffusivities that are each a thousand times smaller than those for case (a). All but case (b) are rotating. Case (a) is a

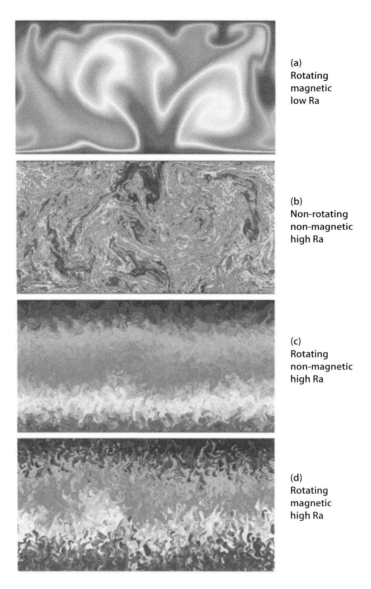

(a)
Rotating
magnetic
low Ra

(b)
Non-rotating
non-magnetic
high Ra

(c)
Rotating
non-magnetic
high Ra

(d)
Rotating
magnetic
high Ra

Figure 13.1 Snapshots of the entropy perturbation in four anelastic simulations of thermal convection in a 2D box with an adiabatic polytropic reference state defined by $n = 1$, $N_\rho = 0.2$, and $Pr = q = 1$ (see *Color Plate* 7 for a color version of this figure). Case (a) has $Ra = 3 \times 10^6$, $Ek = 10^{-4}$, and $Q = 10^4$. The other three cases have $Ra = 3 \times 10^{12}$. Case (b) is nonrotating and nonmagnetic. Cases (c) and (d) are rotating with $Ek = 10^{-9}$. Case (c) is nonmagnetic and case (d) has a vertical background magnetic field with $Q = 10^6$. Reds represent hot fluid, yellows warm fluid, and blues cold fluid. (This material is reproduced from Glatzmaier (2005a) with permission from *Taylor and Francis Group*, LLC, a division of *Informa plc*.)

relatively slow rotator with an Ekman number (Eq. 13.11) of Ek $= 10^{-4}$; cases (c) and (d) rotate a hundred thousand times faster with Ek $= 10^{-9}$. Cases (b) and (c) do not have background magnetic fields. Case (a) has a relatively weak vertical background field with a Chandrasekhar number (Eq. 11.29) of Q $= 10^4$; case (d) has a more intense vertical background field with Q $= 10^6$.

It is clear in Fig. 13.1 that, on the scale of the box, case (a) is laminar and the other three are turbulent. Case (a) is, however, time-dependent; but the upwelling and downwelling thermal plumes are relatively large, typically spanning the entire depth of the box. As is seen in Fig. 13.2, the kinetic energy drops about 25 orders of magnitude from wavenumber 100 to 300. Since the effects of rotation and magnetic field are relatively mild for case (a), the major difference between cases (a) and (b) is the much larger Rayleigh number of case (b), which results in a much larger Reynolds number (Eq. 5.3): Re for case (b) is 10^6; whereas it is 10^3 for case (a). This puts much more kinetic energy into the smaller scales, which experience local shear instabilities. However, since kinetic energy tends to cascade to larger scales in 2D turbulence, there is also considerable energy in the large scales (small mode numbers), as can be seen in part (b) of Fig. 13.2. (Note, both axes in Fig. 13.2 for case (a) are very different from those for the other three cases.) The much smaller viscous and thermal diffusivities of case (b) also cause the boundary layers to be much more shallow than they are in case (a), which is easily seen in Fig. 13.3.

Case (c) is case (b) with rapid rotation. This produces turbulent boundary layers because the plumes that develop from boundary layer instabilities generate vorticity as they rise off the bottom and sink off the top due to the density stratification and to the Coriolis torque in the vorticity equation 13.10. This effect is clearly seen when comparing parts (b) and (c) in Fig. 13.1. Physically, as a fluid parcel rises through density stratification it expands, producing Coriolis forces that cause it to rotate in the opposite sense to the planetary rotation, $\boldsymbol{\Omega}$, i.e., it generates local negative vorticity. Likewise, a sinking fluid parcel contracts, generating positive vorticity (Glatzmaier & Gilman, 1981b). This process removes some of the large-scale kinetic energy by inhibiting plumes that span the entire box; the change in the kinetic energy spectrum is seen when comparing parts (b) and (c) in Fig. 13.2. This process also makes the convective heat transfer less efficient, which therefore requires a superadiabatic temperature gradient (negative entropy gradient) in the bulk of the fluid box, as can be seen when comparing parts (b) and (c) in Fig. 13.3.

Besides the Ekman number, the effect of Coriolis forces on the flow is measured in terms of the *Rossby number*, Ro, the nondimensional ratio of advection effects over Coriolis effects:

$$\mathrm{Ro} = \frac{V}{2\Omega D},$$

where V is a typical flow velocity. Note, Ro = Re Ek. Another convenient nondimensional number is the *convective Rossby number*, Ro_c, the ratio of buoyancy effects over Coriolis effects:

$$\mathrm{Ro}_c = \left(\frac{g\Delta S}{c_p D}\right)^{1/2} / 2\Omega,$$

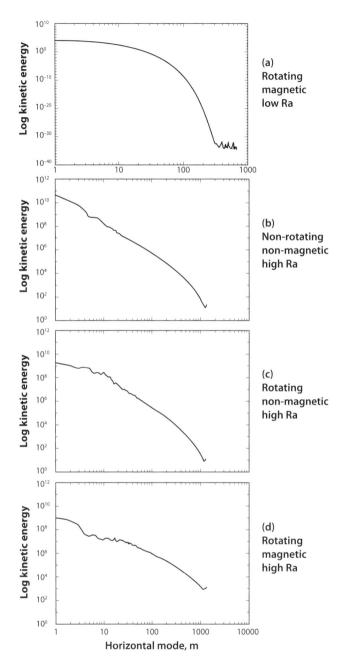

Figure 13.2 The kinetic energy spectra for the four cases illustrated in Fig. 13.1. The log of the volume-averaged kinetic energy is plotted as a function of the log of the horizontal mode number. Note, the axes for case (a) differ from those for the other three cases.

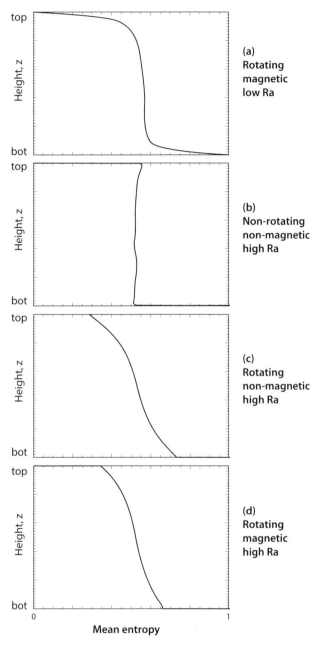

Figure 13.3 The horizontally averaged and time-averaged entropy perturbations for the four
cases illustrated in Fig. 13.1. Entropy is scaled by ΔS and plotted as a function
of height.

which equals $(Ra/Pr)^{1/2}$ Ek. The ratio of magnetic field effects and Coriolis effects can be measured in terms of the *magnetic* Rossby number, Ro_m:

$$Ro_m = \left(\frac{B_o}{(\rho\mu)^{1/2}_{mid} D} \right) / 2\Omega \, ,$$

which equals the ratio of the Alfvén frequency over the Coriolis frequency or the rotational period over the Alfvén travel time across D. The Ro and Ro_c for case (a) are 0.1 and 0.2 and for case (c) are 4×10^{-4} and 2×10^{-3}, respectively, which indicates how much more dominant Coriolis effects are for case (c).

Case (d) is case (c) with an intense vertical background magnetic field. This inhibits horizontal flows, which deform the background field. The resulting strong Lorentz forces provide a restoring force that produces horizontal oscillations, i.e., Alfvén waves. The horizontal oscillation velocities for this case are comparable to the vertical convective velocities. Turbulent boundary layers still exist; but the character of the plumes in case (d) differ from that in case (c) because of these magnetic restoring forces. This is seen when comparing parts (c) and (d) of Fig. 13.2; a somewhat higher kinetic energy is in the small scales for case (d). It is also seen when comparing parts (c) and (d) of Fig. 13.3, with case (d) being a little more efficient convecting heat and so having somewhat less steep entropy gradient in the bulk of the fluid. That is, magnetic forces tend to offset the development of vortices by Coriolis forces. The magnetic Rossby number, Ro_m, for case (d) is 10^{-6}, compared to 10^{-2} for case (a), indicating how much more dominant Coriolis effects are for case (d).

These results are reflected in the values of the Nusselt number (Eq. 5.1a, but here with $c_p \rho_o v_z T$ replaced by $\overline{T} \bar{\rho} v_z S$). Case (a), with its smaller Ra, has the smallest Nusselt number, Nu $= 11$; and case (b), with the large Ra and no rotation, has the largest, Nu $= 175$. Adding rotation makes the convection less efficient; Nu $= 98$ for case (c). Adding magnetic field reduces the rotational hindrance; Nu $= 120$ for case (d).

Now consider another anelastic polytropic rotating (but nonmagnetic) case, one that has a much larger density stratification, $N_\rho = 5$, and a polytropic reference state with $n = 1.5$, i.e., a perfect gas with $\gamma = 5/3$. The Rayleigh, Prandtl, and Ekman numbers are Ra $= 2 \times 10^{12}$, Pr $= 0.1$, and Ek $= 10^{-9}$, respectively, which makes $Ro_c = 5 \times 10^{-3}$ (Glatzmaier, 2005a). The resulting Reynolds and Rossby numbers are 3×10^6 and 3×10^{-3}, respectively. This rapidly rotating case is illustrated in Fig. 13.4 with four snapshots showing how the initial large-scale plumes break down into small-scale turbulence because of shear instabilities. Small-scale but intense vortices, generated by the Coriolis torque described above, slowly migrate through the turbulence. They survive for a relatively long time because of the small viscous and thermal diffusivities represented here by the large Rayleigh and Reynolds numbers.

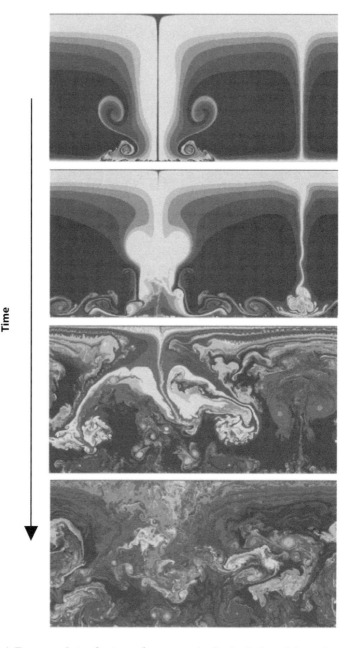

Figure 13.4 Four snapshots of entropy from an anelastic simulation of thermal convection in a 2D box with Ra $= 2 \times 10^{12}$, Pr $= 1/10$, Ek $= 10^{-9}$, and $N_\rho = 5$ (see *Color Plate* 8 for a color version of this figure). Here dark colors represent cold fluid and light colors hot fluid. (This material is reproduced from Glatzmaier (2005a) with permission from *Taylor and Francis Group*, LLC, a division of *Informa plc*.)

13.3 2D ROTATING EQUATORIAL ANNULUS: DIFFERENTIAL ROTATION

Now we assume strictly 2D flow in an annulus rotating about an axis through its center and normal to its plane, the "equatorial plane." We solve the equations in the frame of reference rotating with a constant angular velocity $\boldsymbol{\Omega} = \Omega\hat{z}$ in cylindrical coordinates (r, ϕ, z) with the coordinate axis being the rotation axis. The initial condition is, as usual, no velocity relative to the frame of reference and therefore, with our stress-free boundary conditions, the total integrated angular momentum of the fluid annulus continues to vanish in this rotating frame. We also let $\bar{g}(r)$ represent the sum of the gravitational acceleration and the centrifugal acceleration in this frame.

13.3.1 2D Annulus: Vorticity Waves and Differential Rotation

As discussed in Section 13.2.1, the only modification to the equations is the addition of the Coriolis torque,

$$2\Omega h_\rho v_r \, ,$$

in the vorticity equation 12.88. The Fourier coefficient of this (dimensional) Coriolis torque in terms of the complex streamfunction for mode number m is

$$\frac{2\Omega h_\rho i m \psi_m}{r \bar{\rho}} \, .$$

As for the equatorial box geometry (Section 13.2.1), this term would vanish for this equatorial annulus if the Boussinesq approximation ($h_\rho = 0$) were made.

As discussed in Section 12.4.1, the critical Rayleigh number for thermal convection in a density-stratified fluid depends on the vertical dependences of the reference state variables. Glatzmaier & Gilman (1981a,c) numerically solve for the critical Rayleigh numbers and critical mode numbers for anelastic convection in a 3D spherical shell and describe how the structure and phase propagation of the most unstable mode depend on the chosen reference state and the rotation rate. For the onset of Boussinesq convection in a rapidly rotating sphere the critical Rayleigh number and critical wavenumber go as $\text{Ek}^{-4/3}$ and $\text{Ek}^{-1/3}$, respectively (Roberts, 1968; Busse, 1970); and both increase with Prandtl number (Jones et al., 2000). Glatzmaier & Gilman (1981a,c) find that the critical Rayleigh number and critical wavenumber also increase with density stratification. In addition, they describe how convective cells at onset are tilted in a way that rising fluid has a prograde (i.e., eastward) component of velocity and sinking fluid has a retrograde (westward) component, the opposite of what one might expect based only on the Coriolis forces due to the vertical velocity. This tilt in the convecting flow maintains a perturbation angular velocity relative to the rotating frame of reference, i.e., a *differential rotation*, because of the convergence of Reynolds stress.

Convection in rotating laminar Boussinesq convection (Busse, 1983) is also tilted and so also maintains differential rotation; but the mechanism is quite different compared to that in a rotating density-stratified fluid (Glatzmaier et al., 2009).

The Boussinesq *Taylor-column mechanism* is based on the Proudman-Taylor Theorem, Eq. 13.9. As fluid circulates around a convective Taylor column it needs to expand in planes parallel to the equatorial plane when rising and contract when sinking because the incompressible fluid in the column is confined between the sloping impermeable northern and southern boundaries, which forces the length of the column to be less on the outer side and greater on the side closer to the rotation axis. Note, the sloping boundaries force smaller secondary flows parallel to the rotation axis, in violation of the Proudman-Taylor Theorem. The secondary expansion and contraction flows parallel to the equatorial plane result in Coriolis forces that generate negative local vorticity (i.e., anticyclones) in rising fluid and positive local vorticity (cyclones) in sinking fluid. However, this Boussinesq Taylor-column mechanism requires straight vortex columns, extremely thin because of the small viscosity, spanning continuously from the northern to the southern boundaries without buckling, all within a very turbulent 3D environment, an unlikely scenario.

Let's consider how Coriolis forces cause tilting of convecting cells in a rotating density-stratified fluid and how this maintains differential rotation, which is likely at least part of the reason for the differential rotation observed in our sun and giant planets. The mechanism is based on compressional torque (second term on the right in Eq. 13.8). Consider a buoyant fluid parcel, expanding as it rises through a density-stratified medium. The expansion in the plane normal to the rotation axis produces a Coriolis torque that causes it to rotate (relative to the rotating frame of reference) in the direction opposite to the direction (relative to the inertial frame) of the basic rotation of the planet. That is, this expansion velocity, which is outward from the center of the parcel, is crossed into the angular velocity of the rotating frame, resulting in a Coriolis force tangent to the perimeter of the parcel that produces an angular acceleration in the opposite direction of the angular velocity of the rotating frame. Note that the Coriolis force due to just the vertical rise of the parcel normal to the rotation axis is balanced by part of the pressure gradient force (as it also would be in the Boussinesq model) and so does not generate vorticity; it is the Coriolis torque due to the expansion of the parcel as it rises (i.e., the divergence of the velocity within the equatorial plane) that generates negative vorticity (i.e., an anticyclone) relative to the rotating frame. Likewise, the contraction of a sinking fluid parcel generates positive vorticity (i.e., a cyclone) relative to the rotating frame.

One can also describe this process by defining the potential vorticity of the fluid parcel as the total vorticity (twice the planet's rotation rate plus its local vorticity within the rotating frame) divided by the reference state density and can show that this quantity is approximately conserved as the parcel moves (Ertel, 1972; Glatzmaier & Gilman, 1981b; Glatzmaier et al., 2009). That is, to conserve potential vorticity as the density of a rising fluid parcel decreases, its local vorticity also needs to decrease, and vice versa. The derivation of this theorem includes the nonlinear inertial term in addition to the pressure gradient and Coriolis terms in the momentum equation and therefore is a better approximation for rotating turbulent flows than is the linear steady-state Proudman-Taylor theorem (Eq. 13.9), which neglects the inertial term.

Turbulent convection clearly does not have the traditional large-scale cellular structure seen in laminar convection. Instead, one needs to consider the motion and deformation of small isolated plumes that gain and lose vorticity as they sink and rise, respectively. The *density stratification mechanism* for maintaining differential rotation is based on the resulting tilted flow trajectories, which transport angular momentum from one region to another. It involves a radially dependent phase propagation in longitude of the vorticity pattern, which tilts the plume trajectories in longitude and thus, by the convergence of the nonlinear Reynolds stress, redistributes angular momentum in radius.

How this happens is easier to understand by thinking of a simple series of laminar upflows and downflows as convection cells within the equatorial plane. Positive vorticity peaks in the centers of counterclockwise circulating cells (cyclones) and negative vorticity peaks in the centers of clockwise circulating cells (anticyclones), when viewed in the rotating frame from the north. However, the rate that vorticity is being generated peaks in the upflows and downflows between the cell centers, i.e., 90 degrees out of phase relative to the existing pattern of vorticity.

Since rising fluid generates anticyclonic vorticity on the prograde side of anticyclones and on the retrograde side of cyclones, the phase of this circulation pattern propagates in the prograde direction (Glatzmaier & Gilman, 1981b). This Rossby-like vorticity wave occurs because density decreases with radius. However, the direction the convecting plumes tilt and the resulting type of differential rotation depend on how the phase propagation in longitude varies in radius. The greater the magnitude of the relative change in background density with radius (i.e., the smaller the local density scale height), the greater the phase velocity at the given radius.

Consider the usual case for which the density scale height decreases with radius. In this case, the pattern of circulation propagates eastward faster at greater radii. This causes rising fluid to tilt eastward (i.e., prograde) and sinking fluid to tilt westward (i.e., retrograde). That is, prograde angular momentum is transported toward the top boundary and retrograde angular momentum is transported away from the top boundary. The opposite occurs at the bottom boundary. This convergence of Reynolds stress maintains prograde zonal flow (i.e., the axisymmetric part of the longitudinal velocity) in the outer part of the convection zone and retrograde zonal flow in the inner part. The robustness of this density stratification mechanism for maintaining differential rotation is that it does not require thin straight vortex columns that span from the northern to southern hemispheric boundaries in a laminar environment, as does the Boussinesq Taylor-column mechanism. Instead, this mechanism works locally and in a strongly turbulent environment.

Once this differential rotation is established it provides positive nonlinear feedback. That is, rising and sinking plumes are sheared by the existing differential rotation, which increases the tilt and tightens the spiral flow structure to the degree determined by the turbulent mixing, which attempts to smooth out the spiral shear.

13.3.2 2D Annulus: Simulations

For a simple demonstration of how effective the density stratification mechanism is in maintaining differential rotation consider two cases of anelastic convection in a 2D rotating equatorial annulus (Glatzmaier et al., 2009). Both cases have

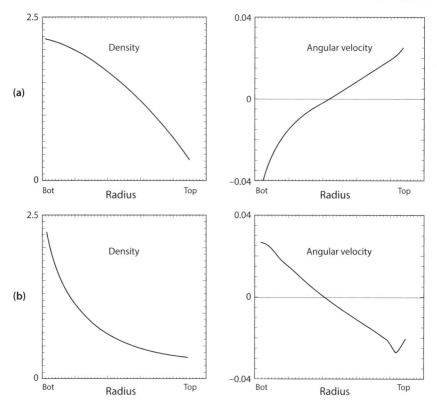

Figure 13.5 Profiles of the density and the resulting angular velocity for the two cases, (a) and (b), of rotating thermal convection in an annulus as described in Section 13.3.2. The density at the bottom is seven times the top density and the angular velocity is relative to the rotating frame and scaled by the frame's rotation rate, Ω.

$Ra = 3 \times 10^{11}$, $Ek = 10^{-8}$, and $Pr = 0.2$; thus the convective Rossby number is $Ro_c = 0.01$, meaning that Coriolis forces are roughly a hundred times greater than buoyancy forces. The boundaries are impermeable, stress-free, and constant entropy. The bottom boundary radius is 21% of the top boundary radius and the density at the bottom boundary is seven times greater than that at the top boundary.

The two cases differ in the radial profile of the reference state density; case (a) has small density scale heights near the top boundary and case (b) has the small scale heights near the bottom boundary, as illustrated in Fig. 13.5. The reference state density profile for case (a) is

$$\bar{\rho}(r) = \rho_{bot}\left(1.040 - 0.897\left(\frac{r}{r_{top}}\right)^2\right)$$

and for case (b) it is

$$\bar{\rho}(r) = \rho_{bot} \left(\frac{r_{bot}}{r} \right)^{1.247} .$$

From these, the gravitational acceleration, $\bar{g}(r)$, for each case can be obtained via Eq. 10.9 and then $\overline{T}(r)$ by integrating the adiabatic temperature gradient, $-\bar{g}/c_p$. We set $\bar{\nu}$, $\bar{\kappa}$, and c_p to constants.

The equations are solved using a Chebyshev-Fourier numerical method (Section 9.4) with a spectral-transform method to calculate the nonlinear terms (Section 10.4) and a second-order semi-implicit time integration scheme (Section 8.2). Both cases use 1365 complex Fourier modes on 4096 longitudinal grid points and 1537 Chebyshev polynomials in radius. Each case was run for more than 2,000,000 numerical time steps, which represents 1000 rotation periods. The resulting fluid velocities correspond to a Reynolds number of about Re $= 10^5$ and a Rossby number of Ro $= 0.01$.

Since there is no flow parallel to the rotation axis, there is no stretching torque (Eq. 13.8); the only torque due to rotation is a compressional torque, the linear part of which is $-2\Omega \nabla \cdot \mathbf{v} = 2\Omega h_\rho v_r$. Recall that the local density scale height is $-h_\rho^{-1}$; therefore, for a given v_r, the local vorticity generation rate is greatest for case (a) near the top boundary and for case (b) near the bottom boundary. As discussed in Section 13.3.1, this causes rising fluid in case (a) to have a prograde velocity component and sinking fluid to have a retrograde component. The resulting tilt transports prograde angular momentum into the region near the top boundary and retrograde angular momentum near the bottom boundary. The convergence of these nonlinear Reynolds stresses maintains a prograde angular velocity (i.e., eastward zonal flow) in the outer region and a retrograde angular velocity in the inner region, both relative to the rotating frame of reference. The opposite differential rotation is maintained in case (b) as illustrated in Figs. 13.5 and 13.6. The perturbation angular velocity in these two cases peaks near the boundaries at 3% to 4% of the angular velocity of the rotating frame of reference, Ω.

The convective velocities in these two simulations are typically a hundred times less than the zonal winds that they maintain. The resulting superposition of nonaxisymmetric convection and the axisymmetric zonal flow is not the typical cellular convection pattern. Instead, as viewed in the rotating frame, the fluid trajectories are wavelike flows in longitude because the fluid parcels move so much faster in longitude than in radius. Since the total angular momentum of the annulus, measured in the rotating frame, remains zero (because of the stress-free boundaries), the angular momentum of the inner region cancels the angular momentum of the outer region. On average, the inner and outer parts of the spiral make one revolution in opposite directions about every 50 planetary rotation periods.

As mentioned in Section 13.3.1, this mechanism has a positive feedback. That is, the established differential rotation also tilts rising and sinking convective plumes, which reinforces the transport of momentum that maintains the particular differential rotation that exists. This feedback mechanism may be even more important than the density stratification mechanism just described since switching the reference state density profile, from case (a) to (b), after the differential rotation has been

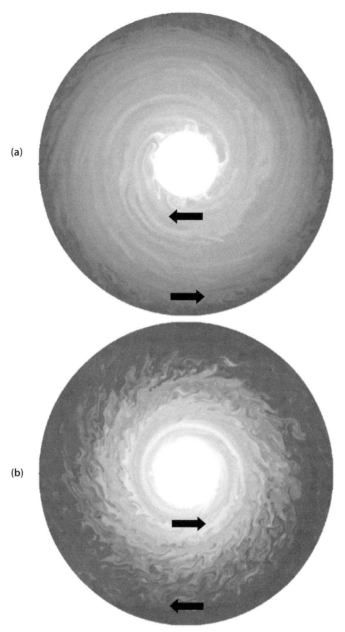

Figure 13.6 Snapshots of the entropy perturbation for the two anelastic simulations, (a) and
(b), of rotating thermal convection in an annulus described in Section 13.3.2
(see *Color Plate* 9 for a color version of this figure). Dark (light) colors represent cold (hot) fluid. The arrows show the angular velocity relative to the
rotating frame, which rotates counterclockwise in the inertial frame. (This material is reproduced from Glatzmaier et al. (2009) with permission from *Taylor
and Francis Group*, LLC, a division of *Informa plc*.)

established does not reverse the differential rotation. This feedback may be the reason differential rotation can be maintained in fairly turbulent rotating Boussinesq convection, for which the Taylor columns are not continuous through the deep interior, when the differential rotation has initially been established within a more laminar convective flow.

This feedback mechanism may have played a role in the establishment of the different banded zonal wind patterns seen on Jupiter and Saturn because of different impact histories these giant gas planets may have experienced. Of course, the different zonal patterns may also be due to the different internal structures these giant planets have and the different tides they experience.

Now consider the 2D annulus case described in Section 12.5.2 and illustrated in Figs. 12.3 and 12.4, which has a greater density stratification, $N_\rho = 3$; but here we set Ra to 10^{11}, Pr to 0.5, and Ek to 2×10^{-8}. The lower Prandtl number usually results in a stronger zonal flow. The higher Rayleigh number here is chosen to drive a comparably turbulent flow since the critical Rayleigh number increases with the rotation rate, Ω, i.e., with decreasing Ekman number.

The results for this case are illustrated in Figs. 13.7 and 13.8. As in case (a) of Figs. 13.5 and 13.6, which also has a density scale height that decreases with radius, this case maintains a differential rotation with much of the inner region being a retrograde zonal flow (blue) and the outer region a prograde flow (yellow). Note, however, that for this case there is also a shallow layer adjacent to the bottom boundary that has a strong prograde zonal flow, presumably due to local boundary layer dynamics. The huge effect of rotation is clearly seen when comparing this strong, persistent, differential rotation with the lack of any significant or persistent differential rotation in the nonrotating case and by comparing the spiral entropy structure in Fig. 13.7 with the plume structure of the nonrotating case in Fig. 12.3.

Another effect of rotation is the enhanced depth of the thermal boundary layers, which is particularly obvious at the bottom boundary when comparing the mean entropy profile in Fig. 13.8 with that of Fig. 12.4. Note that in both cases the mean entropy equals 1 at the bottom boundary ($r = 0.2$). The rotating case (Fig. 13.8) maintains a deep turbulent boundary layer above the very shallow diffusive boundary layer. A similar effect is seen for the 2D box simulations described in Section 13.2.2.

Before moving to 2.5D models, we should mention that considerable progress has been made using 2D anelastic rotating annulus models with a *stable* thermal stratification to study the excitation of internal gravity waves by penetrating convection and how they transport and deposit angular momentum within the stable rotating region. In particular, simulations have been used to study the dynamics of the sun's stable radiative zone below the its convection zone (e.g., Rogers et al., 2008).

13.4 2.5D ROTATING SPHERICAL SHELL: INERTIAL OSCILLATIONS

Recall that a "2.5D spherical-shell model" means that fluid velocity and magnetic field have vector components in all three coordinate directions, radius, colatitude, and longitude; but these vectors and the thermodynamic scalars depend only on

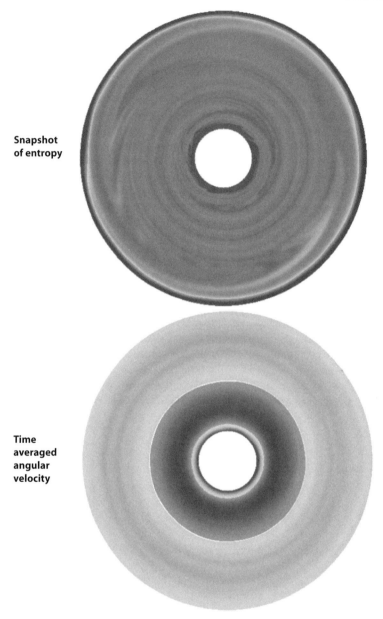

Snapshot of entropy

Time averaged angular velocity

Figure 13.7 (Top) A snapshot of the entropy perturbation for the continuation of the $N_\rho = 3$ simulation illustrated in Fig. 12.3 but with Ra increased to 10^{11}, Pr decreased to 0.5, and now rotating with Ek $= 2 \times 10^{-8}$ (see *Color Plate* 10 for a color version of this figure). Reds represent hot fluid; blues represent cold fluid. (Bottom) Time-averaged differential rotation. Reds and yellows represent counterclockwise angular velocity (i.e., prograde flow) and blues represent clockwise (retrograde) flow, both relative to the counterclockwise rotating frame.

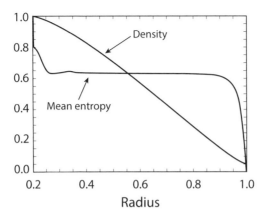

Figure 13.8 The prescribed reference state density and the evolved mean entropy pertur-
          bation maintained by the convection that is illustrated in Fig. 13.7. Both are
          normalized and plotted vs. the normalized radius.

radius and colatitude (and time). The equations and numerical method for a 2.5D
model of convection in a spherical shell, with magnetic field, density stratification,
and rotation, are described in detail in Section 10.6.

Note that if here we let $\bar{g}$ represent the sum of the gravitational and centrifugal
accelerations, as we do for strictly 2D convection in an equatorial plane (Section
13.3), this would be a function of both radius and colatitude, not just a function
of spherical radius, the requirement for our reference state. So here we assume
the centrifugal acceleration is small relative to the gravitational acceleration and
therefore add the centrifugal potential, $\Phi_c = -(\Omega r \sin \theta)^2/2$, to the definition of
the reduced pressure, P (Eq. 12.25). If there were a tidal potential, we could also
add that to P. If, on the other hand, the centrifugal acceleration were not small,
it should be included in the reference state hydrostatic balance (Eq. 12.22), which
would make the reference state pressure and density functions of both radius and
colatitude and would make the shape of the fluid domain be an oblate spheroid.
Here, however, we consider problems for which the centrifugal acceleration is relatively small and therefore include the centrifugal potential in P. Note that the
gradient of P is multiplied by the reference state density, $\bar{\rho}$; but to the next order it
would be multiplied by the density perturbation. The part of this that involves the
prescribed centrifugal and possibly tidal potentials could be included as additional
linear perturbation terms if these potentials were significant. These additional linear
terms would survive the radial component of the curl, and therefore would appear
in the toroidal mass flux equation in addition to the poloidal mass flux equation;
however, here we neglect these terms. We also assume $\mathbf{\Omega}$ is a constant; so there is
no Poincaré force.

13.4.1 2.5D Inertial Oscillations: Linear Analysis in Cartesian Geometry

Before describing nonlinear simulations of rotating convection and inertial oscillations in an axisymmetric spherical shell, we discuss 2.5D linear inertial oscillations

and the corresponding dispersion relation. To keep this simple, in this section we first consider an inviscid incompressible nonmagnetic fluid and work in a local *cartesian* coordinate system as is done in Section 10.5. The angular velocity is $\mathbf{\Omega} = \Omega\hat{z}$. The time-dependent fluid velocity, defined by Eq. 10.40, has components in the x-, y-, and z-directions (Eq. 10.39) of the rotating frame of reference, but is independent of the y-coordinate. The x and z components of the velocity and the y component of the vorticity are defined in terms of the streamfunction, ψ, by Eqs. 2.6 and 2.7; the y component of velocity is v_y (Eq. 10.40) and the x and z components of vorticity are defined by Eqs. 10.41.

We are interested in solutions to the following linear version of the momentum equation (13.6),

$$\frac{\partial \mathbf{v}}{\partial t} = -\nabla P + 2\mathbf{v} \times \mathbf{\Omega} .$$

As usual, we take the curl of this equation to get the linear (incompressible) version of the vorticity equation (13.8),

$$\frac{\partial \boldsymbol{\omega}}{\partial t} = 2\Omega\frac{\partial \mathbf{v}}{\partial z} . \tag{13.12}$$

Next we substitute the above mentioned expressions for the three components of vorticity and velocity into the three components of Eq. 13.12 to get three equations in terms of ψ and v_y and look for propagating plane wave solutions for ψ (Eq. 6.6), and similarly for v_y, in the bulk of the fluid with a frequency, $\bar{\omega}$, and a vector wavenumber, $\mathbf{k} = k_x\hat{x} + k_z\hat{z}$. The x and z components of the vorticity equation both reduce to

$$\bar{\omega}v_y = -2\Omega k_z\psi$$

and the y component gives

$$\bar{\omega}\psi = -2\Omega\frac{k_z}{k^2}v_y ,$$

where $k^2 = |\mathbf{k}|^2 = (k_x^2 + k_z^2)$. Solving these two equations results in the following *dispersion relation* for linear inertial oscillations:

$$\bar{\omega} = \pm 2\Omega\frac{k_z}{|\mathbf{k}|}$$
$$= \pm 2\Omega \cos\theta , \tag{13.13}$$

where the angle θ is the angle between the direction of the vector wavenumber, $\mathbf{k}$ (i.e., the direction of the phase propagation), and the z-direction. The $\pm$ sign means that the inertial oscillation wave propagates in the positive z-direction if $\bar{\omega}$ and k_z are both positive or both negative and in the negative z-direction if not.

As Eq. 6.8 shows for internal gravity waves, the fluid motion in an inertial oscillation is always perpendicular to the direction of the phase propagation, i.e, this is a transverse wave. This provides an explanation for why, according to Eq. 13.13, the maximum inertial oscillation frequency, 2Ω, occurs when the phase of the wave propagates only in the z-direction since then the fluid motions are only in the

x- and y-directions, normal to the rotation axis, and therefore providing the maximum restoring Coriolis forces.

Consider a 2.5D mode for which k_x, k_z, and $\bar{\omega}$ are all positive. Conservation of mass (Eq. 6.8) then says, as it does for internal gravity waves, that when and where v_x is positive v_z is negative, and vice versa. The phase velocity is the velocity of the pattern, and for inertial oscillations it is

$$
\begin{aligned}
\mathbf{c} &= \frac{\bar{\omega}}{|\mathbf{k}|}\hat{\mathbf{k}} \\
&= \frac{2\Omega k_z}{|\mathbf{k}|^3}(k_x\hat{\mathbf{x}} + k_z\hat{\mathbf{z}}),
\end{aligned}
\tag{13.14}
$$

where $\hat{\mathbf{k}} = \mathbf{k}/|\mathbf{k}|$, the unit vector in the direction of the phase propagation. The group velocity, at which the oscillation energy is transported by a superposition of many waves with different frequencies and wavenumbers, is

$$
\begin{aligned}
\mathbf{c}_g &= \frac{\partial\bar{\omega}}{\partial k_x}\hat{\mathbf{x}} + \frac{\partial\bar{\omega}}{\partial k_z}\hat{\mathbf{z}} \\
&= \frac{2\Omega k_x}{|\mathbf{k}|^3}(-k_z\hat{\mathbf{x}} + k_x\hat{\mathbf{z}}).
\end{aligned}
\tag{13.15}
$$

Comparing the vector components of Eqs. 13.14 and 13.15, one can see that the group velocity is perpendicular to the phase velocity, as it is for internal gravity waves (Section 6.1); however, the horizontal components of these two inertial oscillation velocities are in the opposite directions and the vertical components are in the same direction, a relationship opposite to that of the phase and group velocities of internal gravity waves.

Now add buoyancy due to temperature perturbations, i.e., consider a Boussinesq fluid. Again we set $\mathbf{\Omega} = \Omega\hat{\mathbf{z}}$; but now the gravitational acceleration is $\mathbf{g} = -g_o\hat{\mathbf{z}}$ and the reference state temperature gradient is $d\bar{T}/dz > 0$. The linear Boussinesq nondiffusive momentum equation in the rotating frame is

$$
\frac{\partial\mathbf{v}}{\partial t} = -\nabla P + g_o\alpha T\hat{\mathbf{z}} + 2\Omega\mathbf{v}\times\hat{\mathbf{z}}.
\tag{13.16}
$$

The $\hat{\mathbf{y}}$ component of this is

$$
\frac{\partial v_y}{\partial t} = -2\Omega v_x
\tag{13.17}
$$

and vorticity is

$$
\begin{aligned}
\boldsymbol{\omega} &\equiv \nabla\times\mathbf{v} \\
&= -\nabla^2\psi\hat{\mathbf{y}} - \frac{\partial v_y}{\partial z}\hat{\mathbf{x}} + \frac{\partial v_y}{\partial x}\hat{\mathbf{z}}.
\end{aligned}
\tag{13.18}
$$

Taking the curl of Eq. 13.16 and using the Boussinesq mass conservation equation, the vorticity equation is

$$
\frac{\partial\boldsymbol{\omega}}{\partial t} = -g_o\alpha\frac{\partial T}{\partial x}\hat{\mathbf{y}} + 2\Omega\frac{\partial\mathbf{v}}{\partial z}.
$$

The $\hat{y}$ component of this is

$$\frac{\partial \omega_y}{\partial t} = -g_o \alpha \frac{\partial T}{\partial x} + 2\Omega \frac{\partial v_y}{\partial z}, \tag{13.19}$$

which by Eq. 13.18 is also

$$\frac{\partial \omega_y}{\partial t} = -\left(\frac{\partial^3 \psi}{\partial x^2 \partial t} + \frac{\partial^3 \psi}{\partial z^2 \partial t} \right). \tag{13.20}$$

Then, equating Eqs. 13.19 and 13.20, taking another time derivative of this, substituting in Eq. 13.17 and the linear nondiffusive temperature equation 6.1, and assuming a plane wave solution Eq. 6.6 gives the dispersion relation:

$$\bar{\omega}^2 = \left(N^2 k_x^2 + (2\Omega)^2 k_z^2 \right) / k^2, \tag{13.21}$$

where N is the Brunt-Väisälä frequency (Eq. 6.5). This dispersion relation nicely shows how the buoyancy contribution to frequency is maximum and the Coriolis contribution vanishes for a purely horizontally propagating wave, for which the transverse fluid motions are parallel to gravity and the rotation axis. On the other hand, the Coriolis contribution is maximum and the buoyancy contribution vanishes for a purely vertically propagating wave, for which the transverse motions are perpendicular to the rotation axis and gravity.

We can also add a background magnetic field, as discussed in Chapter 11. The dispersion relation for a vertical background field $(B_o \hat{z})$ is

$$\bar{\omega}^2 = \left(N^2 k_x^2 + (2\Omega)^2 k_z^2 + V_A^2 k_z^2 k^2 \right) / k^2 \tag{13.22}$$

and for a horizontal background field $(B_o \hat{x})$ it is

$$\bar{\omega}^2 = \left(N^2 k_x^2 + (2\Omega)^2 k_z^2 + V_A^2 k_x^2 k^2 \right) / k^2, \tag{13.23}$$

where V_A is the Alfvén velocity, Eq. 11.35. Equations 13.22 and 13.23 show how the Alfvén contribution to frequency is maximum when the transverse fluid motions are perpendicular to the background magnetic field, which is the case for a purely vertically propagating wave when the background magnetic field is vertical and for a purely horizontally propagating wave when the background field is horizontal.

This linear analysis of magneto-rotational-gravity waves reveals an additional stability constraint on the numerical time step for 2.5D (and 3D) simulations. That is, to be able to resolve such oscillations in time (within the rotating frame of reference) and so avoid a numerical instability, the numerical time step needs to be considerably smaller than the characteristic time scales:

$$\Delta t \ll 1/\text{MAX}(N, 2\Omega, V_A k),$$

where here k is the largest resolved wavenumber.

As mentioned at the end of Section 11.6, studies of magnetoconvection have also been done in 2.5D with a horizontal background magnetic field in the direction for which no variables depend. For example, an early study by Schubert (1968) examines double-diffusive magnetoconvective instabilities in an axisymmetric sphere with the background field in the ϕ-direction; instabilities occur because buoyancy is partly thermal and partly magnetic, with magnetic diffusivity η being much less than thermal diffusivity κ.

13.4.2 2.5D Inertial Oscillations: Nonlinear Simulations in Spherical Geometry

Here we describe two simulations of 2.5D anelastic convection in a *spherical shell*: a nonrotating nonmagnetic case and a rotating magnetic case. Both have an adiabatic reference state with a polytropic index $n = 1$, the ratio of specific heats $\gamma = 2$, and specific heat capacity at constant pressure $c_p = 3.0 \times 10^8$ ergs/(gm K). The bottom boundary radius is set to 1.4×10^9 cm and the top to 7.0×10^9 cm. The reference state density at the bottom boundary is set to 2.541 gm/cm^3 and at the top boundary to 0.127 gm/cm^3; that is, there are $N_\rho = 3$ density scale heights spanning the convection zone. The Grüneisen parameter (Eqs. 12.78 and 12.79) is assumed constant and is set to $\gamma_G = (\gamma - 1) = 1/n = 1$. The reference state temperature at the bottom boundary is set to 3.868×10^4 K; therefore, with the chosen γ_G and N_ρ, the reference state temperature at the top boundary is 0.194×10^4 K. The mass below the bottom boundary is prescribe to be 2.995×10^{28} gm. The viscous and thermal diffusivities are chosen here to be inversely proportional to the reference state density; this somewhat enhances the amplitude of the fluid velocities in the deeper region where they would otherwise be smaller due to the greater density. The top and bottom boundaries are impermeable (and fixed in space), stress-free, and at constant entropy with the superadiabatic drop in the entropy perturbation across the convection zone set to 3.545 ergs/(gm K). Both cases use 129 Chebyshev levels in radius and 1024 Gaussian colatitudinal grid points with $l_{max} = 682$.

The nonrotating nonmagnetic case has a viscous diffusivity at the bottom boundary of 5×10^8 cm^2/s and a bottom thermal diffusivity of 2×10^9 cm^2/s, which makes the Prandtl number 0.25 and the Rayleigh number (at mid-depth) 4.4×10^5. The numerical time step is set to 3×10^3 s. This case is run for about 500,000 time steps, spanning about 50 simulated years. Fluid velocities get up to about 300 cm/s, a Reynolds number, Re, of several thousand.

This nonrotating nonmagnetic case is illustrated in Fig. 13.9. Examination of the meridional circulation and the entropy in this figure clearly shows a simple flow pattern with cold fluid sinking near the axis in the northern hemisphere, heating up as it flows around the hot bottom boundary, rising as hot fluid in the southern hemisphere, and cooling off as it flows along the cold top boundary. This is the preferred convective circulation pattern for a deep spherical shell (or full fluid sphere) that is not rotating (and not strongly turbulent). The image on the right in this figure shows the zonal flow (i.e., the longitudinal velocity), which is also advected by the meridional circulation; but its kinetic energy decays by about six orders of magnitude during the course of this simulation.

The rotating magnetic case is quite different. First of all, to obtain a similar Reynolds number we need to significantly increase the Rayleigh number to counter the stabilizing effects of rotation. We do this by prescribing much smaller viscous and thermal diffusivities, both of which are inversely proportional to density, as in the nonrotating case. This case has its viscous diffusivity at the bottom boundary set to 2.5×10^6 cm^2/s and its bottom thermal diffusivity set to 2×10^7 cm^2/s, making the Prandtl number 0.125 and the Rayleigh number (at mid-depth) 8.8×10^9. The magnetic diffusivity is constant and set to 7×10^7 cm^2/s. The rotation rate for this

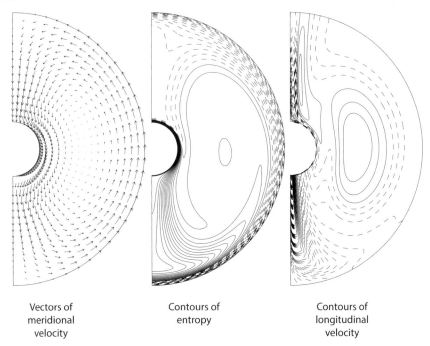

Vectors of   Contours of   Contours of
meridional   entropy   longitudinal
velocity   velocity

Figure 13.9 A snapshot of a 2.5D simulation of thermal convection in a spherical shell that is
nonrotating and nonmagnetic. (Left) Meridional circulation, $(v_r \hat{\boldsymbol{r}} + v_\theta \hat{\boldsymbol{\theta}})$; arrows
indicate the local direction of the fluid velocity, with the length of arrows being
proportional to the velocity amplitudes. (Middle) Entropy perturbation; solid
contours represent fluid warmer (lighter) than the constant adiabatic reference
state entropy; broken contours are colder (heavier) fluid. (Right) The decaying
zonal velocity $(v_\phi \hat{\boldsymbol{\phi}})$; solid contours represent eastward-directed flow and broken contours westward-directed flow, assuming the northern hemisphere is at
the top in this figure.

case is $\Omega = 10^{-6}\,\mathrm{s}^{-1}$, which makes the Ekman number roughly 7×10^{-8} (at mid-
depth). The convective Rossby number is therefore about $\mathrm{Ro}_c = 0.02$, indicating
that Coriolis effects dominate over buoyancy effects. This case has the time step set
to 6×10^4 s and is run for about 1,000,000 numerical steps, spanning about 2000
simulated years. It could be a model of a giant gas planet rotating roughly a hundred
times more slowly than Jupiter and Saturn and with less convective driving. Fluid
velocities get up to about 10 cm/s, a Re of several thousand, which indicates a
degree of turbulence similar to that of the nonrotating case.

A snapshot of this case is illustrated in Figs. 13.10–13.12. The north and south
geographic poles are defined by the direction of the total angular momentum of the
body measured in the inertial frame; by definition, the total angular momentum of
the body vanishes in the rotating frame of reference, which is the frame in which
the equations are solved and the solution is viewed. The eastward, i.e., the prograde,
direction is into the meridian plane in these figures and is the direction of the mean

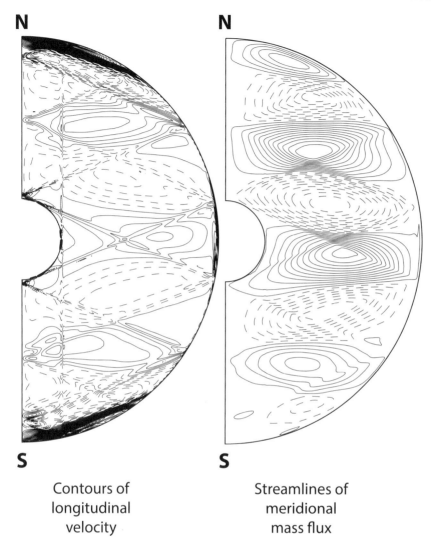

Contours of
longitudinal
velocity

Streamlines of
meridional
mass flux

Figure 13.10 A snapshot of the 2.5D rotating magnetic simulation. (Left) Zonal velocity
($v_\phi \hat{\phi}$); solid contours represent eastward-directed (i.e., prograde) flow and
broken contours westward-directed (retrograde) flow. (Right) Meridional cir-
culation illustrated in terms of streamlines of mass flux, $\bar{\rho}(v_r \hat{r} + v_\theta \hat{\theta})$; solid
streamlines represent clockwise-directed flow, broken are counterclockwise-
directed.

rotational velocity when viewed in the inertial frame. That is, $\boldsymbol{\Omega}$ is everywhere
directed parallel to the rotation axis, from south to north.

The zonal flow (i.e., the axisymmetric longitudinal velocity in the rotating frame)
does not decay in this case but is maintained (mainly) by the Coriolis forces resulting from the meridional circulation. The structures of the meridional and zonal

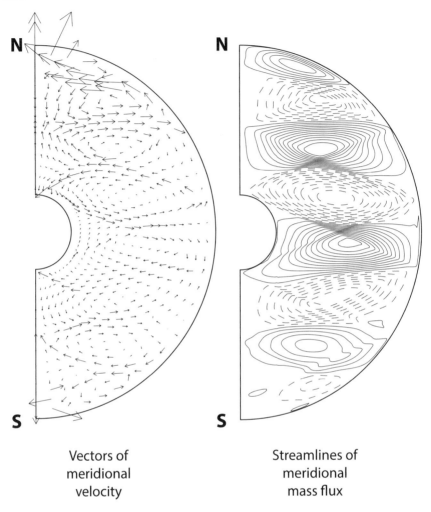

Vectors of
meridional
velocity

Streamlines of
meridional
mass flux

Figure 13.11 The same time step as in Fig. 13.10, here illustrating the meridional circulation
both as streamlines of the meridional mass flux (Eq. 10.94) and as arrows
showing the direction and relative amplitude of the fluid velocity. Note, arrows
are not shown on the bottom and top boundaries because at some places on
these boundaries they are very large.

flows in Fig. 13.10 are strikingly different from those for the nonrotating nonmagnetic case (Fig. 13.9). Strong internal shear flows are maintained along straight
line segments (rays) that crisscross over the meridian plane and "reflect" off the
boundaries. Notice how well the spherical harmonic and Chebyshev polynomial
expansions depict these straight line patterns. The smaller the Ekman number the
more intense and the more narrow the shear layers are. Also notice that the angle of reflection a ray makes with a boundary does not equal its angle of incidence with that boundary. (In the limit of local small-scale *internal gravity waves*

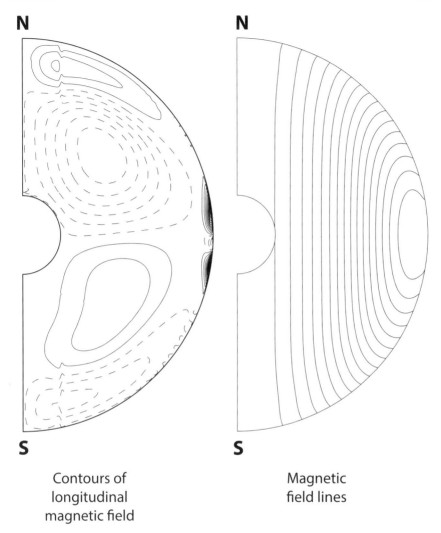

N

S

Contours of
longitudinal
magnetic field

N

S

Magnetic
field lines

Figure 13.12 The same snapshot as in Figs. 13.10 and 13.11, here showing the three components of the magnetic field. (Left) Toroidal magnetic field, i.e., the axisymmetric field directed normal to the meridian plane; solid contours represent eastward-directed field and broken contours westward-directed field.
(Right) Poloidal magnetic field, i.e., the axisymmetric field directed within the
meridian plane; solid field lines represent clockwise-directed field, broken are
counterclockwise directed.

these angles would be approximately equal since the spherical surfaces are locally
perpendicular to gravity; however, for global large-scale gravity waves the reflection of a planewave on a spherical surface alters the shape of the wave.) Instead,
the angle of incidence *relative to the local direction of* Ω for an inertial ray at a

boundary equals the angle of reflection relative to $\boldsymbol{\Omega}$ as predicted for inertial oscillations (Section 13.4.1). As described in that section, the higher the driving frequency for inertial oscillations, $\bar{\omega} \leq 2\Omega$, the closer the direction of the phase propagation, $\mathbf{k}$, is to being parallel to the axis of rotation, $\hat{z}$, and the closer the direction of the group velocity (the rays in Fig. 13.10) is to being perpendicular to the rotation axis. The angle also depends on the radius of the bottom boundary, as seen in Fig. 13.10. A particular ray path is favored if it forms a closed circuit because it is reinforced by the superposition of multiple transits (see Greenspan, 1969; Tilgner, 1999, 2007c). As seen in Figs. 13.10 and 13.11, both meridional and zonal flows are sheared on these rays. Note that only the ray reflections that cause meridional flow along the ray to reverse its direction in *cylindrical* radius cause the Coriolis force (and resulting shear) in longitude to reverse; the reflections off the spherical boundaries that do not reverse the component of the flow in cylindrical radius (those near the poles on both boundaries) do not reverse the Coriolis force. The shear line tangent with the inner-core equator (the "tangent cylinder") is unique; any meridional flow along this tangent cylinder is parallel to the rotation axis and therefore produces no Coriolis force.

As mentioned, the restoring forces for these 2.5D inertial oscillations (Section 13.4.1) are Coriolis forces. That is, flows normal to the axis of rotation in a meridian plane produce Coriolis forces directed in longitude, which then turn the flows into a longitudinal direction, which produce Coriolis forces in the meridian plane that oppose the original meridional flow. At the time of the snapshot in Figs. 13.10 and 13.11 the Coriolis forces resulting from the longitudinal velocity are driving the meridional flow. This can be seen, for example, in the equatorial region where the large-scale eastward zonal flow produces a radially upward Coriolis force that drives the large-scale upward meridional flow there (at this time). The situation is similar at other latitudes. However, at other times the opposite occurs; that is, radially upward flow in the equatorial region produces westward Coriolis forces that drive westward zonal flow. Note, in both cases, the flow that is driven produces Coriolis forces that are in the opposite direction of the original flow and therefore act as restoring forces. However, the time dependence of these simulated flows is not a simple sinusoidal time dependence as described in Section 13.4.1 because of additional forces (buoyancy, Lorentz, and viscous). Occasionally during the simulation the large-scale pattern propagates parallel to the rotation axis on a time scale of about 50 years (compared to the rotation period of 0.2 years) while the pattern of the internal crisscrossed shear layers remains in place.

The magnetic field at this same time step is illustrated in Fig. 13.12: contours of the toroidal field on the left and field lines of the poloidal field on the right. The maximum magnetic field intensity at the time illustrated in Fig. 13.12 is about 15 *gauss*. Because this model is not fully 3D and because we are not maintaining a constant magnetic flux through the boundaries (as we do for the magnetoconvection scenarios of Chapter 11), here the magnetic energy slowly decays (by about 30% during this simulation). Torsional oscillations—i.e., rotating cylindrical fluid shells coaxial with the rotation axis (axisymmetric Taylor cylinders) linked together with magnetic field that provides restoring forces—might occur if the driving frequency of the inertial oscillations were close to the characteristic Alfvén frequency

(Section 12.4.2) of the torsional oscillations. However, for this case the Alfvén frequency is about three orders of magnitude smaller than the characteristic inertial oscillation frequency (2Ω), so the internal shear flows here have little effect on the magnetic field pattern.

An important point to make here is that a 2.5D model is a poor approximation for the axisymmetric part of a 3D model, especially when there is rotation and magnetic fields. The critical contribution to the axisymmetric part of the solution that is missing in a 2.5D model is the axisymmetric part of the nonlinear terms that involve the product of longitudinal mode $m > 0$ with itself. Recall that the square of a Fourier function of mode number m can be written as the sum of a Fourier function of mode $2m$ and one of mode 0 (Eqs. 4.1a,b), where $m = 0$ is the axisymmetric contribution. This contribution by nonlinear terms, including the convergences of the Reynolds and Maxwell stresses, is typically the dominant effect maintaining the axisymmetric part of a 3D solution but is missing in a 2.5D simulation (unless one adds ad hoc forcing terms to the "mean equations"). For example, differential rotation in many cases, like for the sun and giant planets, is maintained *not* by the Coriolis forces due to meridional circulation but by the convergence of Reynolds stress due to the products of nonaxisymmetric velocity components (Section 13.3). In these cases the meridional circulation is mainly maintained by the Coriolis forces resulting from this differential rotation. Also, as mentioned, Cowling's theorem (Cowling, 1934) shows that a *self-consistent* dynamo necessarily produces fully 3D magnetic fields. Therefore, in Sections 13.5 and 13.6 we briefly discuss 3D convection and magnetic field generation in rotating, density-stratified, spherical shells.

13.5 3D ROTATING SPHERICAL SHELL: DYNAMO BENCHMARKS

An electrical conductor moving through a magnetic field generates its own magnetic field (Chapter 11). When the original magnetic field is also maintained by the moving conductor the process is called a "dynamo." When the conductor in a dynamo is a fluid that flows due to gravitational and rotational forces it is called a homogeneous self-sustaining MHD dynamo, or convective dynamo. See Roberts (2007) for a review of the basic theory of convective dynamos. Before describing examples of dynamo simulations, here we discuss two dynamo benchmarks, one Boussinesq and one anelastic.

13.5.1 Boussinesq Dynamo Benchmark

The first benchmark of a convective dynamo for a Boussinesq fluid in a 3D spherical shell was published in 2001 (Christensen et al., 2001; see also Christensen et al., 2009, for the correction of a minor typographical error). Six independently written dynamo codes are compared using cases that have steady-state solutions when observed in a "drifting" frame of reference rotating at a constant angular velocity relative to the rotating frame of reference. Although this type of solution requires a small Rayleigh number (only about a couple times critical), it provides a convenient and precise way to make highly accurate comparisons between codes. All six of the codes employ spherical harmonic expansions; three use Chebyshev expansions in radius and three use finite differences in radius.

Three cases are compared. Case 0 is nonmagnetic. Case 1 is a dynamo with perfectly insulating external regions below and above the convecting zone. Case 2 is also a dynamo with a perfectly insulating external region above the convection zone but with a finitely conducting solid inner core below the convection zone. The solid inner core for case 2 has the same density, electrical conductivity, and magnetic permeability (and therefore same magnetic diffusivity) as that of the fluid in the convection zone.

Since studies of dynamos within terrestrial planetary cores, like the geodynamo, are the main motivation for these codes, the chosen boundary conditions are impermeable and nonslip. Except for the bottom boundary (i.e., the inner core boundary) for case 2, the boundaries are stationary relative to the rotating frame; the solid inner core for case 2 is free to rotate about the axis of the rotating frame in response to viscous and magnetic torques on the inner core boundary. The bottom and top boundaries of the fluid are held at constant temperatures. The magnetic field is matched to a potential field (Section 10.6.4) at both boundaries for case 1 and at the top boundary for case 2. The field is computed within the solid inner core for case 2, accounting for the solid body rotation of the inner core and maintaining the magnetic field and the horizontal component of the electric field continuous at the inner core boundary. Viscous and ohmic heating are neglected.

A slightly different choice of scaling, compared to the one we have been using, is employed in Christensen et al. (2001). If one's code employs a different scaling or uses dimensional variables, care needs to be taken specify the benchmark problem correctly and scale the results according to the paper's scaling choice for comparison. A modified Rayleigh number is defined as

$$\mathrm{Ra} \equiv \frac{\alpha g_o \Delta T D}{\nu \Omega}$$

and is set to 100 for cases 0 and 1 and to 110 for case 2. The gravitational acceleration is proportional to radius (Eq. 10.10); g_o is the value at the top boundary. The Ekman number is defined as in Eq. 13.11 but without the factor of 2 in the denominator; it is set to 10^{-3} for all three cases. The Prandtl number (Eq. 1.19) is set to 1 and for cases 1 and 2 the Roberts number (Eq. 11.27) is set to 5. The radii of the bottom and top boundaries, r_{bot} and r_{top}, are scaled by the depth, $D = r_{top} - r_{bot}$ and set to 7/13 and 20/13, respectively.

The eventual steady-state solution (with the drifting frame) happens to be sensitive to the initial conditions for these low Rayleigh number cases. That is, there are multiple solutions (with different dominant longitudinal mode numbers, m) for these parameters. The benchmark is for a dominant $m = 4$ for the flow; the field is mainly a combination of an axial dipole and an $m = 4$ poloidal component. Check the paper for suggested initial conditions that should converge to the desired solution.

As mentioned, the solutions are steady in a drifting frame, which is prograde relative to the rotating frame for case 0 but retrograde for cases 1 and 2. The constant angular velocity of the solid inner core in case 2 is also retrograde but has an amplitude less than that of the drifting frame. The solutions are symmetric with respect to the equatorial plane; see the paper for a snapshot of the radial components of the flow and field.

The (nondimensional) data compared are the volume-averaged kinetic and magnetic energy and the temperature, the longitudinal velocity, and the colatitudinal magnetic field at mid-depth in the equatorial plane where the radial component of the velocity vanishes and its derivative with respect to longitude is positive. There are four points like this because of the $m = 4$ dominated solution. Calculating these values at one of these points can be done using Newton's method to converge on it each time it is checked. In addition, the angular velocity of the drifting frame and, for case 2, of the solid inner core are also compared among the codes. Care needs to be taken when calculating the angular velocity of the drifting frame to avoid aliasing in the process of converging onto the vanishing radial velocity point mentioned above.

The results of the six basic codes show good agreement. As expected, the fully spectral codes converge onto the correct solution faster than those with finite differences in radius when comparing codes with a comparable number of finite-difference grids and Chebyshev polynomials.

13.5.2 Anelastic Dynamo Benchmark

The first benchmark of a convective dynamo for an anelastic fluid in a 3D spherical shell was published in 2011 (Jones et al., 2011). Four dynamo codes are compared. However, three of the codes were designed or originated from Glatzmaier (1984) and Glatzmaier & Roberts (1996a); these employ spherical harmonic and Chebyshev polynomial expansions. Over the past decade or more these three have been independently modified and parallelized. The fourth code (Jones & Kuzanyan, 2009) also uses spherical harmonic expansions but with a finite-difference method in radius.

Three cases are compared. A *hydrodynamic (nonmagnetic) benchmark* is steady in a frame drifting at a constant angular velocity in the prograde direction relative to the rotating frame of reference. The other two cases are dynamos; one is a *steady dynamo benchmark* that also drifts at a constant rate in the prograde direction and the other is an *unsteady dynamo benchmark* at a higher Rayleigh number. The two steady cases have Rayleigh numbers approximately 1.3 times their critical values; the Rayleigh number for the unsteady dynamo case is roughly three hundred times critical. The unsteady dynamo, the results for which are more difficult to compare among the four codes, was chosen to better test the nonlinear terms, which make greater relative contributions in this higher Rayleigh number case.

Since studies of dynamos within stars, like the sun, and giant gas planets, like Jupiter, are the main motivation for these density-stratified codes, the chosen boundary conditions are impermeable and stress-free. The bottom and top boundaries of the fluid are externally maintained at specific entropies that are constant over each boundary and constant in time; ΔS is the drop in entropy across the depth, $D = r_{top} - r_{bot}$. The regions above and below the convection zone are assumed to be perfect insulators; therefore, the magnetic field is matched to a potential field

(Section 10.6.4) at both boundaries at each time step for the dynamo cases. Viscous and ohmic heating are included; they are needed to maintain the same total heat flow through the bottom and top boundaries for the two steady benchmarks and the same time-averaged values for the unsteady case. For a fully spectral code, the total heat flow through the bottom and top boundaries agree to one part in ten million for the two steady benchmarks.

The fluid is assumed to be a perfect gas (Eqs. 12.3–12.7). The hydrostatic reference state is isentropic and constructed using a simplified polytrope (Section 12.2.3) with polytropic index n; the gravitational acceleration is assumed to be due to only the mass, M_o, below the bottom boundary. The specific heat capacity at constant pressure, c_p, and the ratio of specific heats, γ, are constants such that

$$\gamma \equiv \frac{c_p}{c_v} = \frac{n+1}{n}$$

and the perfect gas constant is

$$R = \frac{p}{\rho T} = \frac{c_p}{n+1}.$$

The three diffusivities, v, κ, and η, are also assumed to be constants. Note, for these benchmark cases the gravitational constant, G, is defined with just three significant places: 6.67×10^{-8} ergs/(gm K).

Jones et al. (2011) express the reference state and perturbation equations in nondimensional forms, using a choice of scaling that differs slightly from what we have been using. Note how they define the Rayleigh and Ekman numbers; they also define a magnetic Prandtl number Pm $\equiv$ Pr q. One could adopt their scaling and the resulting nondimensional equations. If, on the other hand, their scaling is not adopted, care needs to be taken to convert input parameters and output data to the Jones et al. (2011) format for comparison. Alternatively, Jones et al. (2011) provide input parameters and output data in dimensional units for a giant gas planet like Jupiter. The dimensional reference state profiles can easily be calculated using the modified polytropic function, $\Theta_1(r)$, defined in Section 12.2.3.

Here we review the dimensional specifications for the three anelastic benchmark cases using CGS units. For all three cases the reference state polytropic index is set to $n = 2$ and therefore $\gamma = 3/2$. (Note, $n = 1$ would have been more appropriate for Jupiter.) In addition, for all cases

$$r_{bot} = 2.45 \times 10^9\,\text{cm}, \quad r_{top} = 7.00 \times 10^9\,\text{cm}, \quad c_p = 1.05090 \times 10^8\,\text{ergs/(gm s)},$$

$$\Omega = 1.76 \times 10^{-4}\,\text{s}^{-1}, \quad M_o = 1.9 \times 10^{30}\,\text{gm}, \quad \text{and} \quad \rho_{bot} = 1.1\,\text{gm}.$$

For the magnetic cases, the magnetic permeability is $\mu = 4\pi$ and the field is in *gauss*. The reference state temperature at the bottom boundary for the hydrodynamic benchmark is $T_{bot} = 3.48548 \times 10^5$ K and for the two dynamo benchmarks it is $T_{bot} = 4.11829 \times 10^5$ K. The viscous, v, and thermal, κ, diffusivities are both equal to $3.64364 \times 10^{12}\,\text{cm}^2\,\text{s}^{-1}$ for the hydrodynamic benchmark and are both equal to $7.28728 \times 10^{12}\,\text{cm}^2\,\text{s}^{-1}$ for the steady dynamo benchmark. For the unsteady dynamo benchmark, $v = 1.82182 \times 10^{11}\,\text{cm}^2\,\text{s}^{-1}$ and $\kappa = 0.91091 \times 10^{11}\,\text{cm}^2\,\text{s}^{-1}$.

The magnetic diffusivity for the steady dynamo is fifty times smaller than its viscous and thermal diffusivities, $\eta = 1.457456 \times 10^{11}\,\text{cm}^2\,\text{s}^{-1}$; this relatively small value of η makes it possible to obtain a dynamo with a Rayleigh number only slightly larger than critical. For the strongly supercritical unsteady dynamo it is set equal to the thermal diffusivity, $\eta = 0.91091 \times 10^{11}\,\text{cm}^2\,\text{s}^{-1}$. The final parameter needed is the drop in entropy across the depth, which is set to $8.512257 \times 10^5\,\text{ergs}/(\text{gm\,s})$, $7.742683 \times 10^5\,\text{ergs}/(\text{gm\,s})$, and $0.756121 \times 10^5\,\text{ergs}/(\text{gm\,s})$ for the hydrodynamic, steady dynamo, and unsteady dynamo benchmarks, respectively.

The number of density scale heights, N_ρ, across depth D is found via Eq. 12.76; it is 5 for the hydrodynamic benchmark and 3 for the dynamo benchmarks. The time step, Δt, chosen depends on the CFL *stability* condition, which depends on the flow and field amplitudes and the spatial resolution. However, to obtain the accuracy displayed in Jones et al. (2011), the actual time step (for second-order time integration schemes) needs to be several times smaller than the CFL constraint. For example, the time step for the hydrodynamic benchmark with a spatial resolution of 121 radial $\times$ 512 colatitude $\times$ 1024 longitude levels is set to 33 s, for the steady dynamo benchmark with $65 \times 128 \times 256$ levels it is 300 s, and for the unsteady dynamo with $129 \times 256 \times 512$ levels it is 100 s.

Another accuracy issue is discussed in Jones et al. (2011). The total angular momentum of the convection zone should remain zero relative to the rotating frame since it begins as zero and the boundaries are stress-free. However, as discussed in Section 10.3.1 for a 2D annulus, a very slight nonzero value is added at each time step due to numerical truncation, which can grow to a non-negligible value after a million time steps if not corrected. A method suggested, which works well, is to replace the stress-free boundary condition on the degree $l = 1$ parts of the toroidal mass flux scalar, Z_1^m at, say, the top boundary, with a condition that forces the total integrated angular momentum to vanish in the rotating frame of reference. The volume integral over the convection zone of the z component of angular momentum density is proportional to the integral over radius from r_{bot} to r_{top} of $(Z_1^0 r^2)$; the volume integral of the x-component of angular momentum density is proportional to the integral over radius of the real part of $(Z_1^1 r^2)$; and the volume integral of the y component of angular momentum density is proportional to the integral over radius of the imaginary part of $(Z_1^1 r^2)$. Using the formula for the integral of a Chebyshev polynomial over radius, Eq. 9.30, these integral conditions can easily be forced to vanish (instead of the total stress at the top boundary) in the semi-implicit time integration scheme described in Section 8.2. The 2D annulus version of this condition is Eq. 10.24.

Snapshots of the three benchmark solutions are illustrated in Jones et al. (2011). The steady hydrodynamic solution is dominated by longitudinal mode number $m = 19$ and the steady dynamo solution is dominated by $m = 7$, which illustrates how a strong magnetic field (i.e., one with total magnetic energy comparable to total kinetic energy) can increase the dominant length scale relative to that for a nonmagnetic case. The structures for both steady cases are symmetric with respect to the equatorial plane and the patterns drift prograde relative to the rotating frame.

The unsteady benchmark is mildly turbulent. The Rayleigh and Ekman numbers, as we define them (Eqs. 12.89a and 13.11), are 2×10^7 and 2.5×10^{-5}, respectively.

Although these are more extreme than those for the steady benchmarks, they are not extreme enough to produce the Reynolds stresses needed to maintain dominant and banded zonal winds as observed on the surfaces of giant planets like Jupiter and Saturn (Section 13.6.2).

For the steady benchmarks, the same set of data is compared among the codes as is compared in the Boussinesq benchmark (Section 13.5.1), whereas time-averaged energies and luminosities with standard deviations are compared for the unsteady dynamo. The results of the four codes for the three benchmarks, which are provided in Jones et al. (2011), show good agreement.

13.6 3D ROTATING SPHERICAL SHELL: DYNAMO SIMULATIONS

Simulations of the solar dynamo are reviewed by, for example, Miesch (2005) and of planetary dynamos by Stanley & Glatzmaier (2009) and Jones (2011). Here we briefly describe geodynamo and Saturnian dynamo simulations produced with a 3D spherical MHD dynamo code that employs the numerical method outlined in Section 10.6. This code was successfully verified via the Boussinesq benchmark (Section 13.5.1) and the anelastic benchmark (Section 13.5.2). Various versions of this code and others like it have simulated the internal magnetohydrodynamics of stars like the sun that have magnetic cycles of polarity reversals (e.g., Glatzmaier, 1984; Brown et al., 2011) and in much more massive stars that have core convection zones (e.g., Kuhlen et al., 2006). Anelastic convective dynamo simulations have also been used to study the magnetohydrodynamics of the liquid iron-rich cores of terrestrial planets like the Earth (e.g., Glatzmaier & Roberts, 1996a) and the deep hydrogen-helium interiors of giant gas planets like Saturn (e.g., Glatzmaier, 2005b; Jones & Kuzanyan, 2009). Rotating laboratory experiments designed to provide further insight into these stellar and planetary dynamics have also been simulated: nonmagnetic (Hart et al., 1986) and magnetic (Roberts et al., 2010).

13.6.1 Geodynamo Simulations

The mechanism in the Earth's fluid iron-rich core that maintains the geomagnetic field is called the "geodynamo." For a relatively simple discussion of this mechanism see, for example, Glatzmaier & Olson (2005); for much more comprehensive discussions see Gubbins & Herrero-Bervera (2007); Roberts (2007); Jones (2007); Christensen & Wicht (2007); Glatzmaier & Coe (2007).

Here we describe an anelastic geodynamo model and simulation. The anelastic reference state thermodynamic variables (Chapter 12) for the geodynamo model are based on the 1D PREM model of the Earth's interior (Dziewonski & Anderson, 1981). Realistic dimensional values are prescribed for the dimensions of the solid inner core, the fluid outer core, and the "solid" mantle. All three regimes are free to rotate (according to the torques acting them) relative to the frame of reference, which rotates with a period of one day. Convection and magnetic field generation take place within the iron-rich fluid outer core. The field penetrates into the solid inner core, which also has a finite electrical conductivity, and provides a magnetic torque at the inner-core boundary between the inner and outer cores. The magnetic field and the tangential component of the electric field are continuous at this inner-core boundary because this interface is a nonslip impermeable boundary across which the magnetic permeability is continuous. This requires solving for the evolution of the field simultaneously within the fluid outer core and the solid inner core, which at most can undergo "solid-body rotation" relative to the rotating frame of reference (Section 10.6.4; Glatzmaier & Roberts, 1996a). The lower part of the mantle is also electrically conducting, which provides a magnetic torque at the core-mantle boundary. The rest of the mantle and the space out to infinity are considered a perfect electrical insulator; therefore, the magnetic field is a potential field in these regions. Both thermal buoyancy and compositional buoyancy drive convection; separate advective-diffusive equations with different top (core-mantle) boundary conditions govern the perturbations in the light constituent within the fluid and the entropy. At the bottom (inner-core) boundary the solidification of the liquid iron onto the solid core and the accompanying release of latent heat and lighter elements are modeled by defining a boundary condition that makes the local (time-dependent) rate of cooling be proportional to the local diffusive heat and compositional fluxes. This boundary condition is more realistic and less constraining than the traditional constant entropy or constant heat flux boundary conditions. Test simulations (Ogden et al., 2006) attempting to shed light on the question of radiogenic heating in the core have demonstrated that realistic estimates of the amount that might be present produce no significant differences in the simulated magnetic field at the model's surface. Therefore, the heat flow out of the Earth's core is usually assumed to be due to just the cooling of the core, with no radiogenic heating.

Figure 13.13 is a snapshot of the radial component of the magnetic field at the model's core-mantle boundary and at what would be the Earth's surface at two different spatial resolutions: a coarse resolution (up to spherical harmonic degree 12) and a higher resolution (up to degree 95). These are compared to the Earth's field in the year 1980 on both surfaces up to degree 12, which is essentially all that can be detected of the core field at the surface because of the more intense magnetic field in the Earth's crust at degrees higher than 12. The images at the surface are obviously more dipolar dominated and those at the core-mantle boundary have more intense small scales because larger scales decay less with distance from the core (Eq. 10.70). Although the degree-12 images at the core-mantle boundary for the actual and simulated fields are qualitatively similar, the degree-95 simulated image shows much more intense small-scale "core-spots," which likely exist at the Earth's core-mantle boundary but are hidden at the surface due to the crustal magnetic field.

The differential rotation (i.e., zonal flow relative to the rotating frame) is maintained by the Coriolis forces resulting from meridional circulation driven by a warmer region inside the tangent cylinder compared to outside. That is, in this problem, the differential rotation is a *thermal wind* with the region outside the tangent cylinder rotating slowly in retrograde relative to the rotating frame and the region inside the tangent cylinder near the solid inner core rotating in the prograde direction due to fluid sinking toward the rotation axis there. Closer to the surface within the "tangent cylinder" (tangent to the inner-core boundary equator) the fluid

Radial component of the magnetic field

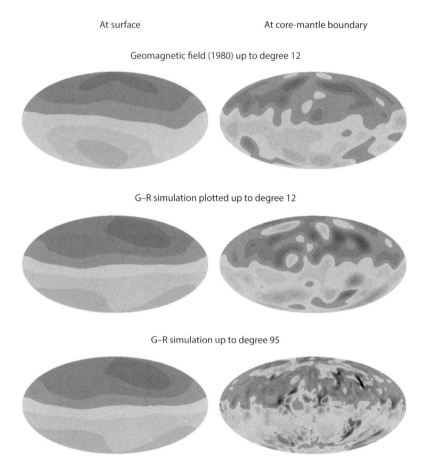

diverges away from the rotation axis, resulting in retrograde zonal flow. In this case the kinetic energy in the differential rotation is relatively small and the total kinetic energy of convection is about a couple thousand times smaller than the

Figure 13.14 A snapshot of the magnetic field maintained in a geodynamo simulation illus-
             trated as magnetic field lines (see *Color Plate* 12 for a color version of this
             figure). Gold field lines are directed outward and blue inward. (This mater-
             ial is reproduced from Glatzmaier & Roberts (1996a) with permission from
             Elsevier Ltd.)

total magnetic energy; that is, this is a strong-field dynamo. Maximum convective
fluid velocities are about 1 cm/s, whereas maximum magnetic field intensities deep
within the core are a few hundred gauss.

A snapshot of the magnetic field maintained in a geodynamo simulation
(Glatzmaier & Roberts, 1996a) is illustrated in Fig. 13.14 using magnetic field
lines; the axis of rotation is "vertical" in the image. The field is a potential field
outside the core-mantle boundary, dominated by an axial dipole. Inside the fluid
core, where the field is generated, the field is much more intense and complicated. The fluid shears the magnetic field along the tangent to the solid inner-core

equator, wrapping the poloidal field into toroidal field and thereby outlining the tangent cylinder.

This field pattern is slowly time-dependent and, on an average of about a couple hundred thousand years, it experiences a spontaneous dipole polarity reversal that typically takes a few thousand years, similar to that seen in the paleomagnetic record. However, the times between reversals vary significantly (unlike the nearly periodic solar dynamo reversal period of 11 years). An example of a geodynamo simulation is illustrated in Fig. 13.15 (Glatzmaier et al., 1999; Glatzmaier & Coe, 2007), showing the magnetic dipole polarity and intensity vs. time, including two spontaneous dipole reversals during the 300,000-year simulation. As in actual paleomagnetic data, the intensity of the dipole decreases significantly during a reversal; however, a reversal does not occur at every intensity minimum. Many spontaneous dipole reversals have occurred in geodynamo simulations (e.g., Olson et al., 2011); each is unique in terms of the evolution of the field during the reversal transition.

The pattern of the fluid flow after the dipole reversal is qualitatively the same as it was before the reversal, including its average direction; the pattern of the magnetic field is also qualitatively the same, but its direction has completely reversed everywhere. The reason for this is that **B** appears in the momentum and energy equations as a quadratic and in the magnetic induction equation as a linear term. Therefore, if **B** at a given time for a given flow pattern satisfies these equations, so does $-$**B**. That is, there are naturally two equally likely attracting polarity solutions. The simulations show that the system is constantly trying to reverse its polarity; only after many attempts do perturbed flows successfully destroy the original field and replace it with a reversed polarity field.

A sequence of four snapshots (a,b,c,d), each separated by 3000 years, during one dipole reversal is illustrated in Fig. 13.16. The pattern of the radial component of the field is shown both at the core-mantle boundary and at what would be the surface of the Earth. The longitudinally averaged field inside the core is also shown, in terms of field lines for the poloidal field (on the left sides of the images) and contours for the toroidal field (on the right). Notice how the reversal appears complete at time (c) when viewed above the core but that it takes another 3000 years for the original field to decay out of the solid inner core and the new polarity to penetrate in.

Another interesting dynamic, but on a much shorter time scale, is the prediction (based of these simulations) that the Earth's solid inner core rotates on average slightly faster than the mantle (Glatzmaier & Roberts, 1995a, 1996b). Simulations that include the gravitational coupling between the inner core and mantle reduce the predicted average rate of eastward drift of the solid inner core to about 0.02 degree longitude per year (Buffett & Glatzmaier, 2000), putting it closer to some seismic analyses of inner-core super rotation. However, this simulated instantaneous angular rotation rate of the inner core varies by typically 0.1 degrees longitude per year with an average period of 75 years.

13.6.2 Saturn Dynamo Simulations

For giant planet simulations we typically construct an adiabatic reference state (Chapter 12) based on a 1D evolutionary model (e.g., Guillot, 1999, 2005) with

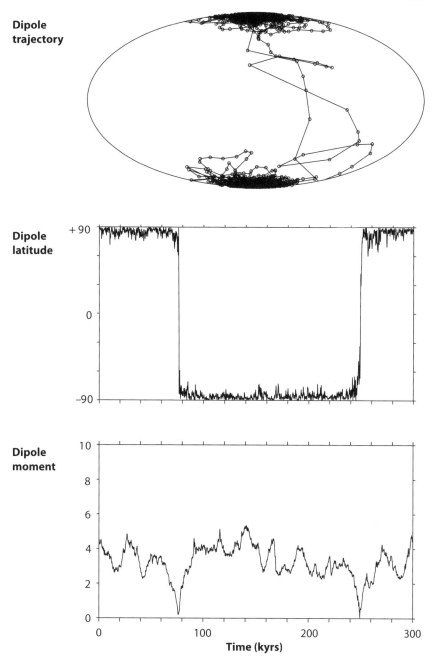

Figure 13.15 A 300,000-year geodynamo simulation showing the evolution of the dipolar
part of the generated magnetic field at the surface of the modeled Earth. (Top)
The trajectory of the magnetic south pole on an equal-area projection of the
entire surface. (Middle) The latitude of the magnetic south pole in degrees.
(Bottom) The magnetic dipole moment in units of 10^{22} Am^2. (This material
is reproduced from Glatzmaier et al. (1999) with permission from the *Nature
Publishing Group*.)

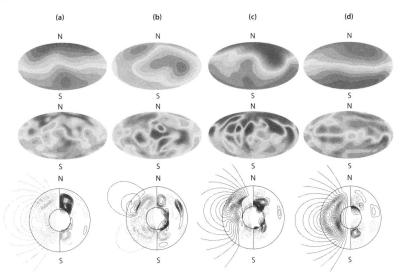

Figure 13.16 A sequence of snapshots, displayed at 3000-year intervals during a spontaneous simulated magnetic dipole reversal (see *Color Plate* 13 for a color version of this figure). (Bottom) The longitudinally averaged magnetic field plotted within the core. The small circle represents the inner-core boundary and
the large circle is the core-mantle boundary. The poloidal field is shown as
magnetic field lines on the left-hand sides of these plots (blue is clockwise and
red is counterclockwise). The toroidal field direction and intensity are represented as contours (not magnetic field lines) on the right-hand sides (red is
eastward and blue is westward). (Middle and Top) The radial component of
the field on the core-mantle boundary and at what would be the surface, plotted as described in Fig. 13.13. (This material is reproduced from Glatzmaier
et al. (1999) with permission from the *Nature Publishing Group*.)

an electrical conductivity (which is inversely proportional to the magnetic diffusivity) in the semiconducting region that increases exponentially with depth based on
analyses by Nellis et al. (1996), Liu et al. (2008), and French et al. (2012). The
turbulent viscous and thermal diffusivities are larger than we would like, due to
computational limitations; therefore, to obtain realistic amplitudes for the fluid velocity and magnetic field, the prescribed luminosity usually needs to be somewhat
larger than observed.

An example of a dynamo simulation that produces a differential rotation similar to that observed on Saturn's *surface* is illustrated in Fig. 13.17 (Glatzmaier,
2005b; Stanley & Glatzmaier, 2009). This figure shows a snapshot of the latitudi-
nally banded longitudinal component of the flow with its broad eastward zonal jet
in the equatorial region. Reds and yellows are eastward-directed flow (relative to
the rotating frame) up to 300 m/s; blues are westward up to 100 m/s. This zonal
wind profile is very different from the westward thermal wind outside the tangent
cylinder in the Earth's fluid core (Section 13.6). The way it is maintained is also different; many small isolated vortices, generated by fluid moving through the density

Simulated banded zonal winds

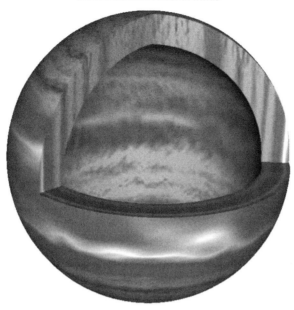

Figure 13.17 A snapshot of the longitudinal flow from a Saturn dynamo simulation (see *Color Plate* 14 for a color version of this figure). Reds and yellows are prograde flow relative to the rotating frame and blues are retrograde. (This material is reproduced from Stanley & Glatzmaier (2009) with permission from *Springer Science + Business Media BV.*)

stratification (Section 13.3.2), fill the convection zone and maintain this differential rotation pattern by the convergence of angular momentum flux (not as a thermal wind).

The banded differential rotation pattern in this simulation is also illustrated in Fig. 13.18. The "constant-on-cylinders" angular velocity *below the surface* is a prediction; a shear flow well below the surface is critical for the maintenance of the dynamo. Reducing the prescribed viscosity in the continuation of this simulation makes the convection more turbulent and causes the high-latitude bands to decay with depth below the surface; a strong shear flow well below the surface remains in the equatorial region.

We have chosen to simulate only the outer 20% in radius in this scenario because another simulation that includes the entire hydrogen-helium interior (Glatzmaier, 2005b) shows the largest kinetic and magnetic energy densities existing in the outer region (Fig. 13.19). In that simulation, convective flows are greatest near the surface where density is small; whereas magnetic induction is most efficient in a relatively narrow layer at roughly 20% of the radius below the surface, where electrical conductivity is large. Dynamo action requires both sufficiently vigorous fluid motions and sufficiently high electrical conductivity to balance the removal of field by

Surface longitudinal winds   **Zonal winds in meridian plane**

Surface radial magnetic field   **Zonal field in meridian plane**

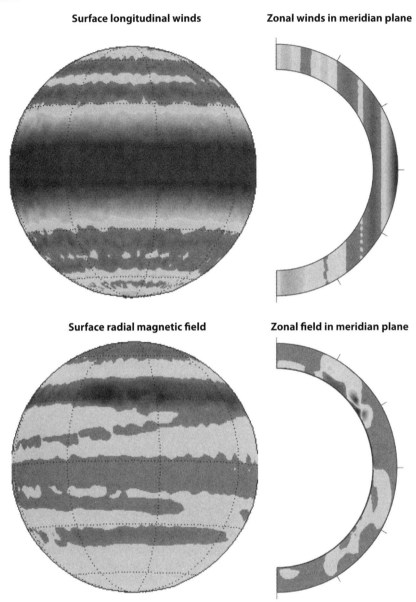

Figure 13.18 A snapshot of a Saturn dynamo simulation, as in Fig. 13.17 (see *Color Plate*
15 for a color version of this figure). (Top, left) The differential rotation (zonal
winds) at the surface; red and yellow represent prograde flow, blues are retrograde. (Top, right) The longitudinal average of the zonal winds below the surface; red and yellow are prograde, blue is retrograde. (Bottom, left) The radial
component of the magnetic field at the surface; yellow represents outward-
directed field, blue is inward. (Bottom, right) The longitudinal average of the
toroidal field below the surface; red and yellow are eastward-directed, blue
is westward. (This material is reproduced from Stanley & Glatzmaier (2009)
with permission from *Springer Science + Business Media BV*.)

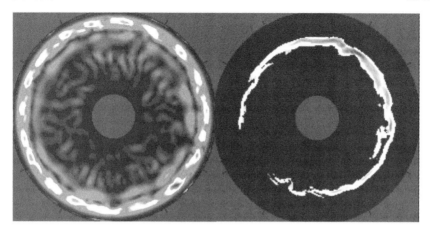

Kinetic energy density   **Magnetic energy density**

Figure 13.19 A snapshot of the kinetic and magnetic energy densities in the equatorial plane of a 3D dynamo simulation of a giant planet like Saturn (see *Color Plate* 16a for a color version of this figure). The kinetic energy is greatest near the surface and decays with depth; the magnetic energy peaks in a narrow layer at roughly 20% of the radius below the surface.

magnetic diffusion. That is, a high local magnetic Reynolds number, Rm, is required, which is the ratio of the RMS fluid velocity, V, at a given depth to the magnetic diffusive velocity at that depth, using a representative length scale, D:

$$\text{Rm} \equiv \frac{V D}{\bar{\eta}} = V D \mu \bar{\sigma} .$$

Here, μ is magnetic permeability (assumed constant), $\bar{\sigma}$ is the electrical conductivity, and $\bar{\eta} = 1/(\mu \bar{\sigma})$ is the magnetic diffusivity. Therefore, although kinetic energy density peaks near the surface, magnetic energy density peaks deeper below the surface in a high-Rm "sweet spot," as seen in Fig. 13.19. The kinetic and magnetic energy densities are both small below this layer.

Plots of the average kinetic and magnetic energy densities vs. radius for the dynamo simulated in just the outer 20% of the radius (as seen in Fig. 13.17) are shown in Fig. 13.20. These show the kinetic energy peaking near the top boundary and magnetic energy peaking near the bottom boundary. The kinetic energy density is about three orders of magnitude greater than the magnetic energy density; so this is a "weak field" dynamo.

Figure 13.18 also illustrates how the radial magnetic field at the surface has a banded structure, especially at high latitude, due to the banded differential rotation. It also shows considerable structure in radius and latitude of the longitudinally averaged toroidal magnetic field, which is also more intense at high latitude.

Besides having strong shear, the fluid flow in a convective dynamo also needs to have sufficient helicity, i.e., the dot product of velocity and vorticity. Helical fluid

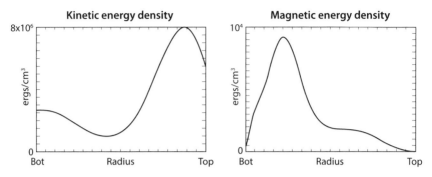

Figure 13.20 Average kinetic and magnetic energy densities plotted vs. radius through the convection zone for the simulated Saturn dynamo illustrated in Fig. 13.17.

motions twist toroidal (east-west) magnetic field into poloidal (meridional) magnetic field while differential rotation shears poloidal field into toroidal field. However, to limit the resulting ohmic dissipation due to this shear flow, poloidal field tends to align parallel to surfaces of the constant-on-cylinders angular velocity, as suggested by Ferraro's isorotation law (Ferraro, 1937). This makes the radial field at the surface significantly larger at high latitudes, as seen in Fig. 13.21. Saturn's observed surface magnetic field is also weaker in the equatorial region between $\pm 45°$ latitude than expected for a pure dipole (Connerney, 1993), possibly for this same reason.

This simulation maintains a magnetic dipole moment (Section 10.6.1) of 1.1×10^{20} Tm^3, somewhat less than Jupiter's (1.5×10^{20} Tm^3) and larger than Saturn's (4.2×10^{18} Tm^3). (Note, 1 Tm$^3 = 10^{10}$ gauss cm^3.) As mentioned in Section 10.6.1, the structure of a potential planetary magnetic field at and beyond the surface is traditionally described in terms of the gauss coefficients of a spherical harmonic expansion. Currently, however, only a very small number of gauss coefficients (typically up to degree 4) have been estimated for the fields of Jupiter and Saturn because the number of flybys is small and NASA's *Galileo* and *Cassini* missions to Jupiter and Saturn, respectively, have not orbited close enough to the planetary surfaces. When better magnetic observations are obtained for Saturn during the NASA *Cassini Solstice* mission and for Jupiter during the NASA *Juno* mission starting in 2016, more detailed comparisons will be made with 3D simulations. These comparisons will, in addition to providing much more detailed measurements of the magnetic field, hopefully provide information, based on gravity measurements, on the depth to which the zonal winds extend below the surface as a function of latitude for these two giant gas planets. Evidence of deep zonal winds would also be obtained if the magnetic fields that will be measured near the surfaces of Jupiter and Saturn have a banded zonal structure correlated with the banded zonal wind pattern (Fig. 13.18) since such a field pattern would need to be maintained deep below the surface where the electrical conductivity is sufficiently high.

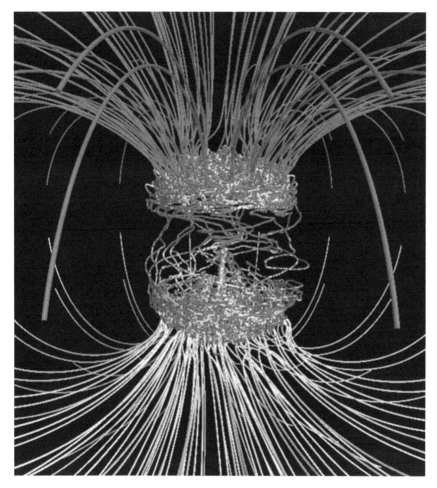

Figure 13.21 A snapshot of the magnetic field, illustrated with magnetic field lines, main-tained in the Saturn dynamo simulation (Fig. 13.17) (see *Color Plate* 16b for a color version of this figure). Gold field lines are directed outward and blue inward. (This material is reproduced from Stanley & Glatzmaier (2009) with permission from *Springer Science + Business Media BV.*)

13.7 CONCLUDING REMARKS

Much progress has been made in our understanding of the internal dynamics of planets and stars since the early 1980s when researchers using 2D local mag-netoconvection codes provided early insight into the magnetohydrodynamics of sunspots. At the same time other researchers using 3D global magnetohydrodynam-ics codes produced the first self-consistent solar dynamo simulations in an attempt to explain the maintenance of the background field assumed in the sunspot simu-lations. "Supercomputers" had become sufficiently powerful at that time to make it possible to run these codes at the spatial and temporal resolutions needed to just barely obtain numerical solutions. Today's computers run parallel versions of these codes with improved numerical methods at resolutions that are better by several orders of magnitude than those of the 1980s.

However, the two basic principles upon which the credibility of simulated results are based are *verification* and *validation*. Verification is checking that the *code* accurately solves the equations that define the model. Validation is checking that the *simulations* accurately represent the real physical problem studied. Validation asks "Are you building the right thing?" and verification asks "Are you building the thing right?" The benchmarks described in Section 13.5, for example, verify codes that successfully produce results that converge to the correct benchmark solutions. However, we still are far from the appropriate parameter regimes needed to validate simulations of convection in planets and stars, i.e., to produce simulations with the strong turbulence that surely exists in the interiors of these bodies. The viscous, thermal, and compositional diffusivities that are currently being used to avoid numerical instabilities are considered "turbulent diffusivities," which serve as crude representations of the transport and mixing done by subgrid-scale turbulence that is not explicitly resolved in the simulations. Besides not representing the correct turbulent structures and transport mechanisms, a problem with using greatly enhanced viscous and thermal diffusivities is that then a simulation usually requires a luminosity (total heat flow) that exceeds the realistic (observed) luminosity for the body in order to drive fluid flows and magnetic fields with realistic amplitudes. As parallel computers and numerical methods improve, solutions will become more realistic by resolving more of the spectrum of scales, allowing the amplitudes of the turbulent diffusivities to be reduced. The diffusivities, however, do not need to be reduced down to the actual molecular values to achieve realistic results over a major part of the spectrum as long as a significant inertial subrange exists within which energy cascades down many orders of magnitude in length scale (i.e., up in wavenumber) before dissipating by "turbulent diffusion."

One test, which the anelastic Saturnian dynamo simulation discussed in Section 13.6.2 already passes, is to obtain patterns and amplitudes of fluid flows and magnetic fields at the surface in qualitative agreement with observations (using the actual dimensions, mean rotation rate, and realistic stratifications of density and electrical conductivity) while maintaining a total (volume integrated) rate of entropy production by ohmic heating less than the *observed* surface luminosity (L_{surf}) times $(1/T_{top} - 1/T_{bot})$, where T_{top} and T_{bot} are the mean temperatures at the top and bottom boundaries, respectively. This constraint comes from equating the rate entropy flows out of the convection zone (at the top) to the rate it flows in (at the bottom) plus the rate it is generated by dissipation within the convection zone (Backus, 1975; Hewitt et al., 1975). Since ohmic dissipation is the major dissipation and (obviously) less than the total dissipation, this constraint can be written as

$$\int J^2/(\bar{\sigma}T)dV \; < \; L_{surf}(1/T_{top} - 1/T_{bot}).$$

Also, since most of the ohmic heating occurs in a relatively thin layer where the magnetic Reynolds number peaks (as discussed in Section 13.6.2), this constraint can be approximated as a constraint on the total rate of ohmic heating:

$$\int J^2/\bar{\sigma}\,dV < L_{surf}T_{peak}(1/T_{top} - 1/T_{bot}).$$

This test should be applied to all convective dynamo simulations as they become better resolved and more turbulent.

Here we end by listing some substantial improvements for planetary and stellar convective dynamo models to which we look forward, many of which are already being developed. Subgrid-scale turbulence models should replace the enhanced turbulent diffusivities by providing additional stresses (in the momentum, energy, and magnetic induction equations) that estimate how the unresolved turbulence disperses (instead of diffuses) energy within the spectrum of length scales (e.g., Buffett, 2003; Chen & Glatzmaier, 2005; Chen & Jones, 2008). The anelastic approximation could be improved, for example, by making the low-Mach-number approximation (e.g., Durran, 1989; Bell et al., 2004; Almgren et al., 2006a; Zingale et al., 2009). Fully compressible global models are needed for problems involving transonic flows like those in the surface layers of stars (where the temperature, and therefore the sound speed, are less than in the deep interior) and in the atmospheres of "hot Jupiters" (giant exoplanets orbiting so close to their stars that the stellar luminosity drives supersonic atmospheric flows). The top boundary in models of stars and giant planets should be extended closer to the visible surface by adding at least a simple treatment of radiative transfer to replace the thermal diffusion approximation there. The impermeable top boundary condition for stars and giant planets should be replaced with a free surface boundary condition. The impermeable bottom boundary should be removed for stars and giant planets so the resulting "full-sphere" model could be used to study convective flow (e.g., Evonuk & Glatzmaier, 2007) or internal gravity wave propagation (e.g., Rogers & Glatzmaier, 2005b) through the center and the resulting feedbacks on nuclear burning for stars (e.g., Almgren et al., 2008; Zingale et al., 2009) and on core erosion for giant planets (e.g., Guillot et al., 2004). The effects of the hydrogen phase transition and helium settling should be simulated in models of giant planets and, likewise, the effects of hydrogen and helium ionization zones in stellar models. It will be interesting to discover how precession of a body's rotation axis (e.g., Tilgner, 2007a; Wu & Roberts, 2009) and a tidal potential due to the body's orbital eccentricity and obliquity (e.g., Tilgner, 2007b; Noir et al., 2009) produce instabilities that modify the way its turbulent convection maintains differential rotation and magnetic field. These and other major model improvements will surely lead to exciting discoveries and better understanding of the internal dynamics of planets and stars.

SUPPLEMENTAL READING

Chandrasekhar (1961)
Greenspan (1969)
Gubbins & Herrero-Bervera (2007)
Olson & Schubert (2007)

EXERCISES

1. *Boussinesq magneto-rotational-gravity wave dispersion relation for a 2.5D box*
   Verify the Boussinesq dispersion relations for 2.5D rotational-gravity waves with a vertical background magnetic field, Eq. 13.22, and with a horizontal background magnetic field, Eq. 13.23.

2. *Conservation of potential vorticity*
   Derive the conservation of potential vorticity theorem. That is, neglect the buoyancy and viscous terms in the vorticity equation, but, unlike the Proudman-Taylor Theorem, retain the full inertial term. Then show that for a 2D equatorial annulus

$$\frac{d}{dt}\left(\frac{2\Omega + \omega}{\bar{\rho}}\right) = 0.$$

That is, the potential vorticity of a fluid parcel, $(2\Omega + \omega)/\bar{\rho}$, remains constant as the parcel moves. In 3D one would need to neglect the local vortex stretching term to arrive at this result. This, however, may not be negligible in a turbulent density-stratified environment.

3. *Perturbation centrifugal force for a 3D spherical fluid shell*
   What additional terms would need to be added to the nonlinear terms in the anelastic equations for convection in a rotating spherical shell, Eqs. 10.74–10.76, if one includes centrifugal force proportional to the density perturbation?

4. *Inertial oscillation dispersion relation for a 2.5D spherical fluid shell with viscosity*
   Derive a dispersion relation for inertial oscillations in a 2.5D spherical shell of an incompressible fluid, as in Eq. 13.13, but with viscous diffusion.

5. *Inertial oscillation dispersion relation for a 2.5D spherical fluid shell with viscosity, density stratification, and buoyancy*
   Derive a dispersion relation for inertial oscillations in a 2.5D spherical shell, as in Eq. 13.13, but for an anelastic fluid with viscous diffusion and buoyancy.

6. *Inertial oscillations in a 3D spherical fluid shell: linear*
   Solve for inertial oscillations, as in Section 13.4.1, but for a 3D spherical shell as a function of the degree l and order m of the spherical harmonic mode when neglecting nonlinear, diffusion, magnetic, and buoyancy terms. That is, take the Eulerian time derivative of Eq. 10.74 and substitute into this Eqs. 10.75 and 10.76 for the time derivatives of Z_{l+1}^m, Z_{l-1}^m, $\partial W_{l+1}^m/\partial r$, and $\partial W_{l-1}^m/\partial r$. Collect the common terms to show how $\partial^2 W_l^m/\partial t^2$ depends on W_{l+2}^m, W_{l-2}^m, their first-order radial derivatives, and $im\,Z_{l+2}^m$ and $im\,Z_{l-1}^m$, with all terms proportional to $(2\Omega)^2$. If one considers only the W_l^m Coriolis term in this expression, what would be the frequency in terms of l and m?

COMPUTATIONAL PROJECTS

1. *Thermal boundary layer thickness*
   Using plots of the mean entropy perturbation as a function of z for a simulations of thermal convection in a box, check how the thickness of the thermal boundary layer depends on Ek for a given Ra, Pr, aspect ratio, and N_ρ.

2. *Anelastic convection in a 2D rotating annulus exciting internal gravity waves in an adjacent stable region*
   Set up a stable thermal stratification below a convectively unstable outer convection zone in a 2D rotating anelastic annulus model in the equatorial plane (see Section 6.2.2 and, for example, Rogers et al. (2006)). This could serve as a 2D simulation of the internal dynamics in a solar like star. Make a movie of the upward propagating internal gravity waves in the stable interior, which are excited by the penetrating convection above. Make another model that simulates a convectively stable outer region above a convectively unstable inner region, which would be a 2D simulation of a star much more massive than the sun.

3. *2D rotating convection in the equatorial plane with vorticity generated by both the compressional torque and a modeled quasi-geostrophic stretching torque*
   Set up a simple model of "quasi-geostrophic" convection starting with a 2D rotating anelastic annulus model in the equatorial plane (Section 13.3). Add a prescribed function of radius to h_ρ in the Coriolis term of the vorticity equation to simulate the additional expansion (contraction) normal to the rotation axis that fluid in Taylor columns parallel to the rotating axis would experience as it moves from (toward) the rotation axis as the distance, H, between the northern and southern boundaries changes, i.e., $H = 2(r_{top}^2 - r^2)^{1/2}$, where r is cylindrical radius. That is, add a prescribed function of radius to the compressional torque term that simulates a stretching torque due to the deformation fluid would be forced to experience if it were circulating in 3D Taylor columns. Compare the effect of this stretching term due to the curved boundaries of a 3D spherical shell to that of the compressional term due to the local density stratification. How many density scale heights spanning the convection zone are needed to have the effect of the compressional torque be comparable to that of the stretching torque?

4. *Anelastic critical Rayleigh number as a function of the Ekman number for a 3D spherical shell*
   Using a linear anelastic code for thermal convection of a polytropic perfect gas in a 3D rotating spherical shell, iterate on the Rayleigh number, as described in Section 3.3, to find the critical Rayleigh number as a function of the spherical harmonic order m (including all degrees l, which are coupled) and a function of the Ekman number, Ek, for a given Prandtl number, Pr, aspect ratio, r_{bot}/r_{top}, polytropic index, n, density stratification, h_ρ, and the amplitude of the initial dipolar magnetic field. Let the $\bar{\rho}\bar{\nu}$ and $\bar{\rho}\bar{\kappa}\bar{T}$ be constants. Then run of few cases to find how Ra_{crit} depends on Pr, r_{bot}/r_{top}, n, and the amplitude of the initial dipolar magnetic field for a given Ek.

5. *Inertial oscillations in a 2.5D spherical fluid shell driven by a time-dependent rotation rate*
Do a series of simulations with a 2.5D spherical-shell model for which the amplitude of the rotation rate equals a constant plus a small, sinusoidally time-dependent perturbation. Study how the amplitude of the inertial oscillation depends on the prescribed frequency of the rotational perturbation (Tilgner, 1999).

6. *Inertial oscillations and internal gravity waves in a 2.5D spherical fluid shell*
Reverse the entropy boundary conditions in a 2.5D rotating spherical-shell model (i.e., make the top boundary hot and the bottom boundary cold) to study how internal gravity waves interact with inertial oscillations when their frequencies are comparable. Note, internal gravity waves reflect off impermeable boundaries with their angle of incidence relative to the local gravitational acceleration being equal to the angle of reflection relative to gravity (Section 6.1); so, for spherical boundaries and a radially directed gravitational acceleration, the incident and reflected angles relative to the spherical boundaries are equal. However, the rays of inertial oscillations reflect off impermeable boundaries with their angle of incidence relative to the local direction of $\mathbf{\Omega}$ at a boundary equal to the angle of reflection relative to $\mathbf{\Omega}$ (Section 13.4.1).

7. *2.5D "intermediate dynamo" simulation*
Produce a 2.5D convection simulation with rotation and magnetic field as described in Section 13.4.2 with the parameters and resolution specified in that section or, alternatively, with less extreme values for the parameters and resolution. After confirming that the field decays (because it is axisymmetric), continue the simulation as an *intermediate dynamo* by adding an "alpha effect" to the poloidal magnetic field equation, 10.77, as is done for kinematic mean field dynamo models, to provide a source for the poloidal magnetic field. This source term is supposed to represent the effect of the local 3D helical flow, which is missing in axisymmetric models but needed to twist toroidal field into poloidal field. (Note that the axisymmetric toroidal field is naturally maintained by differential rotation shearing axisymmetric poloidal field.) The alpha effect is added to the nonlinear induction term in Eq. 10.77, which now has the form

$$\left[\left(\nabla \times (\mathbf{v} \times \mathbf{B} + \alpha B_\phi \hat{\boldsymbol{\phi}})\right)_r\right]_l^m .$$

Let α be defined as suggested by Braginsky & Roberts (1987):

$$\alpha(r, \theta) \equiv 0 \quad \text{for} \quad s \leq 0.8 r_{top}$$

and

$$\alpha(r, \theta) \equiv \left(\frac{20 R_\alpha \eta}{r_{top}^2}\right) z \left(1 - \left(\frac{z}{z_1}\right)^6\right) \sin\left(\pi \left(9 - \frac{10s}{r_{top}}\right)\right)$$

$$\text{for} \quad 0.8 r_{top} < s \leq r_{top} ,$$

where $s = r \sin\theta$, $z = r \cos\theta$, $z_1 = r_{top}(1 - (s/r_{top})^2)^{1/2}$, and R_α is the prescribed nondimensional intensity of the alpha effect. See Fig. 2a in

Glatzmaier & Roberts (1993) for a plot of this prescribed alpha function. Test different values of R_α, from as low as 10 to as high as 10,000, to see what the critical value is for this form of alpha and the set of parameters (mentioned above) you chose, i.e., what value of R_α is needed to just prevent the field from decaying. Also, check how higher values affect the time dependence and spatial structure of the field. Compare you results to those of various cases described in Glatzmaier & Roberts (1993).

8. *Inertial oscillations in a 3D spherical fluid shell: nonlinear*
   Using a nonlinear version of the 3D spherical-shell model, but with no magnetic field or gravitational acceleration, excite at one time step one l, m mode of either W_l^m or Z_l^m with a smooth function in radius that satisfies the boundary conditions. Compare the resulting frequency of the inertial oscillation with that estimated from the *Inertial oscillations in a 3D spherical fluid shell: linear* exercise.

9. *Inertial oscillations in a 3D spherical fluid shell: continuously excited*
   Repeat the *Inertial oscillations in a 3D spherical fluid shell: nonlinear* project but excite one W_l^m continuously with a prescribed frequency. Save the time series of the amplitudes of all the W_l^m at mid-depth and, using a Fourier analysis, plot the dispersion relation, i.e., the energy per mode on an l vs. frequency plot for the m of the exciting mode.

Appendix A

A Tridiagonal Matrix Solver

This is a tridiagonal matrix solver written in Fortran based on LINPACK and LAPACK routines. Note, in this routine the first index of the arrays is 1 and the last is N_z; if one wishes to start the arrays with index 0 and end with $N_z - 1$, all index references in this routine would need to be reduced by 1.

```fortran
subroutine tridi(nz,rhs,sol,sub,dia,sup,wk1,wk2)

real, dimension (1:nz) :: rhs,sol,sub,dia,sup,wk1,wk2
integer :: i,nz

wk1(1)=1./dia(1)
wk2(1)=sup(1)*wk1(1)
do i=2,nz-1
  wk1(i)=1./(dia(i)-sub(i)*wk2(i-1))
  wk2(i)=sup(i)*wk1(i)
enddo
wk1(nz)=1./(dia(nz)-sub(nz)*wk2(nz-1))

sol(1)=rhs(1)*wk1(1)
do i=2,nz
  sol(i)=(rhs(i)-sub(i)*sol(i-1))*wk1(i)
enddo
do i=nz-1,1,-1
  sol(i)=sol(i)-wk2(i)*sol(i+1)
enddo

return
end
```

Appendix B

Making Computer-Graphical Movies

These are Fortran statements that could be added to a computational code to produce a series of movie files, one per snapshot, with the movie snapshot number being part of each filename, i.e., *mov.000001, mov.000002, mov.000003*, etc. Here *istep* is the computational time step count, *imovie* is the number of time steps between movie snapshots, and *imovstart* is the previous movie snapshot count (0 if starting a new movie).

```
      character movfile*10,movstr*6

      if((imovie.gt.0) .and. (mod(istep,imovie).eq.0)) then
        do i=1,nx
          do k=1,nz
            temmov(i,k)=0.
            psimov(i,k)=0.
            do n=0,nn
              temmov(i,k)=temmov(i,k)+tem(k,n)*cosa(n,i)
              psimov(i,k)=psimov(i,k)+psi(k,n)*sina(n,i)
            enddo
          enddo
        enddo
        write(movstr,"(i6)") istep/imovie+imovstart
        do ii=1,6
          if(movstr(ii:ii) .eq. ' ') movstr(ii:ii)='0'
        enddo
        movfile=movfileid//movstr(1:6)
        open(2,file=movfile,form='formatted')
        write(2,22) nx,nz,aspect,ra,pr
22      format(2i7,3(1x,1pe10.3))
        do k=1,nz
          write(2,23) (temmov(i,k),i=1,nx)
23        format(7(1x,1pe10.3))
        enddo
        do k=1,nz
          write(2,23) (psimov(i,k),i=1,nx)
        enddo
        close(2)
      endif
```

This is an *IDL* program that reads a series of movie snapshot files produced as described above and makes an *mpeg* movie of either the temperature perturbation or the streamfunction.

```
pro loaddat,file
common movdat,nx,nz,aspect,tem,psi
if n_elements(file) eq 0 then return
openr,1,file
nx=0L
nz=0L
aspect=0.
ra=0.
pr=0.
readf,1,nx,nz,aspect,ra,pr
tem=fltarr(nx,nz)
psi=fltarr(nx,nz)
readf,1,tem,psi
close,1
free_lun,1
end

pro movie
common movdat,nx,nz,aspect,tem,psi
count=1
read,prompt='starting count number = ',count
iftem=''
read,prompt='plot temperature ? ',iftem
ifpsi=''
if(iftem ne 'y') then read,prompt='plot streamfunction?',$
  ifpsi
file='mov.000000'
nfile=strlen(file)
file=strmid(file,0,nfile-1-(count ge 10)-(count ge 100)-$
  (count ge 1000)-(count ge 10000)-(count ge 100000))+ $
  strtrim(count,2)
loaddat,file
ans=''
if(iftem eq 'y') then begin
  amin=min(tem)
  amax=max(tem)
  read,prompt='want min=min(tem) and max=max(tem)? ', ans
  if(ans eq 'y') then print,'min and max tem=',amin, amax
  if(ans ne 'y') then begin
    read,prompt='min tem = ',amin
    read,prompt='max tem = ',amax
  endif
endif
if(ifpsi eq 'y') then begin
  amin=min(psi)
  amax=max(psi)
  read,prompt='want min=min(psi) and max=max(psi)? ', ans
```

```
   if(ans eq 'y') then print,'min and max psi = ',amin, amax
   if(ans ne 'y') then begin
     read,prompt='min psi = ',amin
     read,prompt='max psi = ',amax
   endif
   nlevs=15
   read,prompt='how many contour levels? ',nlevs
   levs=amin+findgen(nlevs)/(nlevs-1)*(amax-amin)
endif
x=findgen(nx)/(nx-1)*aspect
z=findgen(nz)/(nz-1)
if(iftem ne 'y') then begin
  nxwin=512
  nywin=512
endif else begin
  nxwin=480
  nywin=round(nxwin/aspect)
  loadct,33
  cr=bindgen(256)
  cg=bindgen(256)
  cb=bindgen(256)
  tvlct,cr,cg,cb,/get
  device,true_color=24
endelse
window,1,xsize=nxwin,ysize=nywin,retain=2,/pixmap
mpeg_id=mpeg_open([nxwin,nywin],filename='mov.mpg')
while file_test(file) do begin
  loaddat,file
  if(iftem eq 'y') then begin
    t=congrid(tem,nxwin,nywin,cubic=-0.5,/minus_one)
    t=bytscl(t,min=amin,max=amax)
    t=reverse(t,2)
    y=bytarr(3,nxwin,nywin)
    y(0,*,*)=cr(t(*,*))
    y(1,*,*)=cg(t(*,*))
    y(2,*,*)=cb(t(*,*))
    tv,y,true=1,/order
    write_ppm,file+'.ppm',y
  endif
  if(ifpsi eq 'y') then $
    contour,psi,x,z,thick=1.5,/isotropic,levels=levs, $
      c_linestyle=2*(levs lt 0.),xmargin=[8,8], $
      ymargin=[8,8],xticklen=0.0001,yticklen=0.0001, $
      xminor=1
  mpeg_put,mpeg_id,window=1,frame=count,/order
  print,'count=',count
  count=count+1
  file=strmid(file,0,nfile-1-(count ge 10)-(count ge 100)-$
    (count ge 1000)-(count ge 10000)-(count ge 100000))+ $
    strtrim(count,2)
```

```
endwhile
mpeg_save,mpeg_id
mpeg_close,mpeg_id
if(iftem ne 'y') then begin
  set_plot,'ps'
  device,filename='snap.ps'
  contour,psi,x,z,thick=1.5,/isotropic,levels=levs, $
    c_linestyle=2*(levs lt 0.),xmargin=[8,8], $
    ymargin=[8,8],xticklen=0.0001,yticklen=0.0001,xminor=1
  device,/close_file
  set_plot,'x'
endif
end
movie
end
```

Appendix C

Legendre Functions and the Gaussian Quadrature

This is a Fortran subroutine for computing the value of the normalized associate Legendre function of the first kind for degree $l \geq 0$, order $m \geq 0$, and colatitude θ in radians.

```fortran
subroutine pbar(theta,l,m,p)

real*8 :: theta,s,c,p,p1,p2
integer :: l,m,m1,i,j

s=sin(theta)
c=cos(theta)
p=1./sqrt(2.)
if(m .ne. 0) then
   do i=1,m
      p=sqrt(real(2*i+1)/real(2*i))*s*p
   enddo
endif
if(l .eq. m) return
p1=1.
m1=m+1
do j=m1,l
   p2=p1
   p1=p
   p=2.*sqrt((real(j**2)-0.25)/real(j**2-m**2))*c*p1-
$     sqrt(real((2*j+1)*(j-m-1)*(j+m-1))/
$     real((2*j-3)*(j-m)*(j+m)))*p2
enddo

return
end
```

This is a Fortran subroutine for computing the N_θ ("N") roots of a Legendre polynomial, i.e., the colatitudes (*colat*) in radians at which the Legendre polynomial of degree $l = N_\theta$ and order $m = 0$ vanish. Also calculated are the corresponding N_θ Gaussian weights (gauss), which can be used for a Gaussian quadrature in colatitude.

```fortran
   subroutine gquad(N,colat,gauss)

   integer :: N,l1,l2,l22,l3,k,i
   real :: p,pi,p1,p2,del,co,s,t1,t2,theta
   real, dimension(N) :: colat,gauss

   pi=4.*atan(1.)
   del=pi/real(4*N)
   l1=N+1
   co=real(2*N+3)/real(l1**2)
   p2=1.
   t2=-del
   l2=N/2
   k=1

   do i=1,l2
20  t1=t2
    t2=t1+del
    theta=t2
    call pbar(theta,N,0,p)
    p1=p2
    p2=p
    if((k*p2) .gt. 0.) go to 20
    k=-k
40  s=(t2-t1)/(p2-p1)
    t1=t2
    t2=t2-s*p2
    theta=t2
    call pbar(theta,N,0,p)
    p1=p2
    p2=p
    if(abs(p) .le. 1.e-15) go to 30
    if(p2 .eq. p1) then
      write(6,*) 'sub gquad: zero = ',p,' at i = ',i
      go to 30
    endif
    go to 40
 30  colat(i)=theta
    call pbar(theta,l1,0,p)
    gauss(i)=co*(sin(theta)/p)**2
   enddo

   l22=2*l2
   if(l22 .eq. N) go to 70
   l2=l2+1
   theta=pi/2.
   colat(l2)=theta
   call pbar(theta,l1,0,p)
   gauss(l2)=co/p**2
```

```
70 continue

   l3=l2+1
   do i=l3,N
     colat(i)=pi-colat(N-i+1)
     gauss(i)=gauss(N-i+1)
   enddo

   return
   end
```

Appendix D

Parallel Processing: OpenMP

Here we illustrate the *Open Multiprocessing* (OpenMP) directives used to make a "do loop" parallel (Section 9.5.1), in Fortran. The lines beginning with !$OMP are directives, which would be ignored if the OpenMP library were not loaded with the compiled code.

```
      nthreads=OMP_GET_MAX_THREADS() !nthreads = number of threads
      write(6,*) nthreads,' threads' !prints number of threads

!$OMP PARALLEL DO !begins the parallel region within the code
!$OMP& SHARED(mat,w,wbot,wtop) !lists shared variables
!$OMP& PRIVATE(i,j,tmp) !lists private variables
      do i=1,ni !starts loop over i, divided among threads
         do j=1,nj !loop executed by all threads
            tmp(j)=w(i,j)
         enddo
         if(i .eq. 1) then !if statement not done in parallel
!$OMP CRITICAL !only main thread executes statements between
            tmp(nj)=wtop !CRITCAL and END CRITICAL
            tmp(1)=wbot
!$OMP END CRITICAL
         endif
         call mat(tmp,nj) !a subroutine call all threads execute
         do j=1,nj !loop executed by all threads
            w(i,j)=tmp(j)
         enddo
      enddo !ends the loop over i
!$OMP END PARALLEL DO !ends the parallel region
```

The parallelism here would work best if the total number of iterations for the parallel loop, ni, were evenly divisible by the number of threads, nthreads. The subroutine, mat, the variables, w, wbot, and wtop, and the parameters, ni and nj, were defined outside the parallel region. Each thread (i.e., process) has its own temporary variables, i, j, and tmp, the values of which are not used outside the parallel region.

Appendix E

Parallel Processing: MPI

Here we briefly introduce a few *Message Passing Interface* (MPI) subroutines (Section 9.5.2), in Fortran. The same MPI routines are available in C and C++, with of course slightly different syntax.

The following is the first MPI statement that appears in a (Fortran) code; it loads a file that defines MPI parameters and routines.

```
include mpif.h
```

Then an integer array, here called istatus, needs to be created because it is used by MPI receive commands.

```
integer istatus(MPI_STATUS_SIZE)
```

MPI is initialized with the following subroutine call.

```
call MPI_INIT(ierr)
```

The next subroutine call assigns an ID number to each process within the group being used, called MPI_COMM_WORLD, which can be the only group used. On output, the process ID number, here called myid, will be an integer between 0 and the number of processes minus one, depending on the process. Also, the error check, here called ierr, should equal zero on output.

```
call MPI_COMM_RANK(MPI_COMM_WORLD, myid, ierr)
```

The following subroutine call sets the integer, here called mproc0, to the total number of processes provided for the group MPI_COMM_WORLD, which can be the total for the current job. It is then good to check if mproc0 is actually equal to the number of processes you planned to have for the run, mproc, and to stop if not.

```
call MPI_COMM_SIZE(MPI_COMM_WORLD, mproc0, ierr)
if(mproc0 .ne. mproc) stop
```

One processor, say the one with myid=0, may be designated to do IO. If so, an "if(myid .eq. 0) then" statement should surround the read and write statements. If, for example, after process 0 reads n real words of input data (data(1) to data(n)) from the disk, it can broadcast it to all the other processors with the following command.

```
call MPI_BCAST(data(1),n,MPI_REAL,0,MPI_COMM_WORLD,ierr)
```

Note, if the default word size were 8 bytes, MPI_REAL8 would be used instead of MPI_REAL; or, if the words were integers, MPI_INTEGER would be used. Also, if the words were complex, 2*n would be used instead of n.

The following example illustrates how process 0 could receive data from all the other processes.

```
irecv=0
do iproc=1,mproc-1
  isend=iproc
  itag=iproc
  if(myid .eq. 0) then
    call MPI_RECV(rdata(1,iproc), n, MPI_REAL, isend, itag,
$       MPI_COMM_WORLD, istatus, ierr)
  else if(myid .eq. iproc) then
    call MPI_SEND(sdata(1), n, MPI_REAL, irecv, itag,
$       MPI_COMM_WORLD, ierr)
  endif
enddo
```

This set of commands sends n real words located in memory from sdata(1) to sdata(n) on the processes with myid equal to 1 through mproc-1 to the process with myid equal to 0. Here itag can be any integer one assigns to identify each send. These are called blocking sends and receives because the program does not proceed until the data has been copied from sdata to rdata for each iteration count, iproc.

The next example illustrates how a process can send ns words in its array sdata to the process with myid=irecv and receive nr words into its array rdata from the process with myid=isend.

```
call MPI_SENDRECV(sdata,ns,MPI_REAL, irecv,itags,rdata,
$   nr, MPI_REAL, isend, itagr,MPI_COMM_WORLD,istatus,ierr)
```

A global transpose of the data can be done using the following command, which causes each process to send data to and receive data from all the other processes:

```
call MPI_ALLTOALL(sdata, ns, MPI_REAL, rdata, nr, MPI_REAL,
$   MPI_COMM_WORLD, istatus, ierr)
```

In this case, sdata is dimensioned as (ns,mproc) and rdata as (nr,mproc), where usually ns=nr.

The following is another useful command that takes the maximum of n words in array sdata on each process and sends the maximum of these to the one-word rdata on the process with myid=irecv.

```
call MPI_REDUCE(sdata, rdata, n, MPI_REAL, MPI_MAX, irecv,
$   MPI_COMM_WORLD, ierr)
```

The operation, MPI_MAX, can be replaced with MPI_MIN, MPI_SUM, or MPI_PROD.

Another issue is to keep all processes doing their computations at the right times without some processes getting ahead or behind the others so the computations are done in the correct order. This typically requires the use of the following command each time step, which forces all processes to "check in" at this point before any can proceed.

```
call MPI_BARRIER(MPI_COMM_WORLD, ierr)
```

MPI is terminated by the following command when the job is complete.

```
call MPI_FINALIZE
```

Bibliography

Almgren, A.S., Bell, J.B., Rendleman, C.A., & Zingale, M. 2006a. Low Mach number modeling of type Ia supernovae. I. Hydrodynamics. *Astrophys. J.*, **637**, 922–936.

Almgren, A.S., Bell, J.B., Rendleman, C.A., & Zingale, M. 2006b. Low Mach number modeling of type Ia supernovae. II. Energy evolution. *Astrophys. J.*, **649**, 927–938.

Almgren, A.S., Bell, J.B., Noaka, A., & Zingale, M. 2008. Low Mach number modeling of type Ia supernovae. II. Reactions. *Astrophys. J.*, **684**, 449–470.

Anufriev, A.P., Jones, C.A., & Soward, A.M. 2005. The Boussinesq and anelastic liquid approximations for convection in the Earth's core. *Phys. Earth Planet. Inter.*, **152**, 163–190.

Backus, G.E. 1975. Gross thermodynamics of heat engines in deep interior of Earth. *Proc. Nat. Acad. Sci. USA*, **72**, 1555–1558.

Baines, P.G., & Gill, A.E. 1969. On thermohaline convection with linear gradients. *J. Fluid Mech.*, **37**, 289–306.

Batchelor, G.K. 1953. The conditions for dynamical similarity of motions of a frictionless perfect-gas atmosphere. *Q. J. Roy. Meteor. Soc.*, **79**, 224–235.

Batchelor, G.K. 1967. *An Introduction to Fluid Dynamics*. Cambridge University Press.

Bell, J.B., Day, M.S., Rendleman, C.A., Woosley, S.E., & Zingale, M.A. 2004. Adaptive low Mach number simulations of nuclear flame microphysics. *J. Comp. Phys.*, **195**, 677–694.

Bercovici, D., Schubert, G., Glatzmaier, G.A., & Zebib, A. 1989. Three-dimensional thermal convection in a spherical shell. *J. Fluid Mech.*, **206**, 75–104.

Berkoff, N.A., Kersale, E., & Tobias, S.M. 2010. Comparison of the anelastic approximation with fully compressible equation for linear magnetoconvection and magnetic buoyancy. *Geophys. Astrophys. Fluid Dynam.*, **104**, 545–563.

Boussinesq, J. 1903. *Théorie Analytique de la Chaleur*. Vol. 2. Paris: Gathier-Villars. Pages 157–176.

Boyd, J.P. 2001. *Chebyshev and Fourier Spectral Methods*. Second ed. Dover Publications.

Boyd, J.P. 2011. Comparing seven spectral methods for interpolation and for solving the Poisson equation in a disk: Zernike polynomials, Logan-Shepp ridge polynomials, Chebyshev-Fourier series, cylindrical Robert functions, Bessel-Fourier expansions, square-to-disk conformal mapping and radial basis functions. *J. Comp. Phys.*, **230**, 1408–1438.

Braginsky, S.I. 1964. Magnetohydrodynamics of the Earth's core. *Geomag. Aeron.*, **4**, 698–712.

Braginsky, S.I., & Roberts, P.H. 1987. A model-Z geodynamo. *Geophys. Astrophys. Fluid Dynam.*, **38**, 327–349.

Braginsky, S.I., & Roberts, P.H. 1995. Equations governing convection in earth's core and the geodynamo. *Geophys. Astrophys. Fluid Dynam.*, **79**, 1–97.

Braginsky, S.I., & Roberts, P.H. 2007. Anelastic and Boussinesq approximations. *Pages 11–19 of:* Gubbins, D., & Herrero-Bervera, E. (eds.), *Encyclopedia of Geomagnetism and Paleomagnetism*. Springer-Verlag.

Brandt, A., & Fernando, H.J.S. (eds.). 1995. *Double-Diffusive Convection*. American Geophysical Union.

Brown, B.P., Miesch, M.S., Browning, M.K., Brun, A.S., & Toomre, J. 2011. Magnetic cycles in a convective dynamo simulation of a young solar-type star. *Astrophys. J.*, **731:69**.

Brown, B.P., Vasil, G.M., & Zweibel, E.G. 2012. Energy conservation and gravity waves in sound-proof treatments of stellar interiors. Part I. Anelastic approximations. *Astrophys. J.*, **10**, 1088/0004-637X/756/2/109.

Buffett, B.A. 2003. A comparison of subgrid-scale models for large-eddy simulations of convection in the Earth's core. *Geophys. J. Int.*, **153**, 753–765.

Buffett, B.A. 2007. Core-mantle interactions. *Pages 345–358 of:* Olson, P., & Schubert, G. (eds.), *Treatise on Geophysics: Core Dynamics*. Elsevier.

Buffett, B.A., & Glatzmaier, G.A. 2000. Gravitational braking of inner-core rotation in geodynamo simulations. *Geophys. Res. Lett.*, **27**, 3125–3128.

Bullard, E.C., & Gellman, H. 1954. Homogeneous dynamos and terrestrial magnetism. *Philos. Trans. Roy. Soc. Lond. A*, **247**, 213–278.

Busse, F.H. 1970. Thermal instabilities in rapidly rotating systems. *J. Fluid Mech.*, **44**, 441–460.

Busse, F.H. 1983. A model of mean zonal flows in the major planets. *Geophys. Astrophys. Fluid Dynam.*, **23**, 153–174.

Canuto, C., Hussaini, M.Y., Quarteroni, A., & Zang, T.A. 1988. *Spectral Methods in Fluid Dynamics*. Springer-Verlag.

Carpenter, M.H., & Kennedy, C.A. 1994. *Fourth-Order 2N-Storage Runge-Kutta Schemes*. NASA Langley Research Center. NASA Technical Memorandum 109112.

Chandrasekhar, S. 1961. *Hydrodynamic and Hydromagnetic Stability*. Oxford University Press.

Chapman, B., Jost, G., & van der Pas, R. 2007. *Using OpenMP*. MIT Press.

Charbonnel, C., & Zahn, J.-P. 2007. Thermohaline mixing: a physical mechanism governing the photospheric composition of low-mass giants. *Astron. Astrophys.*, **467**, L15–L18.

Chen, Q., & Glatzmaier, G.A. 2005. Large eddy simulations of two-dimensional turbulent convection in a density-stratified fluid. *Geophys. Astrophys. Fluid Dynam.*, **99**, 355–375.

Chen, Q., & Jones, C.A. 2008. Similarity and dynamic similarity models for large-scale simulations of a rotating convection-driven dynamo. *Geophys. J. Int.*, **172**, 103–114.

Christensen, U.R., & Wicht, J. 2007. Numerical dynamo simulations. *Pages 245–282 of:* Olson, P., & Schubert, G. (eds.), *Treatise on Geophysics: Core Dynamics*. Elsevier.

Christensen, U.R., Aubert, J., Cardin, P., Dormy, E., Gibbons, S., Glatzmaier, G.A., Grote, E., Honkura, Y., Jones, C., Kono, M., Matsushima, M., Sakuraba, A., Takahashi, F., Tilgner, A., Wicht, J., & Zhang, K. 2001. A numerical dynamo benchmark. *Phys. Earth Planet. Inter.*, **128**, 25–34.

Christensen, U.R., Aubert, J., Cardin, P., Dormy, E., Gibbons, S., Glatzmaier, G.A., Grote, E., Honkura, Y., Jones, C., Kono, M., Matsushima, M., Sakuraba, A., Takahashi, F., Tilgner, A., Wicht, J., & Zhang, K. 2009. Erratum to "A numerical dynamo benchmark." *Phys. Earth Planet. Inter.*, **172**, 356.

Clayton, D.D. 1984. *Principles of Stellar Evolution and Nucleosynthesis*. University of Chicago Press.

Connerney, J.E.P. 1993. Magnetic field of the outer planets. *J. Geophys. Res.*, **98**, 18659–18679.

Cook, A.W., & Riley, J.J. 1996. Direct numerical simulation of a turbulent reactive plume on a parallel computer. *J. Comp. Phys.*, **129**, 263–283.

Cooley, J.W., & Tukey, J.W. 1965. An algorithm for the machine calculation of complex Fourier series. *Math. Comp.*, **19**, 297–301.

Courant, R., Friedrichs, K., & Lewy, H. 1928. Über die partiellen Differenzengleichungen der mathematischen Physik. *Math. Annelen*, **100**, 32–74.

Cowling, T.G. 1934. The magnetic field of sunspots. *Mon. Not. Roy. Astron. Soc.*, **94**, 39–48.

Davidson, P.A. 2001. *An Introduction to Magnetohydrodynamics*. Cambridge University Press.

Durran, D. 1989. Improving the anelastic approximation. *J. Atmos. Sci.*, **46**, 1453–1461.

Durran, D.R. 1998. *Numerical Methods for Wave Equations in Geophysical Fluid Dynamics*. Springer-Verlag.

Dziewonski, A.M., & Anderson, D.L. 1981. Preliminary reference Earth model. *Phys. Earth Planet. Inter.*, **25**, 297–356.

Eliasen, E., Machenhauer, B., & Rasmussen, E. 1970. *On a numerical method for integration of the hydrodynamical equations with a spectral representation of the horizontal fields*. Tech. rept. Institut for Teoretisk Meteorologi, University of Copenhagen. Report 2.

Ertel, H. 1972. Ein neuer hydrodynamischer Wirbelsatz. *Meteorolol. Z.*, **59**, 277–281.

Evonuk, M., & Glatzmaier, G.A. 2007. The effects of rotation on deep convection in giant planets with small cores. *Planet. Space Sci.*, **55**, 407–412.

Ferraro, V.C.A. 1937. The non-uniform rotation of the sun and its magnetic field. *Mon. Not. Roy. Astron. Soc.*, **97**, 458–472.

Ferziger, J.H., & Perić, M. 1997. *Computational Methods for Fluid Dynamics*. Springer-Verlag.

Fox, L., & Parker, I.B. 1968. *Chebyshev Polynomials in Numerical Analysis*. Oxford University Press.

French, M., Becker, A., Lorenzen, W., & Nettelmann, N. 2012. Ab initio simulations for material properties along the Jupiter adiabat. *Astrophys. J. Suppl.*, **10**, 1088/0067-0049/202/1/5.

Galloway, D.J., Proctor, M.R.E., & Weiss, N.O. 1978. Magnetic flux ropes and convection. *J. Fluid Mech.*, **87**, 243–261.

Gamet, L., Ducros, F., Nicoud, F., & Poinsot, T. 1999. Compact finite difference schemes on non-uniform meshes. Application to direct numerical simulations of compressible flows. *Int. J. Numer. Meth. Fluids*, **29**, 159–191.

Gerya, T.V. 2010. *Introduction to Numerical Geodynamic Modelling*. Cambridge University Press.

Gilman, P.A., & Glatzmaier, G.A. 1981. Compressible convection in a rotating spherical shell. I. Anelastic equations. *Astrophys. J. Suppl.*, **45**, 335–349.

Ginet, G.P., & Sudan, R.N. 1987. Numerical observations of dynamic behavior in two-dimensional compressible convection. *Phys. Fluids*, **30**, 1667–1677.

Glatzmaier, G.A. 1984. Numerical simulations of stellar convective dynamos. I. The model and method. *J. Comp. Phys.*, **55**, 461–484.

Glatzmaier, G.A. 1988. Numerical simulations of mantle convection: time-dependent, three-dimensional, compressible, spherical shell. *Geophys. Astrophys. Fluid Dynam.*, **43**, 223–264.

Glatzmaier, G.A. 2005a. Planetary and stellar dynamos: Challenges for next generation models. *Pages 331–357 of:* Soward, A.M., Jones, C.A., Hughes, D.W., & Weiss, N.O. (eds.), *Fluid Dynamics and Dynamos in Astrophysics and Geophysics*. Boca Raton, FL: CRC Press.

Glatzmaier, G.A. 2005b. A Saturnian dynamo simulation. *In: AGU Fall Meeting Abtracts*. American Geophysical Union.

Glatzmaier, G.A., & Clune, T.L. 2000. Computational aspects of geodynamo simulations. *Comp. Sci. Eng.*, **2**, 61–67.

Glatzmaier, G.A., & Coe, R.S. 2007. Magnetic polarity reversals in the core. *Pages 283–297 of:* Olson, P., & Schubert, G. (eds), *Treatise on Geophysics: Core Dynamics*. Elsevier.

Glatzmaier, G.A., & Gilman, P.A. 1981a. Compressible convection in a rotating spherical shell. II. A linear anelastic model. *Astrophys. J. Suppl.*, **45**, 351–380.

Glatzmaier, G.A., & Gilman, P.A. 1981b. Compressible convection in a rotating spherical shell. III. Analytic model for compressible vorticity waves. *Astrophys. J. Suppl.*, **45**, 381–388.

Glatzmaier, G.A., & Gilman, P.A. 1981c. Compressible convection in a rotating spherical shell. IV. Effects of viscosity, conductivity, boundary conditions and zone depth. *Astrophys. J. Suppl.*, **47**, 103–116.

Glatzmaier, G.A., & Olson, P. 2005. Probing the geodynamo. *Sci. Am.*, **292**, 50–57.

Glatzmaier, G.A., & Roberts, P.H. 1993. Intermediate dynamo models. *J. Geomag. Geoelectr.*, **45**, 1605–1616.

Glatzmaier, G.A., & Roberts, P.H. 1995a. A three-dimensional convective dynamo solution with rotating and finitely conducting inner core and mantle. *Phys. Earth Planet. Inter.*, **91**, 63–75.

Glatzmaier, G.A., & Roberts, P.H. 1995b. A three-dimensional self-consistent computer simulation of a geomagnetic field reversal. *Nature*, **377**, 203–209.

Glatzmaier, G.A., & Roberts, P.H. 1996a. An anelastic evolutionary geodynamo simulation driven by compositional and thermal convection. *Physica D*, **97**, 81–94.

Glatzmaier, G.A., & Roberts, P.H. 1996b. Rotation and magnetism of Earth's inner core. *Science*, **274**, 1887–1891.

Glatzmaier, G.A., Coe, R.S., Hongre, L., & Roberts, P.H. 1999. The role of the Earth's mantle in controlling the frequency of geomagnetic reversals. *Nature*, **401**, 885–890.

Glatzmaier, G.A., Evonuk, M., & Rogers, T.M. 2009. Differential rotation in giant planets maintained by density-stratified turbulent convection. *Geophys. Astrophys. Fluid Dynam.*, **103**, 31–51.

Gough, D.O. 1969. The anelastic approximation for thermal convection. *J. Atmos. Sci.*, **26**, 448–456.

Greenspan, H.P. 1969. *The Theory of Rotating Fluids*. Cambridge University Press.

Griebel, M., Dornseifer, T., & Neunhoeffer, T. 1967. *Numerical Simulation in Fluid Dynamics: A Practical Introduction*. SIAM.

Gubbins, D., & Herrero-Bervera, E. (eds.). 2007. *Encyclopedia of Geomagnetism and Paleomagnetism*. Springer-Verlag.

Guillot, T. 1999. A comparison of the interiors of Jupiter and Saturn. *Planet. Space Sci.*, **47**, 1183–1200.

Guillot, T. 2005. The interiors of giant planets: models and outstanding questions. *Annu. Rev. Earth Planet. Sci.*, **33**, 493–530.

Guillot, T., Stevenson, D.J., Hubbard, W.B., & Saumon, D. 2004. The interior of Jupiter. *Pages 35–57 of:* Bagenal, F., McKinnon, W., & Dowling, T. (eds.), *Jupiter: The Planet, Satellites, and Magnetosphere*. Cambridge University Press.

Hansen, U., & Yuen, D.A. 1995. Formation of layered structures in double-diffusive convection as applied to the geosciences. *Pages 135–149 of:* Fernando, J., & Brandt, A. (eds.), *Double-Diffusive Convection, Geophysical Monograph 94*. American Geophyiscal Union.

Hart, J.E., Glatzmaier, G.A., & Toomre, J. 1986. Space-laboratory and numerical simulations of thermal convection in a rotating hemispherical shell with radial gravity. *J. Fluid Mech.*, **173**, 519–544.

Hewitt, J.M., Mckenzie, D.P., & Weiss, N.O. 1975. Dissipative heating in convective flows. *J. Fluid Mech.*, **68**, 721–738.

Hubbard, W.B. 1975. Gravitational field of a rotating planet with a polytropic index of unity. *Soviet Astron. A.J.*, **18**, 621–624.

Hughes, D.W., & Weiss, N.O. 1995. Double-diffusive convection with two stabilizing gradients: strange consequences of magnetic buoyancy. *J. Fluid Mech.*, **301**, 383–406.

Ismail-Zadeh, A., & Tackley, P.J. 2010. *Computational Methods for Geodynamics*. Cambridge University Press.

Jackson, D.J. 1998. *Classical Electrodynamics*. Third ed. Wiley.

Jarvis, G.D., & McKenzie, D.P. 1980. Convection in a compressible fluid with infinite Prandtl number. *J. Fluid Mech.*, **96**, 515–583.

Jones, C.A. 2007. Thermal and compositional convection in the outer core. *Pages 131–185 of:* Olson, P., & Schubert, G. (eds.), *Treatise on Geophysics: Core Dynamics*. Elsevier.

Jones, C.A. 2011. Planetary magnetic fields and fluid dynamos. *Annu. Rev. Fluid Mech.*, **43**, 583–614.

Jones, C.A., & Kuzanyan, K.M. 2009. Compressible convection in the deep atmospheres of giant planets. *Icarus*, **204**, 227–238.

Jones, C.A., Soward, A.M., & Mussa, A. I. 2000. The onset of thermal convection in a rapidly rotating sphere. *J. Fluid Mech.*, **405**, 157–179.

Jones, C.A., Boronski, P., Brun, A.S., Glatzmaier, G.A., Gastine, T., Miesch, M.S., & Wicht, J. 2011. Anelastic convection-driven dynamo benchmarks. *Icarus*, **216**, 120–135.

Kuhlen, M., Woosley, S.E., & Glatzmaier, G.A. 2006. Carbon ignition of Type Ia supernovae. II. A three-dimensional numerical model. *Astrophys. J.*, **640**, 407–416.

Kundu, P.K., & Cohen, I.M. 2008. *Fluid Mechanics*. Fourth ed. Academic Press.

Lantz, S.R. 1992. *Dynamical Behavior of Magnetic Fields in a Stratified, Convecting Fluid Layer*. Ph.D. thesis, Cornell University, Ithaca, NY.

Lantz, S.R., & Fan, Y. 1999. Anelastic magnetohydrodynamic equations for modeling solar and stellar convection zones. *Astrophys. J. Suppl.*, **121**, 247–264.

Ledoux, P. 1947. Stellar models with convection and with discontinuity of the mean molecular weight. *Astrophys. J.*, **105**, 305–321.

Lele, S.K. 1992. Compact finite difference schemes with spectral-like resolution. *J. Comp. Phys.*, **103**, 16–42.

Lessani, B., & Papalexandris, M.V. 2006. Time-accurate calculation of variable density flows with strong temperature gradients and combustion. *J. Comp. Phys.*, **212**, 218–246.

Liao, W. 2008. An implicit fourth-order compact finite difference scheme for one-dimensional Burger's equation. *Appl. Math. Comput.*, **206**, 755–764.

Lin, D.J., Bayliss, A., & Taam, R.E. 2006. Low Mach number modeling of type I X-ray burst deflagrations. *Astrophys. J.*, **653**, 545–557.

Liu, J., Goldreich, P.M., & Stevenson, D.J. 2008. Constraints on deep-seated zonal winds inside Jupiter and Saturn. *Icarus*, **196**, 506–517.

Livermore, P.W., Jones, C.A., & Worland, S.J. 2007. Spectral radial basis functions for full sphere computations. *J. Comp. Phys.*, **227**, 1209–1224.

Majda, A., & Sethian, J. 1985. The derivation and numerical solution of the equations for zero Mach number combustion. *Combust. Sci. Tech.*, **42**, 185–205.

Merryfield, W.J. 1995. Hydrodynamics of semiconvection. *Astrophys. J.*, **444**, 318–337.

Miesch, M.S. 2005. Large-scale dynamics of the convection zone and tachocline. *www.livingreviews.org*, **lrsp-2005-1**, 1–138.

Moore, D.R., Peckover, R.S., & Weiss, N.O. 1973. Difference methods for time-dependent two-dimensional convection. *Comp. Phys. Comm.*, **6**, 198–220.

Nellis, W.J., Weir, S.T., & Michell, A.C. 1996. Metalization and electrical conductivity of hydrogen in Jupiter. *Science*, **273**, 936–938.

Noir, J., Hemmerlin, F., Wicht, J., Baca, S.M., & Aurnou, J.M. 2009. An experimental and numerical study of librationally driven flow in planetary cores and subsurface oceans. *Phys. Earth Planet. Inter.*, **173**, 141–152.

Ogden, D.E., Glatzmaier, G.A., & Coe, R.S. 2006. The effects of different parameter regimes in geodynamo simulations. *Geophys. Astrophys. Fluid Dynam.*, **100**, 107–120.

Ogura, Y., & Phillips, N.A. 1962. Scale analysis of deep and shallow convection in the atmosphere. *J. Atmos. Sci.*, **19**, 173–179.

Olson, P., & Schubert, G. (eds.). 2007. *Treatise on Geophysics: Core Dynamics*. Elsevier.

Olson, P.L., Glatzmaier, G.A., & Coe, R.S. 2011. Complex polarity reversals in a geodynamo model. *Earth Planet. Sci. Lett.*, **304**, 168–179.

Orszag, S.A. 1970. Transform method for the calculation of vector-coupled sums: application to the spectral form of the vorticity equation. *J. Atmos. Sci.*, **27**, 890–895.

Orszag, S.A. 1971a. On the elimination of aliasing in finite-difference schemes by filtering high-wavenumber components. *J. Atmos. Sci.*, **28**, 1074.

Orszag, S.A. 1971b. Numerical simulation of incompressible flows within simple boundaries: accuracy. *J. Fluid Mech.*, **49**, 75–112.

Patankar, S.V. 1980. *Numerical Heat Transfer and Fluid Flow*. Taylor and Francis.

Peyret, R. 2002. *Spectral Methods for Incompressible Viscous Flow*. Springer-Verlag.

Phillips, N.A. 1959. An example of non-linear computational instability. *Pages 501–504 of:* Bolin, B. (ed), *The Atmosphere and the Sea in Motion*. Rockefeller Institute Press and Oxford University Press.

Piacsek, S.A., & Toomre, J. 1980. Nonlinear evolution and structure of salt fingers. *Pages 193–219 of:* Nihoul, J.C.J. (ed.), *Marine Turbulence*. Vol. 28. Elsevier.

Plourde, F., Pham, M.V., Kim, S.D., & Balachandar, S. 2008. Direct numerical simulations of a rapidly expanding thermal plume: structure and entrainment interaction. *J. Fluid Mech.*, **604**, 99–123.

Press, W.H., Flannery, B.P., Teukolsky, S.A., & Vetterling, W.T. 1992. *Numerical Recipes: The Art of Scientific Computing*. Cambridge University Press.

Proudman, J. 1916. On the motion of solids in a liquid possessing vorticity. *Proc. R. Soc. Lond. A*, **92**, 408–424.

Radko, T. 2008. The double-diffusive modon. *J. Fluid Mech.*, **609**, 59–85.

Rai, M.M., & Moin, P. 1991. Direct simulations of turbulent flow using finite difference schemes. *J. Comp. Phys.*, **96**, 15–53.

Rappaport, S., Verbunt, F., & Joss, P.C. 1983. A new technique for calculations of binary stellar evolution, with application to magnetic braking. *Astrophys. J.*, **275**, 713–731.

Rayleigh, Lord. 1916. On convection currents in a horizontal layer of fluid, when the higher temperature is on the under side. *Philosophical Magazine*, **32**, 529–546.

Rempel, M., Schüssler, M., & Knölker, M. 2009. Radiative magnetohydrodynamic simulation of sunspot structure. *Astrophys. J.*, **691**, 640–649.

Roberts, G.O. 1972. Dynamo action of fluid motions with two-dimensional periodicity. *Philos. Trans. R. Soc. Lond. A*, **271**, 411–454.

Roberts, P.H. 1967. *An Introduction to Magnetohydrodynamics*. London: Longmans.

Roberts, P.H. 1968. On the thermal instability of a rotating fluid sphere containing heat sources. *Phil. Trans. R. Soc. Lond. A*, **263**, 93–173.

Roberts, P.H. 2007. Theory of the geodynamo. *Pages 67–105 of:* Olson, P., & Schubert, G. (eds.), *Treatise on Geophysics: Core Dynamics*. Elsevier.

Roberts, P.H., & Glatzmaier, G.A. 2000. Geodynamo theory and simulations. *Rev. Mod. Phys.*, **72**, 1081–1124.

Roberts, P.H., Glatzmaier, G.A., & Clune, T.L. 2010. Numerical simulation of a spherical dynamo excited by a flow of von Karmon type. *Geophys. Astrophys. Fluid Dynam.*, **104**, 202–220.

Rogers, T.M., & Glatzmaier, G.A. 2005a. Penetrative convection within the anelastic approximation. *Astrophys. J.*, **620**, 432–441.

Rogers, T.M., & Glatzmaier, G.A. 2005b. Gravity waves in the Sun. *Mon. Not. Roy. Astron. Soc.*, **364**, 1135–1146.

Rogers, T.M., Glatzmaier, G.A., & Woosley, S.E. 2003. Simulations of two-dimensional turbulent convection in a density-stratified fluid. *Phys. Rev. E.*, **67**, 026315-1–6.

Rogers, T.M., Glatzmaier, G.A., & Jones, C.A. 2006. Numerical simulations of penetration and overshoot in the sun. *Astrophys. J.*, **653**, 766–773.

Rogers, T.M., MacGregor, K.B., & Glatzmaier, G.A. 2008. Non-linear dynamics of gravity wave driven flows in the solar radiative interior. *Mon. Not. Roy. Astron. Soc.*, **387**, 616–630.

Rosenblum, E., Garaud, P., Traxler, A., & Stellmach, S. 2011. Turbulent mixing and layer formation in double-diffusive convection: Three-dimensional numerical simulations and theory. *Astrophys. J.*, **doi:10.1088/0004-637X/731/1/66**.

Schmitt, R.W. 1994. Double diffusion in oceanography. *Annu. Rev. Fluid Mech.*, **26**, 255–285.

Schubert, G. 1968. The stability of toroidal magnetic fields in stellar interiors. *Astrophys. J.*, **151**, 1099–1110.

Schubert, G., Bercovici, D., & Glatzmaier, G.A. 1990. Mantle dynamics in Mars and Venus: influence of an immobile lithosphere on three dimensional mantle convection. *J. Geophys. Res.*, **95**, 14105–14129.

Schubert, G., Turcotte, D.L., & Olson, P. 2001. *Mantle Convection in the Earth and Planets*. Cambridge University Press.

Slingerland, R., & Kump, L. 2011. *Mathematical Modeling of Earth's Dynamical Systems*. Princeton University Press.

Snir, M., Otto, S., Huss-Lederman, S., Walker, D., & Dongarra, J. 1998. *MPI—The Complete Reference: Volume 1, The MPI Core*. MIT Press.

Spiegel, E.A., & Veronis, G. 1960. On the Boussinesq approximation for a compressible fluid. *Astrophys. J.*, **131**, 442–447. Correction, **135**, 655–656.

Spiegel, E.A., & Weiss, N.O. 1982. Magnetic buoyancy and the Boussinesq approximation. *Geophys. Astrophys. Fluid Dynam.*, **22**, 219–234.

Stanley, S., & Glatzmaier, G.A. 2009. Dynamo models for planets other than Earth. *Space Sci. Rev.*, **doi:10.1007/s11214–009–9573-y**.

Stellmach, S., Traxler, A., Garaud, P., Brummell, N., & Radko, T. 2011. Dynamics of fingering convection. Part 2. The formation of thermohaline staircases. *J. Fluid Mech.*, **677**, 554–571.

Stern, M.E. 1960. The salt fountain and thermohaline convection. *Tellus*, **12**, 172–175.

Stern, M.E., Radko, T., & Simeonov, J. 2001. Salt fingers in an unbounded thermocline. *J. Mar. Res.*, **59**, 355–390.

Stevenson, D.J., & Salpeter, E.E. 1977. The dynamics and helium distribution in hydrogen-helium fluid planets. *Astrophys. J. Suppl.*, **35**, 239–261.

Takahashi, F. 2012. Implementation of a high-order combined compact difference scheme in problems of thermally driven convection and dynamo in rotating spherical shells. *Geophys. Astrophys. Fluid Dynam.*, **106**, 231–249.

Taylor, G. 1917. Motion of solids in fluids when the flow is not irrotational. *Proc. Roy. Soc. Lond. A*, **93**, 99–113.

Tilgner, A. 1999. Driven inertial oscillations in spherical shells. *Phys. Rev. E*, **59**, 1789–1794.

Tilgner, A. 2007a. Kinematic dynamos with precession driven flow in a sphere. *Geophys. Astrophys. Fluid Dynam.*, **101**, 1–9.

Tilgner, A. 2007b. Zonal wind driven by inertial modes. *Phys. Rev. Lett.*, **99**, 194501.

Tilgner, A. 2007c. Rotational dynamics of the core. *Pages 207–243 of:* Olson, P., & Schubert, G. (eds.), *Treatise on Geophysics: Core Dynamics*. Elsevier.

Veronis, G. 1965. On finite amplitude instability in thermohaline convection. *J. Marine Res.*, **23**, 1–17.

Veronis, G. 1968. Effect of a stabilizing gradient of solute on thermal convection. *J. Fluid Mech.*, **34**, 315–336.

Weiss, N.O. 1966. The expulsion of magnetic flux by eddies. *Proc. Roy. Soc. A.*, **293**, 310–328.

Weiss, N.O. 1981a. Convection in an imposed magnetic field. Part 1. The development of nonlinear convection. *J. Fluid Mech.*, **108**, 247–272.

Weiss, N.O. 1981b. Convection in an imposed magnetic field. Part 2. The dynamical regime. *J. Fluid Mech.*, **108**, 273–289.

Wu, C.-C., & Roberts, P.H. 2009. On a dynamo driven by topographic precession. *Geophys. Astrophys. Fluid Dynam.*, **103**, 467–501.

Yu, S.T., Hultgren, L.S., & Liu, N.S. 1994. Direct calculations of waves in fluid flows using a high-order compact difference scheme. *AIAA J.*, **32**, 1766–1773.

Zingale, M., Almgren, A.S., Bell, J.B., Nonaka, A., & Woosley, S.E. 2009. Low Mach number modeling of type Ia supernovae. IV. White dwarf convection. *Astrophys. J.*, **704**, 196–210.

Index

Prandtl number. *See* Nondimensional numbers

Predictor-corrector schemes. *See* Time integration schemes

Proudman-Taylor Theorem, 232, 242, 279

Pseudo-incompressible approximation. *See* Equation sets

Pseudo-spectral method. *See* Solution methods

Radiative heat flux. *See* Heat flux

Rayleigh-Bénard convection. *See* Convection

Rayleigh damping, 65, 116

Rayleigh number. *See* Nondimensional numbers

Rayleigh-Taylor instability, 81

Reynolds number. *See* Nondimensional numbers

Reynolds stress, 8, 121, 161, 241, 243, 259

Roberts number. *See* Nondimensional numbers

Rossby number. *See* Nondimensional numbers

Rotation, 229–281

Runge-Kutta schemes. *See* Time integration schemes

Salt-fingering. *See* Convection

Schwarzschild criterion. *See* Stratifications

Self-gravity, 199

Semiconvection. *See* Convection

Semi-implicit schemes. *See* Time integration schemes

Solar convection. *See* Convection

Solar dynamo. *See* Dynamos

Solid inner core: Finitely conducting, 155, 156, 260, 265; Insulating, 155, 161, 260; Latent heat, 265; Perfectly conducting, 155; Rotating, 156, 260, 265

Solution methods: Chebyshev collocation, 102–108, 119, 126, 127, 149–153; Chebyshev-Fourier, 102–108, 119, 129, 245; Chebyshev-spherical-harmonic, 135, 149, 152, 153, 157–159, 162, 259–277; Direct coupled-equation matrix, 107, 125–127, 152, 153; Finite-difference, 21–23, 43, 112; Fourier spectral, 19–29, 71, 117–119, 124–126; Galerkin, 35–39, 177–179, 185, 186; Influence matrix, 127–129, 152, 153; Poloidal-toroidal decomposition, 140–149, 159; Pseudo-spectral, 131; Spectral-transform, 130–132, 139–145, 149–153, 162; Vorticity-streamfunction decomposition, 28, 59, 176, 181, 214–216, 220, 221, 232, 233, 250–252

Solvers: Fast Chebyshev transform, 108; Fast Fourier transform (FFT), 108, 111, 131, 139, 151, 163; Fourier transform, 19, 40,

51, 67, 131, 132; Gaussian quadrature, 139, 140, 151, 289, 290; Legendre functions, 137–143, 288; Poisson, 24, 25, 88, 92, 101, 107; Tridiagonal, 24, 25, 86, 88, 92, 99, 283

Sound waves. *See* Waves

Spectral aliasing, 132, 140

Spectral-transform method. *See* Solution methods

Spherical harmonic divergence and curl, 144

Spherical harmonic expansions, 136–145, 158–159

Spherical harmonic truncations, 140, 144, 145

Staircase profiles, 76–78

Stellar convection. *See* Convection

Stellar dynamos. *See* Dynamos

Stratifications: Adiabatic, 197; Compositionally stable, 68, 72–74; Compositionally unstable, 68–72, 265; Convectively stable, 3–5, 48, 59–66, 68–80, 197; Convectively unstable, 3–5, 43–48; Density, 193–228; Ledoux criterion, 73; Schwarzschild criterion, 4–6, 73; Subadiabatic, 4, 5, 59–66, 69–72, 76–77, 197, 202, 220–222; Superadiabatic, 5, 6, 72–74, 77–78

Stress-free boundary conditions. *See* Boundary conditions

Subadiabatic stratification. *See* Stratifications

Subgrid-scale turbulence. *See* Diffusivities

Sunspots, 169, 187

Superadiabatic stratification. *See* Stratifications

Taylor columns, 232, 242, 247, 279

Taylor number. *See* Nondimensional numbers

Taylor-Proudman Theorem. *See* Proudman-Taylor Theorem

Thermal convection. *See* Convection

Thermal diffusivity. *See* Diffusivities

Time integration schemes: Adams-Bashforth, 23, 85, 90, 91; Adams-Moulton, 90, 91; Crank-Nicolson, 87–91; Implicit vs. explicit, 22; Predictor-corrector, 89, 90; Runge-Kutta, 52, 53, 79, 85, 86; Semi-implicit, 87–91

Time step constraints: CFL, 40–42, 62, 87, 88, 91, 97, 157, 177, 193, 194, 263; Diffusive, 23, 42, 62, 71, 87, 88, 97

Torsional oscillations. *See* Waves

Tridiagonal solver. *See* Solvers

Verification and validation, 276

Viscous diffusivity. *See* Diffusivities

Viscous heating, 8, 9, 12, 92, 145, 151, 179, 203, 218

Viscous stress, 7, 8, 12, 199

Visualization. *See* Postprocessing